A Math-Based Writing System for Engineers

Brad Henderson

A Math-Based Writing System for Engineers

Sentence Algebra & Document Algorithms

Springer

Brad Henderson
University of California, Davis
Davis, CA, USA

ISBN 978-3-030-10754-3 ISBN 978-3-030-10756-7 (eBook)
https://doi.org/10.1007/978-3-030-10756-7

Cover illustration by Steven Morse

This Springer imprint is published by the registered company Springer Nature Switzerland AG
The registered company address is: Gewerbestrasse 11, 6330 Cham, Switzerland

for my father, Professor Jerald M. "Jerry" Henderson, Ph.D., P.E., who once told me, "With a degree in engineering, you can go on in life to accomplish almost anything your mind and heart desire."

Preface

My first exposure to engineering writing being taught by an engineer—within the context of technical content and culture and with fervent maverick passion—was at the University of California, Irvine, several decades ago.

I had earned my B.S. in mechanical engineering from Cal Poly, San Luis Obispo, worked as a design engineer in the burgeoning aerospace industry of the early 1980s, and experienced a moment of "enlightenment" which invited me to exit the corporate world and embrace a pent-up bohemian desire to become a famous novelist. I pursued a graduate degree in creative writing from University of Southern California and worked furiously on several novels destined (I thought) for publication by a big New York publisher and critical acclaim. Eventually, I had another giant epiphany: fame and fortune as a novelist wasn't anywhere in sight, and I needed an actual job that paid the bills. That's when a friend of a friend told me there was a professor at UC Irvine who was looking for an engineer who could write.

Frederick Sawyer taught a popular course called "Engineering Communication in the Professional World" (ENG 190W). A renowned expert in chemical engineering, Fred had developed the course for the university's College of Engineering. The class was affiliated with UC Irvine's relatively new Writing-Across-the-Curriculum (WAC) program, and nearly every engineering undergraduate wanted to take ENG 190W. It fulfilled the four-unit upper-division writing requirement with an engineering course number rather than the alternatives, which at the time were limited to "Advanced Composition" or a writing-intensive, upper-division literature class taught by the English Department.

Fred was looking for someone with the engineering and writing skills to teach the course exactly as he had designed it. As a part of my instructor's training, he required that I observe him teach every class meeting for an entire quarter for a section of Engineering 190W. During the first week of class, Fred had each student identify a newspaper or magazine article that was aimed at a technical audience and another aimed at a nontechnical audience and calculate "readability" using the Fog Index equation, as well as with some graphical analysis tools. Throughout the rest of Fred's lectures and assignments, there was the same emphasis on clear-cut, objective methods—time-tested recipes for writing workplace documents such as reports and proposals and practical coaching on public speaking that emphasized basic public speaking mechanics. "Make your voice audible and clean," he would say, "because after all, it's a sound wave, people!"

I joined the faculty at UC Irvine, and for 3 years, I taught Fred's specialized engineering writing class, as well another he had designed for physics, chemistry, and math majors. I felt like I had truly found a way to combine my skills and background in engineering and writing into something useful. Then, in the early 1990s, fate and family obligations took me to the Pacific Northwest, where I landed a job with Hewlett-Packard's Inkjet Division's corporate headquarters.

I worked as a technical education specialist, focused on hard-skills training. Newly hired technicians and engineers had a lot to learn about HP's equipment and proprietary technologies before they became fully productive. HP provided me with an opportunity to work with a truly brilliant team of design, manufacturing, research and development, and management engineers doing their jobs within a high-tech company that operated in global partnership with satellite manufacturing centers in Singapore, Puerto Rico, Ireland, Spain, and beyond. My expertise in engineering communication was a key factor to success at HP.

Yet, my experience at UC Irvine left a persistent pull toward teaching. When a job at UC Davis became available for a writing instructor with expertise in engineering and science writing, I returned to my roots—my father and grandfather had taught engineering here—and for the past 15+ years, that's been my full-time gig. I have carried on the tradition of Fred Sawyer's brand of engineer-to-engineer writing instruction with what I now call "math-based writing." Over time, I developed several new components—sentence algebra, sentence optimization, and document algorithms.

Developing Math-Based Engineering Writing

The idea of using graphics to teach engineering writing came from a seed planted during my years teaching with Fred at UC Irvine. However, structural flowcharts of document cores were something that I first developed when I was assigned to teach some sections of advanced essay writing. This class covered the standard suite of college composition forms—evaluation, position, compare-and-contrast papers, and so on. I observed that my students' papers often didn't develop coherently—not because of sentence-level blunders but because the students didn't understand how to propel the development of their topics logically and compellingly, from thesis to resolution.

It occurred to me to develop flowcharts of the various essays' rhetorical structures as a teaching tool. Many students found them helpful. I began to use these flowcharts in my engineering writing classes, where they were an even better fit. I renamed them "document algorithms" to align with the language of engineering students.

The most frequent major mistake I observe engineering students making is missing (or misplaced) pieces in their documents' structures. For instance, they might not state a project objective at the beginning of a project report because it seems "obvious" (when it isn't), or fail to share their decision-making process and criteria for drawing conclusions and making recommendations, again because "it's obvious." Document algorithms help writers avoid these pitfalls.

The sentence optimization process covered in Part II is designed to help engineering writers develop a sentence-level writing style that is concise, clear, correct, and appropriate in tone. Fred's approach was to have students read Strunk and White's *The Elements of Style* and practice using "SCC" exercises: *s*implify, *c*larify, and *c*orrect flawed sentences. I've tried using Strunk and White's classic book, but my students find it hard to connect with.

My system teaches students how to troubleshoot, repair, and avoid common sentence errors (nothing original, really) in a way that's user-friendly to engineers—same usual 20 or so errors but taught using math-based engineering language and engineering content.

Sentence Algebra: The Rules and Logic of Grammar

When it comes to writing, I believe that you need to know how to manufacture the building blocks (sentences and sentence streams) before you construct buildings (documents). Not everyone agrees with me; nor does every engineering student and professional need focused training in sentence algebra. My book delivers a stem-to-stern writing system in three major parts that can be studied individually, with the basics presented first.

Before I went to graduate school in creative writing, I experienced a pang of imposter syndrome about my ability to talk intelligently about sentence-level craft with professors and peers. *What's a participle? What's a relative clause?* I had received praise for being a skilled writer, but my lifetime of holistic writing education hadn't given me a complete understanding of how those little machines called sentences operated behind the panels. I realized that much of my writing style was based on intuition and mimicry, and that I did not possess the same core-skill rigor in writing that I possessed in my math-based technical skill set. I wanted both.

I took a specialized grammar review class for adults taught by a professor in Medieval Studies at Long Beach State University, and everything popped into place. From that point on, I was able to understand how the words and word groups making up sentences functioned and interacted between initial capital letter and terminal punctuation mark. I don't think everybody needs to know grammar, but if you're a "how things work" person like me, it's satisfying, comforting, and empowering to understand it.

Over the years, I have tried to get my engineering writing students interested in grammar. Some students wince when I so much as utter the word. Most of them write serviceable sentences and are interested chiefly in learning how to write documents, improve writing style, and avoid errors that cost them a lot of points when I'm grading papers. Few are interested in "grammar," in and by itself.

In the back of my mind, I kept thinking I'd like to find a way to make grammar palatable to engineers. I began tinkering with replacing the eight parts of speech with algebraic variables and plugging them into "equations" representing the five architectures for building basic sentences. It's a rudimentary system, but it's a way in. And that's what I'm including in this book as the Part I chapters, along with the Part II chapters on avoiding common errors and maximizing sentence style, and the Part III chapters on writing what I believe to be five essential short-form documents that engineers need to know how to write.

How to Use This Book for Teaching and Self Study

For college instructors interested in using this book to teach engineering students, here is what has worked for me. For several years, I have used Parts II and III as the principal textbook for my engineering writing classes at UC Davis. During the 10 weeks of our quarters, I have the students read all of the chapters in those two parts and write a short-form version of all five essential documents that I cover in document algorithms. I also assign graded editing exercises tied to the material covered in the book's lessons on sentence optimization.

Individual chapters from the Part III section also work well for just-in-time educational modules integrated into engineering classes that require reports. For instance, capstone design series classes often require students to write proposals, project status reports, and final project reports—which are covered in Chaps. 15–17. There are many ways to use individual chapters from this book for just-in-time training applications—at universities and in industry.

I developed and taught "STEM Grammar," an experimental grammar course for engineers and other STEM majors, using Part I as the principal textbook, along with selected chapters from Part II. The class stirred a bit of controversy at my university over where a STEM grammar class belongs—Linguistics, English, Rhetoric and Composition, Engineering, or elsewhere—and frankly, some people wondered if the class was legit. During the class, I observed the students to be engaged and enthusiastic. At the end, my instructor evaluation was sufficiently positive. I'm looking forward to refining this class and offering it again.

If you are an engineering student or industry professional interested in improving your writing skill set through self-study, I recommend that you read the entire book. However, how and in what sequence you address the book's chapters is up to you and your individual needs. If you are already a highly skilled sentence-level writer, you can, of course, skip directly to Part III and check out my document algorithms.

I am hoping that many readers will benefit from this book. The math-based writing system is not the only way to develop engineering writing skills, but it does provide a novel alternative. Keep in mind that this book pioneers the system's version 1.0 release. I look forward to version 2.0 and future updates, and I welcome your suggestions for improvement.

I believe that engineering writing and doing engineering are inextricably linked. I also believe that an engineering education is an excellent foundation for all types of writing. I have published in multiple genres—fiction, poetry, creative nonfiction, and technical writing. One of my greatest honors was being selected to be featured in Tom Moran's innovative book *Engineers Write!* (IEEE Press) as one of the 12 "literary engineers" writing in the USA. Too many people stereotype engineers as inept writers. Not true. We are capable of high performance in all areas of our jobs. Write on, engineers. Write on.

Woodland, CA, USA
February 2019

Brad Henderson

Acknowledgments

Writing this book was not a solo effort. A number of fine people provided help, support, and keen insights along the way. I will mention some of them here by name. To all of the others (and there are many), please know that this acknowledgment segment in spirit includes you.

The top slot on my list of names belongs to my partner in love and in life, Sharon Campbell Knox. Sharon is a highly educated, talented writer and critic. She never holds back when she provides feedback to me about my work. Six years ago, I wrote several exploratory chapters to try to find my way into this book. They attempted to articulate a new method for teaching engineers how to build grammatical sentences using math metaphors, equations, and symbols as tools. A bit nervous, because the material was in some ways even more "out there" than my poetry, I went ahead and shared the pages with Sharon. Her response was immediate and decisive. She said, "This is the book you need to write." Sometimes writers need a little assurance to bring a new project out of hiding and get it rolling. She provided that.

After I wrote more pages and developed a book proposal, I sought a publisher. Several felt my book was interesting, but too risky to market. My acquiring editor at Springer, Michael Luby, was more optimistic. He saw sufficient promise in the book's math-based writing system and offered me a book contract. My friend, mentor, and fellow Springer author, Michael Alley, was also very instrumental during the manuscript placement phase of this project. He championed my book's concept, and he helped me to refine, perfect, and properly aim my manuscript pitch. For those of you who don't know who Mr. Alley is—when it comes to engineering and science communication (both oral and written)—he is the equivalent of an international rockstar.

After that, I had to take my exploratory chapters and proposal and develop the book into a complete manuscript. Initially, I naively envisioned putting the burners on and getting Springer my final manuscript in 12 months, ideally sooner. Instead, developing the book's content took over 4 years. This was a long-haul, large-scale book project, and a lot of it involved "front-end" design and prototyping of my new invention. I soon realized that I needed to bring in additional help, enter George Hayhoe, professor emeritus of technical communication at Mercer University School of Engineering and editor-in-chief of the *IEEE Transactions on Professional Communication*. George agreed to join my team as my book's private developmental editor. George is amazing—another rockstar like Michael Alley. I could not have gotten this book written without George's help. And even if I would've gotten it done, its quality would have suffered.

Along with George (and Sharon), there were some other insiders who worked with me during the entire arc of my book's development. These were my engineering writing students at UC Davis, where I teach as a continuing lecturer for the University Writing Program (UWP). Academic quarter after quarter, class after class, I test-taught versions of my book's chapters in progress. My students' input was both candid and brilliant. Their enthusiasm for engineering writing and prodigious talent as young writers was extraordinary. Thank you, UC Davis engineering students. You are the best. "Document algorithms" came to be because of you.

My project also benefitted tremendously from students who served on my book project team as engineering writing interns. These students assisted me with all aspects of the project but were particularly helpful with the development of the book's math-based grammar chapters on "sentence algebra." These interns need to be recognized by name: Bernadette Azizkhan, Antonio Chaj, Diana Chen, Yuanxian Chen, Katelyn Cooper, Arancha Ducaud, Ian Gordon, Samin Mohammadi Kamangar, Marie-Pierre Kippen, Gabriella Lahti, Ross Lai, Claire Loncarich, Maureen Njuguna, Richard Quigley, Nicolette Sarmiento, Joel Schmierer, Georgina Serrano, Jack Taylor, Evan Widjaja, and Jenna Wooster.

There are several other students I would like to extend special thanks: Matthew Gabel for helping me develop the illustrative sketch of the avionics valve in Chap. 16; Kristy Perano for writing the tech-to-nontech brief that I adapted for use as the sample document ("effective version") in Chap. 18; Sean Alling, Chungho Cheng, Michael Durkee, and Joshua Manderville for devising the strain gauge installation scenario that I adapted for use in the Chap. 19 job aid example; and Amy Freitas for writing the long-form project report that I adapted for use as the sample document in Chap. 20.

Thanks, as well, to graphic artists, Pearl Skelton and Steven Morse, who created the signature look of the graphics featured in this book. Thank you to subject matter experts Carl Luckenbill, Mike Hill, and Ransom Stephens, who reviewed selected chapters from this book for technical accuracy. I also want to graciously acknowledge Brinda Megasyamalan (project coordinator), Gomathi Mohanarangan (project manager), and the production team at Springer Nature, SPi, India, for their patience, persistence, and above-and-beyond effort during the book's production process. Thank you, in general, to my father, Jerry Henderson, and my granddad, Milton Henderson, for being engineers and inspiring me with engineering in my blood; and thanks as well to my son, Silas Henderson, for carrying on the tradition and choosing to pursue an engineering education. And finally, thanks to Dr. Frederick G. Sawyer, who gave me my first job teaching engineering writing at the University of California, Irvine, several decades ago. Fred was my first professional mentor. He showed me how to teach engineering writing like an engineer, for engineers. Fred's lessons were transformative. I have never forgotten them.

Contents

Chapter 1
Introduction to the Math-Based Writing System

Learning Objectives

- *Understand conceptually what is meant by "math-based writing system."*
- *Know the book's main purpose and target audience, as well as the rationale behind the book's modularized, variable-entry-point structure.*
- *Know the three levels of instruction that this book offers to engineering writers and the key elements of each level.*

No matter where you are from around the globe, if you're an engineer, it's almost certain that you can read, write, and speak about engineering content using numbers, symbols, and equations. For this reason, math can be thought of as a universal language among engineers. Is it possible to use this common-ground language of numbers, symbols, and equations as a touchstone or analog for teaching engineers how to write in a word-based language such as English? This book says *yes*.

Like most natural systems, whether abstract, organic, or inert, there is an architecture of order underlying language. At least since Aristotle, rhetoricians in the West have documented how sentences and paragraphs join together to record and communicate logical arguments, patterned descriptions, and archetypal stories. For centuries, grammarians have defined common anatomies for sentences, identified how words function within them, and determined how specific words should be arranged to achieve acceptable constructions. In recent decades, writing experts have defined and advanced a set of generic principles for good writing, which almost invariably apply to engineering writing. The math-based writing system that this book proposes embraces these architects and anatomists of language, and does not argue with them.

What's different about this book's system is that it examines the structure and operation of the English language—its building blocks (words and sentences) and buildings (documents)—using math-based thinking (see Fig. 1.1). It uses one language to help teach another—something not all that extraordinary given that, on the flip side, we use word-based language as a medium for discussing math concepts and doing "word problems" in math classes. In engineering and scientific discourse, we oftentimes toggle back and forth between equations and words, as well.

B. Henderson, *A Math-Based Writing System for Engineers*,
https://doi.org/10.1007/978-3-030-10756-7_1

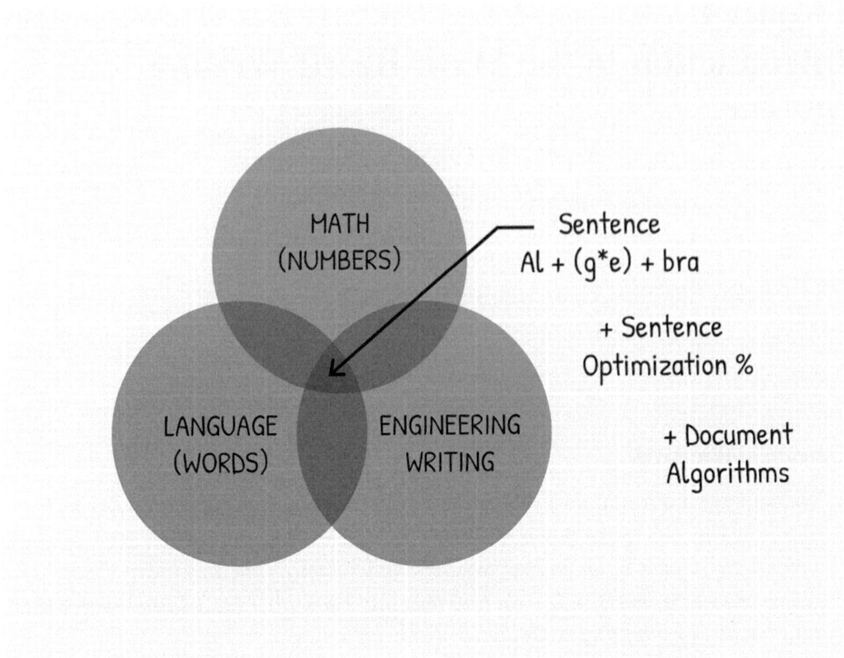

Fig. 1.1 The domains and intersections of math-based writing for engineers

Instead of approaching writing through the lens of traditional writing instruction, this book's system uses symbols, equations, graphics, and flowcharts to help engineers understand and master the writing component of their jobs and to assist engineering students in integrating applied writing into their professional preparation. For engineers who view writing as a largely subjective, qualitative endeavor, the system may provide a fresh perspective that allows them to approach writing in an objective, quantitative fashion similar to the way that they approach other engineering tasks. Also, all of the book's examples feature engineering content.

The principal audience for this book is engineering professionals—from students and novices preparing for initial success to seasoned experts who work in industry and want to refine and develop their engineering writing skill set. This book may also be useful to international professionals and students working to master written English as a work language, as well as professionals in other STEM disciplines.

Some readers will benefit from reading the book from beginning to end. Others may choose to access only those chapters of instruction they need—when they need them.

- Engineers who seek to develop a deeper, more extensive understanding of how sentences operate at the equation level (like little message-carrying machines) can begin with the Part I chapters on *Sentence Algebra*. This alternate approach to English grammar uses algebraic variables to represent the parts of speech (N = noun, V = verb, X = pronoun, and so on) and math-based equations to represent sentences (N_s + V = "Computers calculate").

- Those of you who already know the sentence-building fundamentals might choose to turn to the Part II chapters on *Sentence Optimization*. Here you will find lessons on how to troubleshoot, repair, and avoid common sentence errors that occur in applied workplace writing, as well as how to develop an optimal writing style.
- Expert sentence engineers who are ready for "calculus-level" writing can choose to skip to Part III on *Document Algorithms*, which provides best practices and flowcharted blueprints for producing effective proposals, project updates, reports, and so forth. The Part III chapters also focus principally on documents that are short in length and digital in format to meet the needs of professionals writing in today's global, fast-paced work environment.

Additionally, the book's structure models best practices in instructional systems design. Each chapter begins with learning objectives, ends with a review of how the objectives were met, and assigns the reader optional action items designed to be quick to complete and easy to thread into an engineer's busy workday. The action items include a variety of mini thought tasks to further understanding of concepts and mini "do" tasks to test drive application techniques.

A few things that this book does not do: it does not teach generic technical writing for aspiring professional technical writers. Nor does it discuss technical communication theory or techniques for writing academic articles, reviews, and dissertations. There are many good books that already cover that terrain. This book was written by an engineer for engineers and other math-based thinkers who seek further development and refinement in the "writing part" of their jobs.

Action Items

α	Create a new electronic document (e.g., Microsoft Word file, Google Docs, etc.), and title it, "An Analysis of My Engineering Writing." As a first entry in this document, type a single paragraph (keep it short, 200 words or fewer) that discusses what you like versus what you don't like about the writing you have had to do in school and at work prior to encountering this book. After completing this mini-task, save the file. You needn't show this paragraph to anyone. But please keep it for future review and reflection.
α	Retrieve the document that you created for action item one ("Analysis of My Engineering Writing"). Add a second paragraph (also 200 words or fewer), in which you do a self-analysis of your workplace writing ability that considers your strengths and weaknesses as an engineering writer. Also, as best as you can, rate your *current writing skill level overall* using a percent scale, where 0% is none (unlikely) and 100% is ultimate mastery. Also determine a desired rating for you to achieve in the future. Record both ratings (now and future) at the end of your second paragraph. For example, "And overall, my self-appraised skill level as an engineering writer scores 80% now; whereas, I would like it to be 95% (or better)." Once again, you needn't show this document to anyone. But do keep it for future review and reflection.

(continued)

α Consider the following terms for various types of writing: "engineering writing," "science/scientific writing," "business writing," "technical writing," "journalism," and "literary writing." Use Google to look up a definition of each one of these terms. What comes up? You will probably discover multiple posts and mostly consistent definitions for all of these types of writing except one—"engineering writing"—because "engineering writing" is typically not thought of as a discrete genre of writing. This book suggests a definition. But more importantly, what do you think? Henceforth, in your engineering career, I challenge *you* to define what engineering writing is—more specifically, what excellent engineering writing is—by example of the documents that *you* write and circulate as a professional.

Recap

Key Learnings/Takeaways

- "Math-Based Writing" is an instructional system that teaches writing using math concepts and symbols, math-based thinking, and an engineering context. Because equations and symbols are a universal language for encoding mathematical concepts among engineers, the system uses that language to help engineers learn to write correctly in English.
- The book's chief purpose is to teach engineers and other math-based thinkers how to write successfully on the job, and the book's primary audience is professional engineers (beginners through seasoned experts). Its secondary audiences are engineering students and students and professionals in other STEM disciplines.
- This book is designed as a variable-entry-point series of chapters. Some readers will benefit from reading the book from beginning to end. Others will choose to access only the chapters they need, when they need them.
- This book's system applies the three math-based perspectives to the writing of engineering document and arranges the chapters into three parts providing three levels of instruction:
  - *Part I—Sentence Algebra* provides engineering writers with an understanding of how sentences are designed and built.
  - *Part II—Sentence Optimization* presents a set of quality control techniques for applied sentence-level writing in engineering documents.
  - *Part III—Document Algorithms* characterizes key document structures using blueprints (flowcharts) known as document algorithms.
- Part III chapters also focus principally on documents that are short in length and digital in format to meet the needs of professionals writing in today's global, fast-paced work environment.
- This book does not teach generic technical writing, technical communication theory, or techniques for writing academic articles, reviews, and dissertations. It was written by an engineer for engineers and other math-based thinkers who seek further development and refinement in the "writing part" of their jobs.

Part I
Sentence Algebra

Chapter 2
Part I Primer: The Elements of Sentence Algebra

Learning Objectives

- *Understand the eight functional roles that words and word groups can play in a sentence equation and the sentence algebra variables that represent those eight functions (traditional grammar's eight parts of speech).*
- *Understand "spark" and the way that it enables words to join together to form sentences.*
- *Understand what a basic sentence is and the five basic sentence types you can build with the variables.*
- *Understand what an advanced sentence is.*
- *Know the basic coding conventions that guide sentence algebra equations.*

Part I Chapters 2 through 8 present the "grammar" of math-based writing that I call *sentence algebra*. In Part I, we will use the sentence algebra system to investigate sentences as if they were little message-carrying machines that can be modeled with functional equations. In these equations, up to eight different algebraic variables represent the machines' components. The equations themselves record how each machine's components are arranged and interconnected so that they operate together to accomplish the work of message encoding (by writers), transmission (via documents), and communication (to readers, who then decode the message text). After you complete the Part I chapters, when you write a sentence and see it on the page or screen or when you read a sentence written by someone else, you should be able to understand technically how that sentence works beneath its surface.

Once acquired, most of the time when you are writing on the job, your sentence algebra skill and knowledge will reside in your subconscious and implicitly (automatically) inform and guide your sentence-level writing. This is true of any grammar system, of course. When you've mastered sentence algebra, however, on those occasions when you are called upon to troubleshoot and repair malfunctioning sentences in documents, or fine-tune a stream of sentence text so that it performs optimally, your sentence algebra training will allow you to approach these tasks as a math-based-thinking sentence engineer—confident, competent, and armed with a

B. Henderson, *A Math-Based Writing System for Engineers*,
https://doi.org/10.1007/978-3-030-10756-7_2

full set of sentence engineering tools and schematics for a wide variety of sentence types, basic through advanced.

Eight Functions, Eight Variables

As illustrated in Fig. 2.1, the sentence algebra system begins with words and the eight functional roles that words can play in a sentence, traditional grammar's eight parts of speech, and assigns each role an algebraic variable. Here are the eight variables—each partnered with a simple definition and several sample words that can function as that variable:

Variable	**Definition**	**Examples**
N (noun)	Names a person, place, or thing	*Jeff, company, product*
V (verb)	Expresses action	*Invent, test, sell*
X (pronoun)	Stands in for a noun	*Her, that, it*
M_n (adjective)	Modifies a noun	*Successful, marginal, heavy*
M_v (adverb)	Modifies a verb (or sometimes an adjective or another adverb)	*Often, quickly, carefully*
L (preposition)	Begins a phrase and expresses a relationship to another word or element	*In, on, between*
C (conjunction)	Connects words, phrases, and sentence equations	*And, or, although*
I (interjection)	Expresses emotion as a word	*Oh, yeah, wow*

Our system further divides the eight variables into two main categories: the *core variables*, N, V, and X, and the *accessory variables*, M_n, M_v, L, C, and I. The core variables create basic sentence cores, which for now you can think of as a basic sentence's stripped-down engine. The other variables further accessorize a basic sentence to enrich its core function with additional details and enable connections to and between advanced sentence structures. The first seven variables listed above (N through C) are used regularly in engineering writing. The last variable listed (I) is not. For completeness, our lessons include a brief segment on interjections in Chap. 4. Subsequent chapters in Part I, as well as in Parts II and III, do not.

The chapters on the sentence algebra variables also discuss a variety of features and properties that impact how words playing the eight functional roles are written in text sentences. These characteristics do not all impact sentence algebra coding and equations, but they must be observed and applied when writing grammatically correct sentences for inclusion in engineering documents.

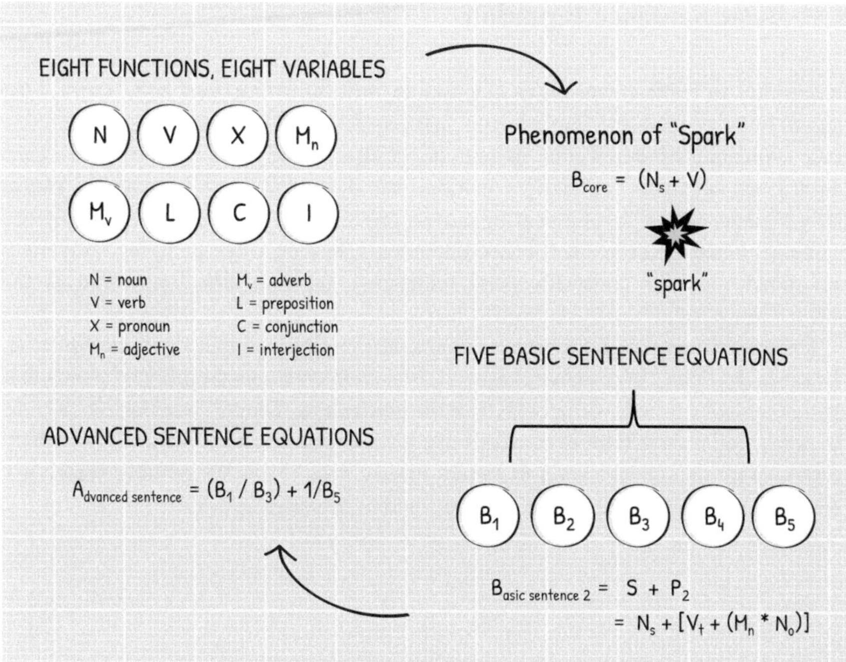

Fig. 2.1 Sentence algebra overview: eight word-function variables, "spark," five basic sentence equations, and advanced equations

"Spark"

Outside of sentence equations, individual words (whether written as text or as a variable) encode individual units of meaning. For further study of individual words, you can consult a dictionary. Along with each entry's definition(s), these databases catalog each word entry's spelling, part(s) of speech, usage(s), and more. Take the word "plane," for example. What does it signify? A flying machine, a flat surface, a tool? A standard online dictionary such as *Merriam-Webster.com* indicates that the word "plane" can function as a noun (N), adjective (M_n), or verb (V) and lists over a dozen definitions related to aeronautics, architecture, geometry, horticulture, and the fine arts [1].

Despite their impressive versatility, words by themselves do little more than assert *dictionary definitions*. Thus, the sentence algebra system considers individual words to be *cognitively static*. To enable words to become *cognitively dynamic* and act out their dictionary roles in useful communications, we plug words into sentence equations. And at the core of each one of these equations that functions as a sentence (or message-carrying machine), there are two essential components: a noun subject (N_S) joined together with a verb (V). This construction, $N_S + V$, is known as a *basic clause*, and it produces a synergistic reaction that this book's system refers to as the phenomenon of "spark." "Spark" allows words to advance their cognitive potential—from static definitions into word streams that encode human thoughts.

Five Basic Sentence Equations

In addition to eight consistently identifiable word-unit functions and "spark," the natural order beneath the English language also reveals five consistently identifiable basic word-unit arrangements, or patterns, that scaffold *five basic sentence equations*. Each of these five basic equations has a single "spark"-producing clause ($N_s + V$) at its center. As engineering writers, we regularly use the five basic sentence equations when we create text-based documents. In fact, when combined, inverted, and transformed (to form advanced sentences), the five equations serve as the underpinning for all sentence-level constructions in the English language.

The master template for a basic sentence (B) is a short, simple algebraic equation, "B = S + P." The "S" variable, known as the *subject*, is an assembly of words that always includes a noun subject (N_s) or the equivalent. The "P" variable, known as the *predicate*, is an assembly of words that always includes a *main verb* (V)—which partners with the subject noun to form a basic clause ($N_s + V$). A basic sentence's predicate can contain other core variables and accessories as well.

Here is the algebraic template for a basic sentence:

$$\mathbf{B_n = S + P_n}$$

where B_n = one byte of human thought
S = subject (which always includes N_s or equivalent)
P_n = predicate (which always includes a main V)
$n = 1, 2, 3, 4$, or 5

As we progress through Part I, we will analyze each of the five scenarios for "P_n = predicate" and the five resultant B_n—B_1 through B_5. The chapters will demonstrate coding examples using each basic equation and also feature a lengthier example at the end of Chap. 6 demonstrating the coding process of a paragraph, which contains all five basic sentence types. As a preview of things to come, Fig. 2.2 illustrates the analysis of a text sentence that reveals a B_2 sentence equation underneath. We will discuss this example further and refer back to it in this chapter's final segment on coding conventions.

Advanced Sentence Equations

Advanced sentences bear similarities to basic sentences, yet they are typically designed to encode more content, capture more specificity, and model more intricate and multifaceted sentence equations. The basic framework underneath an *advanced sentence equation* is always one or more basic sentences or basic sentence variations or combinations. Because of this framework, advanced sentences always contain at least one subject noun (or equivalent) and one main verb pairing ($N_s + V$), which in

original sentence (text)

The robotic arm inside the chamber safely repositions the radioactive sample.

scanned

N_s	$LP(M_n)$	M_v	V_t	M_n	N_o
The robotic arm	inside the chamber	safely	repositions	the radioactive	sample.

core formula

$$B_2 \;=\; S + P_2 \;=\; N_s \;+\; (V_t \;+\; N_o)$$

coded (full equation)

$$(N_s * LP(M_n)) \;+\; (M_v * V_t) \;+\; (M_n * N_o)$$

Fig. 2.2 An example of using sentence algebra to scan, code, and understand the inner workings of a sentence

turn creates at least one instance of spark. Advanced sentences, unlike basic sentences, can also contain more than one core spark-producing N_s + V pair.

Chapters 7 and 8 on advanced sentences discuss a variety of advanced sentence structures and components including:

- Compound and complex sentences that join basic sentences together with conjunctions (C) in various combinations
- Sentence inversions ($1/B_n$) and variations such as passive voice, questions, commands, and so on
- Multi-word subassemblies such as relative clauses (RC) and verbal phrases (VP) that writers can use to further accessorize sentences

There is also a lengthier, more intricate coding example at the end of Chap. 8 that analyzes a paragraph containing a variety of advanced sentence structures.

Coding Conventions

When I first developed the sentence algebra system, I considered what I had learned about grammar in my kindergarten through college classes, my many years of experience speaking and writing in English, my math-based engineering education, my curiosity about language and the transcription of thought, my observations (not original to me) that there was natural order and intrinsic logic built into English, and, finally, the premise that English grammar can be described using mathematical symbols, metaphors, variables, and equations (also not original to me). After carefully thinking about all of these things, I eventually synthesized them into an instructional tool—a method for investigating and characterizing how sentences work.

This tool is customized for math-based-thinking engineering writers (people like me). The sentence algebra system is a math analog, a hybrid blend. It is not pure "1 + 1 always equals 2" math. To get the system to work, repeatably and reproducibly, you have to apply a set of sentence algebra conventions (system-specific rules). I will cover some of these conventions, later on, when they become necessary and relevant in the chapters on advanced equations. However, the system's basic conventions are best covered up front. They apply to all of this book's sentence algebra lessons, from this point onward.

The Basic Coding Conventions

- *Code left-to-right, the same as you read and write*: When coding a sentence as a sentence algebra equation, the left-to-right sequencing of variables in the equation should match the text's original structure, that is, the original left-to-right arrangement of words in the sentence.
- *Use the plus sign (+) and asterisk (*) to signal basic operations.* To connect sequences of core variables (Ns, Xs, and Vs) to form basic equations, use the plus sign (+); to add specificity to core variables with modifiers (such as M_n or M_v), use the asterisk (*) as a specificity multiplier.
- *Use parentheses and brackets to organize terms.* For example, we can organize the earlier Fig. 2.2 example to highlight the B_2 equation underneath. $B_2 = [S] + [P_2] = [N_s * LP(M_n)] + [(M_v * V_t) + (M_n * N_o)]$.
- *Identify and code top-level functional components in equations.* When you are scanning and coding a sentence, identify and code its top-level functional components, even though further disassembly is often possible. For instance, disassembling a prepositional phrase, an "LP" for short, into its sub-parts can introduce another layer of interesting analysis. But if the phrase's chief role in a sentence is to modify a noun or verb, coding it as a prepositional phrase playing the role of modifier, $LP(M_n)$ adjective or $LP(M_v)$ adverb, is usually sufficient for applied sentence engineering. This is why, in the Fig. 2.2 example, the prepositional phrase "inside the chamber" is coded as "$LP(M_n)$", rather than "$(L + (m_n * N_o))$."
- *In sentence equations, usually consider articles ("the," "a," and "an") to be invisible (elliptical), and do not code them with variables.* Thus, the word "the" is not explicitly coded in the Fig. 2.2 equation, even though it operates as a specialized type of adjective. As we will see in Chap. 4, when necessary, we represent articles with the special variable "m_n".
- *In basic and advanced sentence equations, omit the text version's initial capital letter and terminal punctuation mark.*

- *Omit the text version's internal punctuation marks inside basic sentences*. These internal punctuation marks insert "process points" into basic sentence structures and include commas setting off introductory words, phrases, or clauses and also nested elements inserted between two commas or two dashes. As we will see in Chap. 7, however, sentence algebra equations do not omit punctuation marks such as semicolons ";" used to connect basic sentences to form advanced sentence structures.
- *Apply the Basic Math Laws selectively to sentence algebra equations.*
  - The *Commutative Law does* apply (selectively) to *specificity multiplication* within sentences, especially when it comes to M_v (adverbs). For example, the basic sentence containing the subject pronoun "she," main verb "wins," and adverb "often" can be written correctly in two different ways: "She often wins" [coded as $X_s + (M_v * V)$] or "She wins often" [coded as $X_s + (V * M_v)$]. On the other hand, for this chapter's scan and code example (Fig. 2.2), a similar manipulation does not apply. "The robotic arm … *safely* repositions the radioactive sample" works. "The robotic arm … repositions *safely* the radioactive sample" is an awkward construction.
  - The *Commutative Law* generally does *not* apply to sentence algebra *addition of core elements* (with or without accessorizing) in equations. For example, the spark-producing pair "$N_s + V$" is not functionally equivalent to "$V + N_s$." As applied to our running example, something like $(M_v * V_t) + (N_s * LP(M_n)) + (M_n * N_o)$ or "Safely repositions the robotic arm inside the chamber the radioactive sample" *doesn't work.*
  - The *Associative Law* often *does* apply, since it involves organization mainly (with parentheses and brackets), not operations.
  - The *Distributive Law* generally does *not* apply.

You will learn more about where and when the Basic Math Laws do and do not apply in Chaps. 5–8. (Note: The Review Box below provides a quick refresher on the Commutative, Associative, and Distributive rules.)

Review Box
The Basic (Rudimentary) Math Laws for Manipulating Equations

Commutative Law:	$a + b = b + a$
	$a \times b = b \times a$
Associative Law:	$(a + b) + c = a + (b + c)$
	$(a \times b) \times c = a \times (b \times c)$
Distributive Law:	$a \times (b + c) = a \times b + a \times c$

Some Additional Resources

One final note before moving on to Chap. 3 and examining sentence algebra's core variables in detail: As you proceed, you may come across a grammar question for which you find no sufficient answer within this text. Just as a professional engineer might sometimes need to consult an engineering theory and principles handbook—these days, usually a digital source available online—the engineering writer can often look up and find answers to emergent grammar questions using an online dictionary or online grammar guidebook. Here are several good ones.

Grammar, Purdue Online Writing Lab (OWL), Purdue University, Indiana, U.S.A.
https://owl.purdue.edu/owl/general_writing/grammar/index.html

Tutorials on the Essence of Grammar, Punctuation, and Usage in Scientific Writing; Writing Guidelines for Engineering and Science; Pennsylvania State University, Pennsylvania, U.S.A.
https://www.craftofscientificwriting.com/exercises.html

Aggie Grammar Guide, University of California, Davis, U.S.A.
https://tutoring.ucdavis.edu/agg

Action Items

α	Can you write a sentence that contains words playing the roles of the *seven* sentence algebra variables regularly used in engineering writing—N, V, X, M_n, M_v, L, and C. Give this task a try. To make it easier, first examine the benchmark example (see below), and then create your own. *We installed the new O-ring into the water widget assembly, and this action immediately stopped the leakage.* *where:* $N =$ *"O-ring,"* $V =$ *"installed,"* $X =$ *"we,"* $M_n =$ *"new,"* $M_v =$ *"immediately,"* $L =$ *"into,"* $C =$ *"and"*
α	After completing the first action item, try writing another sentence using words that function *only* as Ns (nouns) and Vs (verbs). Next, expand that sentence, or create another sentence, so it includes accessory word types such as M_n (adjectives), M_v (adverbs), and so on.
α	Make a list of five to ten typical grammar obstacles and unknowns that you recall encountering in your writing before encountering this book. An example of an obstacle/unknown might be uncertainty whether a long, intricate sentence is grammatically complete or, with regard to personal pronouns, whether a pronoun should be "I" or "me"—"we" or "us." Later, after reviewing the basic sentence algebra chapters (Chaps. 3–6), revisit your list of grammar obstacles/unknowns. Have any items on the list been resolved? Later still, after you've reviewed the advanced sentence algebra chapters (Chaps. 7 and 8), consult the list again. Most likely, the number of unresolved/unanswered items will have significantly decreased (hopefully to 0) by the time you complete this book's Part I training on sentence algebra.

Recap

Key Learnings/Takeaways

- There are eight functional roles that words can play in a sentence equation. Sentence algebra assigns each role a variable: N = noun, V = verb, X = pronoun, M_n = adjective, M_v = adverb, L = preposition, C = conjunction, and I = interjection.
- The first seven of these variables regularly apply in engineering writing; one, I = interjection, does not.
- Words by themselves record dictionary definitions, and to be useful in communicating messages, they must be plugged into sentence equations that are built around at least one spark-generating clause—that is, a noun subject paired with a main verb, $N_s + V$.
- In this book's system, "spark" is the reaction that occurs when a basic clause forms.
- In the sentence algebra system, there are five basic sentence formulas for expanding a basic clause into a sentence-level unit of thought—and these five structures are assemblies made out of words that can play eight functional roles.
- The equation template for a basic sentence is $B_n = S + P_n$, where B_n = a sentence-level unit of thought; S = subject (which always includes an N_s or equivalent); P_n = predicate (which always includes a main V); and n = 1, 2, 3, 4, or 5.
- The basic sentences are named B_1, B_2, B_3, B_4, and B_5.
- The following basic sentence algebra conventions impact how sentences code into sentence equations:
  - *Code left-to-right, same as you read and write.*
  - *Use the plus sign (+) and asterisk (*) to signal basic operations.*
  - *Use parentheses and brackets to organize terms.*
  - *Identify and code top-level functional components in sentence equations.*
  - *In sentence equations, usually consider the articles ("the," "a," and "an") elliptical (invisible), and do not code them with variables.*
  - *In basic and advanced sentence equations, omit the text version's typographic start and stop signals.*
  - *Omit internal punctuation marks inside individual basic sentences.*
  - *Apply the Basic Math Laws selectively to sentence algebra equations.*
- An advanced sentence is one or more basic sentences or basic sentence variations or combinations. Like a basic sentence, an advanced sentence's communication purpose is to encode and express a dynamic unit of thought—although an advanced sentence's level of specificity and encoded content can, by design, be more intricate and multifaceted than one expressed by a basic sentence.

Reference

1. Plane. Merriam-Webster.com. https://www.merriam-webster.com/dictionary/plane. Accessed 1 Nov 2018

Chapter 3
The Core Variables N, V, and X

Learning Objectives

- *Know the core sentence algebra variables for noun, verb, and pronoun and how words and word groups representing these variables function in sentences.*
- *Know the key features and properties of nouns, verbs, and pronouns that guide their correct coding in sentence algebra equations and correct usage in text sentences.*
- *Understand how to move a verb through a complete conjugation sequence.*
- *Know the relationship between a pronoun and its antecedent.*

This chapter covers what you need to know about the sentence algebra variables N, V, and X to understand how they function in sentence equations. We classify them (see Fig. 3.1) as the *core variables* because noun and verb (N + V) pairs are the central spark-producing elements of English sentences. We also consider pronouns (Xs) to be core variables because pronoun and verb pairings (X + V) can also operate as sentence cores.

N = Noun

In sentence algebra, "N" represents a word or group of words that names or identifies a *person, place, or thing*. A noun can also represent a set of persons, places, or things. Thus, when an N appears in a sentence equation, it could represent any of the following:

Person(s)	**Place(s)**	**Thing(s)**
manager	conference rooms	truss
engineers	New York City College	HP Inkjet Products Division
design team	Iowa	plasma deposition station

B. Henderson, *A Math-Based Writing System for Engineers*,
https://doi.org/10.1007/978-3-030-10756-7_3

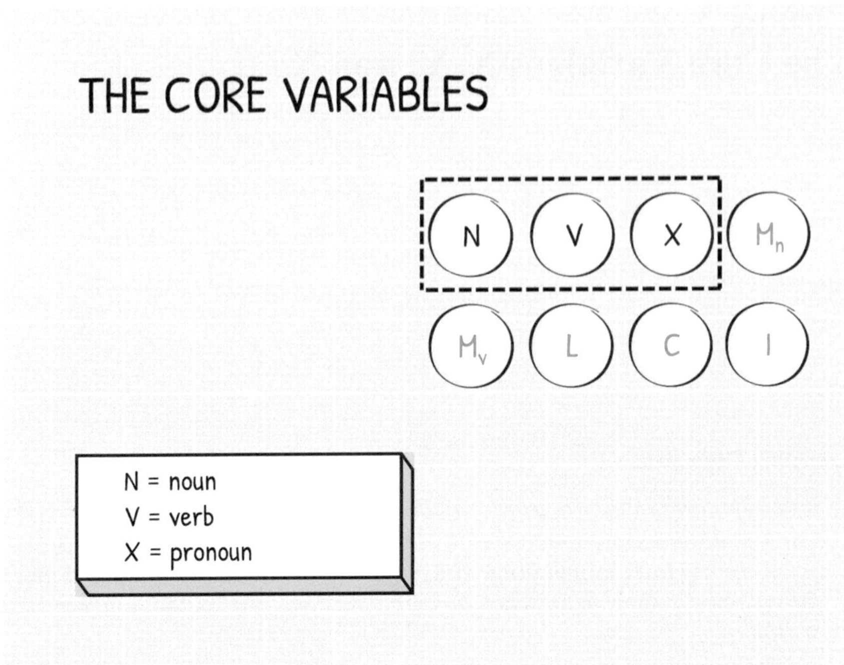

Fig. 3.1 The core variables representing three out of eight total parts of speech

Aside from a noun's principal function, nouns have other features and properties that further impact how we use them and how we write them in text sentences.

Features and Properties of Nouns

Compound Nouns

In the examples listed above, four of the nine nouns consist of one word, and the other five nouns consist of more than one word. The multi-word nouns ("conference rooms," "design review," "New York City College," etc.) are known as *compound nouns*. Sometimes it is debatable whether a multi-word cluster is a compound noun or a combination of adjectives and nouns. For instance, is "plasma deposition station" a compound noun, or is "plasma" an adjective modifying "deposition station" (a two-word compound noun)? Another interpretation might be that "plasma" and "deposition" are both adjectives modifying "station." When two or more words form a cohesive unit that functions as a single noun, our system prefers the simplest interpretation and assigns the unit a single N.

Number

Nouns have the *property of number*. This property expresses quantity and is independent of how many words form a noun. There are two noun numbers: $n = 1$ (singular) or $n > 1$ (plural). In most cases, a *singular noun* word(s) does not end in "s"; when that same noun represents a count more than one, it adds an "s" (for instance, one battery, two anode**s**, one-and-a-half volt**s**.) In the examples at the beginning of this segment, "engineers" and "conference rooms" demonstrate *plural nouns* following this simple rule.

With number, there are occasional irregularities, especially for English nouns borrowed from Greek or Latin. For instance, the plural of "formula" can be written as either "formulas" or "formulae"; other exceptions include "criterion/criteria," "nucleus/nuclei," and "memorandum/memoranda." Some singular nouns not of Latin derivation already end with an "s," such as with "truss," "bus," and "dress." In this case, for plural you add "es": "truss/trusses," "bus/buses," "dress/dresses." Also, when a noun ends with a "y," as in "battery." The plural form deletes the "y" and adds "ies," as in "batteries."

Collective

Another exception is that some nouns represent a collection of items that are *not individually counted* and are instead viewed as a whole set. An example of this type of noun is "equipment." There is no such word as "equipments." Whether an equipment set contains 1, 2, or 53 pieces, it is still called "equipment." Words like "equipment," "furniture," and also "research" are *collective nouns*. (Note: For any given noun, if you are in doubt about how to spell its singular versus plural form, or whether it is a collective noun form that doesn't take an "s," consult a dictionary.)

Proper vs. Common Nouns

Specific nouns, or *proper nouns*, begin with a capital letter. The rest, *common nouns*, do not. Capitalizing the first letter of a proper noun tags the noun as a word that represents the specific name of a specific person, place, or thing. For example, it is "Intel Corporation" not "intel corporation" or "Intel corporation," and "California" not "california."

On the other hand, the name of a generic person, place, or thing is a common noun and not capitalized, even when it names a specific example of that noun. For example, civil engineering students learn about the stress and strain that can be placed upon a "concrete pillar" not a "Concrete Pillar." And even if your college diploma says "B.S. in Mechanical Engineering," your personal bio should read, "I earned a bachelor's degree in mechanical engineering." There is no capitalization because in this case you are describing a type of degree, not the degree title.

V = Verb

A verb is a word or group of words that *expresses action.* Some verbs can also express "states of being" or function as auxiliary or "helping" verbs. In sentence algebra, a verb takes the symbol "V." Along with Ns (and Xs standing in for nouns), we know that Vs are one of two key components at the nucleus of every sentence equation.

Below are lists of four Ns (nouns), four Vs (verbs), and four N + V pairs, with Ns in regular font and Vs in *italic*. The left and middle columns present static words (dictionary listings). The right column combines entries from the first two columns to display functional cores of sentences—each expressing a unit of dynamic human thought (a little message).

N (static)	**V (static)**	**N + V (dynamic)**
engine	*rev*	engine *revs*
computer	*compile*	computer *compiled*
bridge	*support*	bridge *has supported*
circuit	*transmit*	circuit *is transmitting*

As the above examples illustrate, the writing of a verb word can change its spelling/form when that verb goes from dictionary listing to "live" operation in an N + V pair—for example, *rev/revs*, *compile/compiled*, *support/has supported*, and *transmit/is transmitting*. These changes are known as *verb conjugations*, and they indicate when the verb's action occurs, the number of people or things performing the verb's action, and more.

Features and Properties of Verbs

Verb Tense: The Dimension of Time (When?)

In text sentences, the way a verb is written can change as a result of *when* the verb's action takes place. This expression of time frame is known as *verb tense*. As you will see in the upcoming examples, tense can change a verb's form/spelling in a variety of ways, ranging from adding an "s," "ed," or "ing" to the verb's end, to placing other (auxiliary) words such as "will," "have," and "is" in front of the verb.

Simple Tense: The most common verb tense is *simple tense*. In simple tense, the different forms/spellings or conjugations of a verb distinguish whether the verb's action occurs in the general (simple) present, past, or future. The following sentence illustrates general *past/present/future tense* (three choices) for V = "finish":

The engineer *finished/finishes/will finish* the project's four phases.

Beyond this simple approach to time frame, there are several *complex tenses* that distinguish how and when a verb's action takes place *within* a general time frame—(simple) past, present, or future.

Perfect Tense: An example of a complex tense is *perfect tense*, which specifies action that has been completed at some point within one of the three main time frames—(simple) past, present, or future.

To construct perfect tense, you combine a principal (root) verb ending in "ed" (for regular verbs) with the helping (auxiliary) verb "have" (various forms—"had," "has," "have," and so on). Here is a sentence that shows three possible conjugations for the principal (root) verb "finished" in *past perfect/present perfect/future perfect tense*:

The engineer *had finished/has finished/will have finished* phase II of the project.

Instead of documenting that the engineer completed (or completes or will complete) all four phases of the project sometime in the general past (or general present or future), the perfect tense pinpoints the engineer's work status more specifically. It reports that the engineer's work has progressed to an interim point in time (in the general past, present, or future) occurring *after* the milestone of completing phase II and *prior to* the completion of subsequent phases (e.g., phases III and IV).

Progressive Tense: Another complex tense that offers time frame-within-time frame capability is *progressive tense*. We construct this verb form by partnering the helping verb "be" (presenting as "is," "are," "was," and so forth) with a principal (root) verb ending in "ing." Instead of recording action that begins and ends within one of the three general time windows, progressive tense expresses action that is *ongoing* within one of the general time stretches. For example (again using V = "finish," this time written with the "ing" ending), here is another version of the example sentence displaying *past progressive/present progressive/future progressive* tense:

The engineer *was finishing/is finishing/will be finishing* phase II of the project.

For further review and consolidation of our discussion about verb tense and type, consider Table 3.1. Each horizontal listing represents a sample verb conjugation for V = {calibrate, build, analyze, design}, with each V conjugated either in *simple tense* (rows 1 and 2), *perfect tense* (row 3), or *progressive tense* (row 4) within each of the three general time frames—past (column 1), present (column 2), and future (column 3). For all N + V pairings, we will use the noun, N = "engineers." One listing, V = "build" (row 2), features an "irregular verb" (further explained soon). The other three listings display regular verbs.

Table 3.1 Sampling of simple and complex verb tenses for four verbs

	Column	1	2	3
Row	Tense	Past	Present	Future
1	***Simple***	[Engineers] calibrated	[Engineers] calibrate	[Engineers] will calibrate
2	***Simple (irregular verb)***	[Engineers] built	[Engineers] build	[Engineers] will build
3	***Perfect***	[Engineers] had analyzed	[Engineers] have analyzed	[Engineers] will have analyzed
4	***Progressive***	[Engineers] were designing	[Engineers] are designing	[Engineers] will be designing

The above sampling displays verb conjugations taking shape in three different ways:

1. As the verb's "root-word" dictionary listing alone (e.g., "calibrate," row 1, column 2) or alone and prefaced by "will" ("will calibrate," row 1, column 3)
2. As an alternate spelling of the verb's root-word listing alone, involving the addition of the suffix "ed" or an irregular variation (e.g., "calibrated," row 1, column 1, and "built," row 2, column 1)
3. As a root-word verb (or an alternate spelling of the root word involving either the suffix "ed" or "ing") combined with one or more auxiliary or helping verbs—such as "have," "had," "are," "will be," and so forth—for example, "have analyzed" (row 3, column 2) and "are designing" (row 4, column 2)

Irregular Verbs

In Table 3.1's sample conjugations, one respelling of a root verb does not incorporate the suffix "ed" and is *irregular*. That word is "built" (row 2, column 1). In twenty-first-century English, there is no such word as "builded." There are more than 600 irregular verbs in English, about 50 of which are most commonly used. Irregular verbs come in a variety of forms, such as make/made/making; teach/taught/teaching; understand/understood/understanding; and write/wrote/written/writing [1].

The *dictionary entry* for all regular and irregular verbs indexes the verb's root form. Immediately beside the root, there are usually pronunciation guidelines and a description of function(s), and then the next item listed is the verb's simple past tense spelling—the "ed" version for a regular verb. If the verb is irregular, its simple past tense listing shows the alternate spelling. Most dictionaries also post the verb's "ing" version after simple past. Occasionally, an irregular verb's entry lists four verb forms rather than three—for instance, *write/wrote/written/writing*.

Take a look at the sample dictionary entries (just the initial listings without definitions, examples, synonyms, and so on) shown below for the regular verb, "calibrate," and two irregular verbs, "build" and "write":

Sample Dictionary Entries

Calibrate [kal-*uh*-breyt] [2]
verb (use with object), **calibrated, calibrating.**

Build [bild] [3]
verb (use with object), **built** or (Archaic) builded**; building.**

Write [rahyt] [4]
verb (use with object), **wrote** or (Archaic) writ**; written** or (Archaic) writ**; writing.**

The sample dictionary entries for "calibrate" and "build" display three verb spellings—(1) root form, (2) simple past tense (second), and "ing" form (third). These root forms and simple past forms (one regular and one irregular) appear in Table 3.1's conjugation samples.

Table 3.1 also displays a sample past perfect tense conjugation for "analyze," combining the helping verb "have" (had/have/will have) with the "ed" (simple past) form "analyzed." In addition, Table 3.1 features a sample progressive tense conjugation for "design," combining the helping verb "be" (was/is/will be) with the "ing" form "designing." If you were to look up dictionary entries for "analyze" and "design," you would find that they, too, list three spellings for each verb.

However, as shown above, our third sample dictionary entry for "write" (irregular verb) lists four different spellings—(1) "write" (root verb), (2) "wrote," (simple past tense), (3) "written" (additional spelling), and (4) "writing" ("ing" form). Thus, take note: for some irregular verbs like this one, the recipe for *simple tense conjugation* continues to use spelling (2) "wrote," as in:

Correct

The engineer ***wrote****/writes/will write* an excellent project report.

Yet the recipe for the *perfect tense conjugation* uses an additional spelling (3), in this case "written," instead of spelling (2), as in:

Correct

The engineer *had* ***written****/has* ***written****/will have* ***written*** an excellent project report.

When we examine advanced sentence equations including verbal phrases, you will learn more about additional verb forms and irregularities. For now, let's move on to further explore the fundamental features and properties of verbs.

Verb Number

In N + V constructions, just as the noun's spelling adjusts as its count goes from singular to plural, count also impacts the verb. Specifically, for an N + V pair, if the count of the activating noun N is one ($n = 1$), then the count of the verb V must also be one (singular N pairs with singular V). Conversely, if the count of N is greater than one ($n > 1$), the count of V must be $n > 1$ (plural N pairs with plural V).

As you may recall, the spelling adjustment for plural nouns sometimes adds an "s," as in "engineers" (plural). When the activating noun is singular, the spelling adjustment for a singular verb is just the opposite and sometimes adds an "s." For example, if Table 3.1 had featured the singular noun "engineer" rather than the plural "engineers" in its conjugation samples, the row 1, column 2 entry would change to "[the engineer] calibrates" rather than "[engineers] calibrate."

Point of View (POV) or "Person"

In addition to tense and number, a final attribute that impacts verb conjugations is *point of view (POV)* or "*person*." POV characterizes the *stance* of the action doer(s) and action observer(s) relative to the action's stage, that is, where the verb action takes place. There are three main points of view (POVs):

1. Someone or something doing an action firsthand onstage ("first person"): "I am reading."
2. Someone onstage directly observing an action being done secondhand by another person, place, or thing onstage ("second person"): "You are reading."
3. Someone offstage indirectly observing an action being done thirdhand by another person, place, or thing onstage ("third person"): "She is reading."

The three main POVs also have singular and plural cases; thus, there are *six POVs* total. For instance, for a singular person, the three stances regarding action are *I* did the action, *you* did the action, or *she/he (or it)* did the action. Three additional POVs apply to the plural case. For multiple persons, or people, the stances are *we* did the action, *you* (you all) did the action, or *they* did the action. POV follows the same logic for place- and thing-naming nouns.

Verb Conjugation Matrices

Though not a regular part of applied engineering writing, verb conjugation matrices are useful learning tools for examining how a specific verb changes its form/spelling as it cycles through a full array of application scenarios—varying tense, number, and POV. For further (optional) review of the verb features and properties covered in the V = Verb segment, I invite you to explore the matrices in Figs. 3.2, 3.3, 3.4, 3.5, and 3.6.

POV	Singular past/present/future	Plural past/present/future
first-person	I integrated/integrate/will integrate	we integrated/integrate/will integrate
second-person	you [1 person] integrated/integrate/ will integrate	you [>1 person] integrated/integrate/ will integrate
third-person	s/he integrated/integrates/will integrate	they integrated/integrate/will integrate

Fig. 3.2 Conjugation of "integrate" (regular verb) for all instances of simple tense. Note: All five matrices use the shorthand "s/he" to stand in for the third person singular doer—which could be "she," "he," "it," or, otherwise, depending on personal preference

POV	Singular past/present/future	Plural past/present/future
first-person	I unwound/unwind/will unwind	we unwound/unwind/will unwind
second-person	you [1 person] unwound/unwind/ will unwind	you [>1 person] unwound/unwind/ will unwind
third-person	s/he unwound/unwinds/will unwind	they unwound/unwind/will unwind

Fig. 3.3 Conjugation matrix of "unwind" (irregular verb) for all instances of simple tense

POV	Singular past/present/future	Plural past/present/future
first-person	I had dampened/have dampened/will have dampened	we had dampened/have dampened/will have dampened
second-person	you [1 person] had dampened/have dampened/will have dampened	you [>1 person] had dampened/have dampened/will have dampened
third-person	s/he had dampened/has dampened/will have dampened	they had dampened/have dampened/will have dampened

Fig. 3.4 Conjugation of "dampen" (regular verb) for all instances of perfect tense

POV	Singular past/present/future	Plural past/present/future
first-person	I had driven/have driven/will have driven	we had driven/have drive/will have driven
second-person	you [1 person] had driven/have driven/will have driven	you [>1 person] had driven/have driven/will have driven
third-person	s/he had driven/has driven/will have driven	they had driven/have driven/will have driven

Fig. 3.5 Conjugation of "drive" (irregular verb) for all instances of perfect tense

POV	Singular past/present/future	Plural past/present/future
first-person	I was amplifying/am amplifying/will be amplifying	we were amplifying/are amplifying/will be amplifying
second-person	you [1 person] were amplifying/are amplifying/will be amplifying	you [>1 person] were amplifying/are amplifying/will be amplifying
third-person	s/he was amplifying/is amplifying/ will be amplifying	they were amplifying/are amplifying/will be amplifying

Fig. 3.6 Conjugation of "amplify" (regular verb) for all instances of progressive tense

The above matrices present complete conjugations for five example verbs V = {integrate, unwind, dampen, drive, amplify}. Each matrix features a specific verb type (regular or irregular) and tense type (simple, perfect, or progressive) covered in the V = Verb segment. For convenience and brevity, each verb conjugation matrix uses a placeholder term (X) for the generic actor noun that travels through an array of scenarios with the matrix's verb (V). Here, X presents as a variety of *personal pronouns*—I, you, she, we, they, and so on—which are a key topic of discussion in Chap. 3's final segment on the third core variable.

Note that Fig. 3.5's perfect tense conjugation of the irregular verb "drive" features the alternate form "driven," instead of its simple past form "drove," in its various combinations with the auxiliary verb "have." Like the verb "write" (discussed earlier), it is an irregular verb with four possible forms/spellings ("drive"/"drove"/"driven"/"driving"), instead of the more common three (as is the case, for instance, of Fig. 3.3's irregular verb—"unwind"/"unwound"/"unwinding").

Also note that the example set does not include an irregular verb conjugation for progressive tense because the "ing" form of an irregular verb is never irregular. In progressive tense, regular verbs and irregular verbs conjugate the same way.

X = Pronoun

A *pronoun* is a word that *stands in* for a person, place, or thing—that is, a noun—in sentence equations. Since pronouns serve as *placeholders* or substitutes for nouns, they are an ideal choice for bearing the label of "X" in sentence algebra.

There are five types of pronouns. Ultimately, this book will address them all, but here we focus on the *personal pronouns*, which already made a brief appearance in this chapter's V=Verb segment. Personal pronouns stand in for a person or persons in the form of "I," "you," "she," "him," "his," "we," "they," "us," etc. They can also refer to a place or thing in the form of "it." The rules that govern personal pronouns can also apply to the other four types: *relative*, *interrogative*, *demonstrative*, and *indefinite* pronouns. Examples of these would be "*who*," "*what*," "*this*," and "*someone*," respectively.

No matter what type, the basic purpose and function of a pronoun is the same: a pronoun is a convenient, universally recognized, easy-to-use "nickname" for a noun. For all types of pronouns, the term for the noun being replaced is *antecedent*.

Features and Properties of Pronouns

POV and Number

Both point of view (POV) and the property of number impact how we write pronouns. Let's look at a conjugation matrix for the verb V = "iterate" as it moves through a conjugation sequence of pronoun and verb (X + V) pairings. This time, let's focus on how the conjugation changes **X**. See Fig. 3.7.

In the below matrix, notice that as POV shifts from first person to second person to third person in the singular column, the personal pronoun changes form, from "I" to "you" to "s/he." (Recall that "s/he" is shorthand for "she," "he," "it," or, otherwise, depending on personal preference.) Likewise, as number shifts and we move from the singular (= 1 person) to plural ($\geq$ 2 persons) column, form changes from "we" to "you" to "they."

POV	**Singular [1 person]** past/present/future	**Plural [≥ 2 persons]** past/present/future
1st-person	***I*** iterated/iterate/will iterate	***we*** iterated/iterate/will iterate
2nd-person	***you*** iterated/iterate/will iterate	***you*** iterated/iterate/will iterate
3rd-person	**s/he** iterated/iterates/will iterate	***they*** iterated/iterate/will iterate

Fig. 3.7 Conjugation of V = "iterate" for all cases of simple tense, point of view (POV), and pronoun number in the X + V pair

Gender

In addition to POV and number, *gender* also impacts the form of a pronoun in action. Gender can reflect the biological sex (or non-sex) of a person or thing. Tradition and/or culture can also dictate gender for a pronoun and thus also its form/spelling. For example, when sailors refer to a ship, they often use the personal pronoun "she." Relative to culture and profession, when you use personal pronouns, it is important to avoid sexist nuances. For example, when you are making a general reference to a role or position that can be assumed by either a man or woman—for instance, a "task force chairperson"—take care to acknowledge both genders by using inclusive terms such as "he or she," "she or he," or, better yet, use the plural forms "you" or "they."

In twenty-first-century society, we also acknowledge transitional and alternative genders—people who sometimes prefer personal pronouns other than "he/him" or "she/her." As an engineering writer and professional, you need to stay aware of this. Often when addressing a workplace audience, your best (and most respectful) choice is to remain gender neutral. You may have noticed that in this book, I choose to refer to my readers either indirectly as "engineer writers" (third person) or directly as "you" (second person).

Case

Finally, along with number, POV, and gender, the other functional role impacting a pronoun's form is *case*—or, the role a pronoun assumes in a sentence equation relative to other variables. Nouns have the property of case, too, but case does not impact how we write nouns in sentence equations like it does when we write pronouns.

There are three cases:

1. *Subjective*: The pronoun is an instigator or agent of action—the subject variable (located before the verb) in the equation X + V, where X substitutes for a subject N. The subjective case pronouns are *I*, *you*, *he*, *she*, *it*, *we*, *you*, and *they*. An example of a subjective case pronoun would be, "*They* excelled."

   Note: Sentence algebra notation adds a subscript "s" for "subject" to designate a subjective case pronoun or noun, as in $X_s + V$ or $N_s + V$.
2. *Objective*: The pronoun is the recipient of action—the object variable (located after the verb) in the equation (N + V) + X, where X substitutes for the second N, known as the object N. The objective pronouns are *me*, *you*, *him*, *her*, *it*, *us*, *you*, and *them*. An example of an objective case pronoun would be, "The company hired *her*."

   Note: Sentence algebra notation adds a subscript "o" for "object" to designate an objective case pronoun or noun, as in $(N_s + V) + X_o$. or $(N_s + V) + N_o$.
3. *Possessive*: The pronoun locates itself immediately *before* a noun to indicate "possession" or "ownership" associated with the noun immediately next to it—possessive pronoun on the left and "owned" noun on the right. The possessive pronouns are *my*, *his*, *her*, *its*, *our*, *your*, and *their*. An example of a possessive case pronoun would be, "Fiona liked *her* job."

   Note: Sentence algebra notation adds a single apostrophe (') to the possessive pronoun or noun and connects the possessive element to the noun it "owns" with a hyphen (-), as in X'-N or N'-N.

Reconciling Pronouns in Action: Matching Xs with Ns

Let's now examine a sequence of six sentences containing several pronouns. Each sentence displays as a sentence algebra equation on the left and word (text) equation on the right. As you read through the sentences below, try to identify the antecedent (that is, the specific noun the pronoun replaces) for each pronoun that you encounter. In addition, note the pronouns' positions in sentences, and consider how the property of case—subjective, objective, or possessive—impacts each pronoun's form/spelling when written as a word. To assist you in correctly matching Xs with Ns in the example sentences, I have underlined all nouns that function as antecedents, and I have *italicized* all corresponding pronouns.

(1)	$N_s + V + N_o$	Sheila attended college.
(2)	$X_s + V + N_o$	*She* studied wafer-fab technology.
(3)	$N_s + V + X_o$	Hewlett-Packard hired *her*.
(4)	$N_s + V + N_o$	The job targeted IC chip production rate.
(5)	$N_s + V + X_o$	Sheila improved *it*.
(6)	$X'\text{-}N_s + V + X_o$	*Her* boss promoted *her*.

In the above progression, the first pronoun you encounter is X_S = "She" in sentence two. Two nouns in sentence one, "Sheila" and "college," precede the pronoun. Since the pronoun in question refers to a female person, not a thing, we can deduce that the antecedent is "Sheila" not "college." In sentence three, "Sheila" continues to be the antecedent of the next pronoun, "X_o" = "her." Here, since case changes from subjective to objective, the correct form/spelling is "her," not "she."

The next pronoun appears in line five. X_O = "it" refers to a singular thing, not a person. Several nouns precede this pronoun—first, "Sheila" (restated again) in sentence five; next, the compound noun, "IC chip production rate," in line four; and then, prior to that, the regular noun, "job." Although "Sheila" is the noun that is nearest upstream, the logical antecedent for "it" is the compound noun representing a singular, inanimate object, N_O = "IC chip production rate."

Finally, in line six, the pronoun "her" appears twice. The first instance is in the two-variable pairing "X'-N_s". The pairing describes an employee-manager relationship between someone (female) and the "boss" (whose gender is not revealed). The next "her" in line six refers to the same person (female) whom the boss (whose gender is not revealed) promotes. The logical noun antecedent match for both of these Xs is N = "Sheila" in line five.

As the above example demonstrates, when a reader encounters a pronoun "in action" in a sentence equation, the reader has to decode the meaning of that X based on the nearest logical choice of an N upstream. When there are multiple choices, best choice is signaled by both *position* (nearest noun) and *context* (number, gender, and meaning of nearby nouns as presented in the text). Thus, as an engineering writer, you must be vigilant not only in selecting and using proper pronoun forms/spellings but also in arranging pronouns and their associated antecedents within sentence streams so that your readers can decode X variables with ease and accuracy. We will learn more about optimal usage of pronouns in applied engineering writing in Part II.

Action Items

α	Consider this scenario: You are a naval engineer stationed on an aircraft carrier. You are writing an email to the ship's chief engineer to report recent inspection findings. Based on what you have learned in this chapter about common vs. proper nouns, determine which of the two underlined choices in the following sentence would best represent what you intend the sentence to mean: *My team discovered a possible hairline fracture in one of the main propeller shafts/Main Propeller Shafts.* (Hint: "Main propeller shafts" is the generic name of multiple things in a set. One of these things (a type of shaft) may have a hairline fracture. Regardless of ship type and size, "main propeller shafts" are common components in engine-driven ships–so are "anchors," not "Anchors.")

(continued)

α	Consider the two verbs "lie" and "lay." "Lie" is the root-word dictionary listing for a verb that expresses the action of someone or something being in a horizontal resting position (e.g., *Pencils lie on the desktop*). On the other hand, "lay" is the root-word dictionary listing for a verb that expresses the action of moving someone or something down and into place (e.g., *The building crew lay conduit in the trench*). On a piece of scratch paper, draw a 3 row by 2 column table. Then use this table to conjugate *both* "lie" and "lay" in simple tense. You will find duplicate spellings in some of your table's cells. Based on what you've just learned in this chapter and action item, you can now explain why writers often mix up and misuse these two verbs in action. Do the same exercise for two more verbs: "affect" and "effect." This time, try using the two words as verbs (Vs) in sentences. If possible, try using them as nouns (Ns) in sentences as well. What potential errors might result if the "functional roles" of "affect" and "effect" get mixed-up in sentences?
α	Based **not** on a grammar checker program, your intuition, or advice from a colleague, but rather from your own functional understanding of how nouns, verbs, and pronouns operate in a sentence, determine which of the two choices in the following sentence is correct and why: *Bui Aerospace's Control Systems Team includes Arianna Thompson and I/me, as well as Tim Merrick.*

Recap

Key Learnings/Takeaways

- Sentence algebra considers the three variables N, V, and X—which represent three of the eight total parts of speech—to be the system's *core variables*. The system classifies these three variables as core elements because noun and verb (N + V) pairs and pronoun and verb (X + V) pairs are the central spark-producing engines of English sentences.
- In sentence equations, the function of an N is to name/identify a person, place, or thing (or set of persons, places, and things).
- The following additional features and properties are associated with nouns:
  - A noun can be a single word, or it can be a group of words known as a *compound noun*.
  - Nouns have the *property of number*. This property expresses quantity and is independent of how many words form a noun. There are two noun numbers: *singular* ($n = 1$) or *plural* ($n > 1$). In most cases, a singular noun word(s) does not end in "s"; when that same noun represents a count of more than one, it adds an "s."
  - Some nouns represent a collection of items that are not individually counted; these are *collective nouns*.
  - Specific nouns, or *proper nouns*, begin with a capital letter. The rest, *common nouns*, do not. Proper nouns represent the name of a specific person, place, or thing. Common nouns represent generic persons, places, or things.

- A V = verb is a word or group of words that expresses action. Some verbs can also express "states of being" or function as auxiliary or "helping" verbs.
- When paired with an activating noun (or pronoun) in a text sentence, a verb's form/spelling can change. These changes, known as *verb conjugations*, indicate who or what is performing the verb's action, when the action occurs, and more.
- The following additional features and properties are associated with verbs at work in a sentence equation (i.e., paired with an activating N or X):

  – The way a verb is written changes in response to when the verb's action takes place. This expression of time frame is known as *verb tense*, of which there are several types:

    Simple tense locates action in the general (simple) past, present, or future.
    Perfect tense establishes the action has taken place and concluded within the three general time frames (simple past, present, and future). To form this tense, the conjugations combine the root (main) verb's "ed" spelling or alternative spelling with the helping verb "have."
    Progressive tense signals the verb's actions are ongoing within the three general time frames. To form this tense, the conjugations combine the root (main) verb's "ing" spelling with the helping verb "be."

  – There are two *verb types—regular* (root-word dictionary listing for simple past tense adds an "ed" suffix) and *irregular* (root-word dictionary listing for simple past tense is an alternate spelling).
  – Verbs also have *the property of number*. A verb's count—*singular* or *plural*—is determined by the count of the activating noun (or pronoun) with which the verb is paired.
  – In addition to tense, type, and number, a final attribute that impacts verb conjugations is *point of view (POV)*, or "*person*." POV characterizes the stance of the action doer(s) and action observer(s) relative to the action's *stage* (where the verb action takes place). There are three main points of view (POVs), and the three main POVs also have singular and plural cases:

    "*First person*": someone or something doing an action firsthand onstage: "I"/"We" am/are reading."
    "*Second person*": someone onstage directly observing an action being done secondhand by another person, place, or thing onstage: "You"/"You [all]" are reading."
    "*Third person*": someone offstage indirectly observing an action being done thirdhand by another person, place, or thing onstage: "She"/"We" is/are reading."

- The sentence algebra variable for a *pronoun* is *X*. A pronoun's function in a sentence equation is to stand in for a noun. The noun that a pronoun stands in for is its *antecedent*.
- There are five types of pronouns. This chapter focuses on the *personal pronouns*—"I," "you," "she," "him," "his," "we," "they," "us," "it," etc. The other

four pronoun types are *relative*, *interrogative*, *demonstrative*, and *indefinite* pronouns. Examples of these would be "who," "what," "this," and "someone," respectively.

- The following additional features and properties are associated with pronouns:
  - Like nouns, pronouns change form based on *number (singular or plural)*; and, like verbs, pronouns change form based on *point of view (POV)*.
  - The property of *case* also affects how pronouns form.

    Subjective case characterizes a pronoun positioned on the left side of a verb and functioning as a performer of the verb's action.

    Objective case characterizes a pronoun positioned on the right side of a verb and functioning as the receiver (object) of the verb's action

    Possessive case characterizes side-by-side positioning, wherein the pronoun on the left "owns" the noun situated immediately after it on the right.

  - Additionally, the property of *gender* affects a pronoun's form/spelling. Biological sex, as well as culture and/or tradition, can dictate gender classification (male, female, neutral, and so on) for a pronoun.
- The property of case impacts *sentence algebra notation* for both pronouns and nouns:
  - To designate a subjective case pronoun or noun, we add a subscript "s" for "subject," as in X_s or N_s.
  - To designate an objective case pronoun or noun, we add a subscript "o" for "object," as in X_o or N_o.
  - To indicate possession, we add a single apostrophe (') to the possessive pronoun or noun and also connect the possessive element to the noun it "owns" with a hyphen (-), as in X'-N or N'-N.

References

1. 50 most commonly used irregular verbs. Esl-lounge.com. http://www.esl-lounge.com/reference/grammar-reference-most-common-irregular-verb-list.php. Accessed 15 July 2018
2. Calibrate. Dictionary.com. Dictionary.com Unabridged. Random House, Inc. http://dictionary.com/browse/calibrate. Accessed 15 July 2018
3. Build. Dictionary.com. Dictionary.com Unabridged. Random House, Inc. http://dictionary.com/browse/build. Accessed 15 July 2018
4. Write. Dictionary.com. Dictionary.com Unabridged. Random House, Inc. https://www.dictionary.com/browse/go. Accessed 15 July 2018

Chapter 4
The Accessory Variables: M_n, M_v, L, C, and I

Learning Objectives

- *Know the additional sentence algebra variables—adjective, adverb, preposition, conjunction, and interjection—and why they are classified as "accessory" rather than "core" variables.*
- *Know the key features of adjectives, adverbs, prepositions, conjunctions, and interjections that guide their correct coding in sentence algebra equations, as well as the standard rules and conventions for correct usage and punctuation of words representing these variables in text sentences.*

This chapter completes our discussion of the eight fundamental building blocks of English sentences or eight parts of speech. In Chap. 3, we examined N = nouns, V = verbs, and X = pronouns and the ways that these essential components pair together—either as N_s + V or X_s + V—to form spark-producing cores of sentences. Along with adding more nouns, verbs, and pronouns to sentence cores, we can also expand sentence equations into more complex structures using five "accessory" components—M_n = adjectives, M_v = adverbs, L = prepositions, C = conjunctions, and I = interjections—as shown in Fig. 4.1.

As writers, we use these additional components to further accessorize basic sentence machines. We can add modifier words and word groups to increase the specificity of sentence-level messages (add depth), and we can join words and word groups together to expand the complexity of sentence structures (add breadth). As with nouns, verbs, and pronouns, the additional variables also have a variety of features that affect how they're coded in sentence equations, as well as how they're written and punctuated in text sentences.

B. Henderson, *A Math-Based Writing System for Engineers*,
https://doi.org/10.1007/978-3-030-10756-7_4

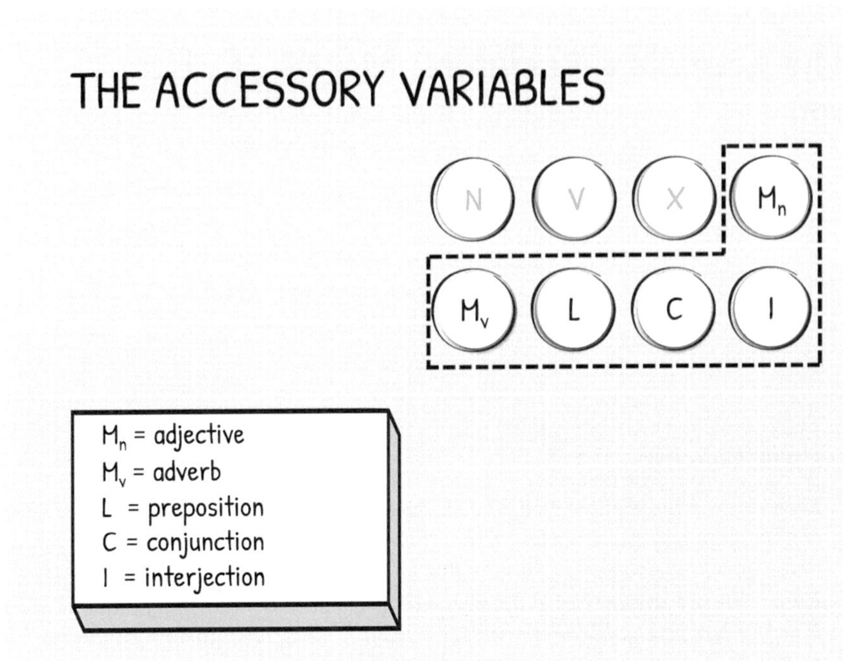

Fig. 4.1 The accessory variables representing five out of eight total parts of speech

M_n = Adjective

In excellent engineering writing, concision is an attribute of elegance. The "wow" factor comes from the information that sentence streams transfer from writer to reader, not from fancy stylistics. Our system's math analog embraces this attribute. Many great scientific equations require only a few variables—for example, $F = M * A$, $I = V/R$, and of course $E = M * C^2$. On the other hand, many scientific equations simply cannot be written accurately and precisely if they are constrained to a quota of two or three variables. The same holds true for sentence equations.

For instance, consider what might happen if a marine engineer were to focus predominantly on word count rather than precision when writing the following procedural step for replacing a critical relief valve in a nuclear submarine's ballast system:

To reattach the relief valve's flange,
install *several* bolts. **non-specific**

To reattach the relief valve's flange,
install *eight 4-cm-diameter, 20-cm-long, class-Z titanium* bolts. **specific**

In the above two versions, the procedural description that specifies *number*, *size*, and *type* of bolts instead of using the economical yet vague term *several* will most likely provide a safer, better-performing outcome if the valve fails and the sub's mechanic needs to replace it.

In either case, the descriptive words before the noun "bolts" function as *adjectives*—single words or word groups that *modify a noun*. The sentence algebra variable for adjective is "M_n"—"M" for modifier and "n" for noun. Since adjectives work as *specificity multipliers*, sentence algebra equations link adjective and noun pairings together with the multiplication symbol, "*". For example:

$M_n{}^*N$ = one adjective modifying one noun ("*tall* building")

$M_n{}^*M_n{}^*N$ = two adjectives modifying one noun ("*sturdy suspension* bridge")

Let's consider a variety of adjectives partnered with nouns. The *italicized* words and groups of words listed below are the *adjectives*; the non-italicized words and word groups function as nouns.

state-of-the-art process	*rough* surface finish
an erratic, faint signal	*frequent hands-on* PLC practice
big, green, plastic solvent tray	*nanoscale* capacitor
the rebar foundation grid	*a* rebar foundation grid
small (smaller or smallest) anomaly	*good (better or best)* fiscal quarter

The list above reveals several important features of adjectives in action.

Four Features of Adjectives in Action

1. *Hyphenating Compound Adjectives*: The first feature to notice in the preceding adjective-noun pairs is that most of them are $M_n * N$ structures, consisting of either a one-word adjective like "rough" or a multi-word, hyphenated cluster like "state-of-the-art" placed before a noun. These clusters function collectively as single *compound adjectives*. Hyphenating compound adjectives is a punctuation convention to help readers distinguish them from compound nouns in strands of text, since compound nouns, by opposite convention, are usually not hyphenated.

   Sometimes with compound words, both adjectives and nouns alike, there is room for interpretation. Case in point: should the column 2, line 1 entry be written "*rough-surface* finish" instead of "*rough* surface finish"? Technically, it could be either, although the latter is probably more common. In gray-area scenarios such as these, a good engineering writer must be decisive: you must choose how you would like the reader to decode a multi-word strand and then encode it—in this case, hyphenate it—accordingly.
2. *Punctuating Multiple Adjectives*: A second feature to note in the above adjective-noun pairings is that two of the nouns are modified with *multiple adjectives* that are separated by *serial commas*. A serial comma inserts (immediately after a

serial adjective) a very brief processing point that cues the reader to recognize and process a serial adjective's individual impact on the noun it modifies.

For instance, the column 1, line 2 example shows a singular noun modified by a series of two serial adjectives separated by a comma—"an *erratic, faint* signal." (Note: The "an" in front of "erratic" will be explained soon (4).) The column 1, line 3 example features a compound noun modified by a string of three *serial adjectives* separated by commas—"*big, green, plastic* solvent tray."

An alternative case is multiple adjectives with no commas, as displayed in column 2, line 2. Here, the compound noun "PLC practice" is modified by two adjectives ("frequent" and "hands-on) with no commas in between. Instead of a one-at-a-time process, this construction cues the reader to process the two-word word cluster as a *collective adjective* exerting collective impact on the noun it modifies—"*frequent hands-on*" (collective adjective) with "PLC practice" (compound noun). Note: In either case, serial or collective, sentence algebra equations code multiple adjectives the same way, "(M_n * M_n * N)," ignoring the commas.

So which practice is best when writing text that includes multiple adjectives—comma-separated series or continuous stream? The answer depends upon how you, the writer, want your reader(s) to decode that particular strand of text—series or stream.

3. *Degree of Comparison*: A third feature to observe about adjectives in action is *degree of comparison*. Columns 1 and column 2 line 5 show the adjectives "small" and "good," respectively, each cascading through three progressive degrees of comparison: *positive*, *comparative*, and *superlative*.

   "Small" showcases a regular progression that advances by adding "-er" and then "-est" to the adjective's root form. A regular progression like this can also advance using the words "more" (or "less") and "most" (or "least)—for example, "rigorous," "more rigorous," and "most rigorous." On the other hand, "good" showcases an irregular progression that advances with non-regular spellings just like an irregular verb. Note: Adjectives with irregular comparisons (just like irregular verbs) are relatively few but frequently used. As usual, a dictionary is the best source for clarifying questions about regular and irregular words.

4. *Articles*: The fourth feature revealed in the example display is that sentence algebra considers the *articles*—"the," "a," and "an"—to be a special type of M_n. The earlier example (displaying a variety of adjectives partnered with nouns) italicizes the three articles working as noun modifiers. The *definite article*, "the," denotes that a noun refers to a specific object in a set; the *indefinite articles* "a" and "an" both indicate that a noun refers to a non-specific object in a set. Thus, "*the* rebar foundation grid" (column 1, row 4) refers to a specific rebar foundation grid selected from a field of choices. "*A* rebar foundation grid" (column 2, row 4) means any rebar foundation grid (doesn't matter which) selected. The indefinite article "an" in "*an* erratic, faint signal" (column 1, row 2) works similarly and is used before a noun that begins with a vowel or a vowel sound.

   Although articles serve a specialized adjective function in text sentences and are essential to include wherever they are applicable, they are generally not coded in first-order sentence algebra equations (i.e., they are not assigned an "M_n"). For example, "*an erratic, faint* signal" would code as "$M_n * M_n * N$"—an M_n for "erratic" and another "M_n" for "faint," "N" for "signal," but no M_n for the leading "an." For both "*a* rebar foundation grid" and "*the* rebar foundation grid" (each a case of one article before one compound noun), the coding would, in both cases, be "N" for just the compound noun, "rebar foundation grid," with coding for the articles omitted. Note: When articles do appear in sentence algebra equations, the variable that represents them is "m_n."

In general, articles don't impact the basic syntax or core variable arrangements of the math models that this system identifies underneath English language sentences. Thus, leaving out articles simplifies a sentence algebra equation without affecting the equation's purpose of revealing the sentence's functional schematic.

Three Special Conventions for Articles

Despite their limited use in sentence algebra equations, articles cannot be considered invisible when you are building an engineering document out of text. When going from schematics to fully operational machines, articles must appear where they are needed to complete sentence-level messages and fulfill formal language conventions.

Beyond all adjectives' general features, there are three special conventions for articles that engineering writers should know and apply:

1. An article placed before another adjective(s) is never followed by a comma. That is why the "an" before "erratic" in the column 1, row 2 entry ("an erratic, faint signal") takes no comma even though the next adjective in the series, "erratic" before "faint," does.
2. For indefinite articles, whether to use "a" versus "an" depends case-by-case upon the partner noun. If the partner noun begins with a consonant sound, use "a." If the noun begins with a vowel sound, not necessarily vowel letter, use "an." For example, "a c̲athode" ("c" aloud is the consonant sound "cah"), "an a̲node" ("a" aloud is the vowel sound "ae"), and "an N-type silicone" ("N" aloud is the vowel sound "en," even though "n" is a consonant).
3. Some nouns are not compatible with articles at all. In this book, for example, I write about "sentence algebra," not "a sentence algebra" or "the sentence algebra." This convention is governed by *conventional phrasing*—that is, this system's term for native-speaker preferences and accepted idioms that influence correct phrasing in formal English. In Part II, Sentence Optimization, Chap. 12 further discusses articles and conventional phrasing. If you are multilingual and seek advice on how to improve your phrasing skills over time, here is where you will find it.

M_v = Adverb

There is a second type of modifier function that a word or word group can assume in a sentence—the role of *adverb*. In sentence algebra, we denote adverbs with the variable "M_v"—"M" as in modifier, "v" as in verb.

While adjectives add specificity to nouns, the main function of adverbs is to add specificity to an action expressed by a verb. Adverbs can also add specificity to adjectives and other adverbs. Oftentimes, this increase in specificity is driven in purpose by the journalist's prompts—"who?" "what?" "when?" "where?" "why?" and "how?"—or, as they are referred to in this book's system, "5W + H."

For centuries, the 5W + H prompts have helped journalists to flesh out articles and news stories to an appropriate level of detail and completeness while avoiding unnecessary tangents and fluff. I advise you, too, to consider and apply the 5W + H prompts, wherever applicable, when you are optimizing the specificity of descriptions and actions in your engineering writing.

Below is a series of six examples which plug a noun, a verb, and an adverb into the equation N + (V * M_v). All of the nouns (Ns) are multi-word clusters functioning as compound nouns. All verbs (Vs) in the examples are partnered with italicized adverbs (M_vs) answering 5W + H prompts (displayed on the right, also italicized):

the air pressure dropped *suddenly*	(dropped *how?*)
an Excimer laser ablates *well*	(ablates *how?*)
the micro orifice will clog *soon*	(will clog *when?*)
a network hub failed *catastrophically*	(failed *how?*)
the synthetic O-ring degraded *more rapidly*	(degraded by *what* degree?)
the company's market value has increased *significantly*	(has increased to *what* degree?)

Looking at the above adverbs in action, you can observe that adverbs frequently end in "-ly." Four of the six examples display this feature: "suddenly," "catastrophically," "rapidly," and "significantly." Two, on the other hand, do not: "well" and "soon." Also, just like adjectives, you should also note that most adverbs have the capacity to convey three degrees of comparison—*positive*, *comparative*, and *superlative*. For instance, in the fifth example, a writer could swap in either "rapidly," "more rapidly," or "most rapidly" for the M_v = adverb, depending on which degree best fits the purpose of that text.

In addition, you should note that sometimes adverbs add specificity to adjectives, for instance, "M_v * M_n * N" as in "very heavy strut." Most of the time in engineering writing, however, it is best to use adjectives that do not require modification. For example, it would be better to use a precise adjective and write "2.3-ton strut" rather than the less precise adverb-adjective combination, "very heavy strut."

L = Preposition

The next sentence algebra variable—*preposition*, represented by "L"—has a three-fold function:

1. It leads ("L") a cluster of words that centers around a noun or pronoun object and typically shows a relationship.
2. It acts like a flexible hinge that enables the cluster to attach to another sentence variable functioning as a noun, pronoun, or verb.
3. Once attached, the L-led cluster modifies the noun, pronoun, or verb unit as an adjective or adverb, respectively.

Below is a sampling of common prepositions. Most are short, single words. However, as shown in the last row, prepositions can also be compound structures of two or three words.

of	in	to	for
with	on	at	from
by	about	as	into
like	through	after	over
between	under	against	without
aside from	instead of	in accordance with	except for

By themselves, prepositions don't mean much. They are construction words rather than content words. However, when an L operates as leader and commander of a specialized left-to-right word sequence known as a *prepositional phrase*—the previously mentioned "cluster," involving at least two terms, an L plus an N or X—the result is a useful and versatile construction.

In a prepositional phrase (or "LP" in sentence algebra shorthand), the noun or pronoun following the preposition assumes an object role (N_o or X_o). The resulting net equation is "L+ (N_o or X_o)." Each of the following six examples is a prepositional phrase built using the preceding formula. To help distinguish the equation underneath each phrase's text, the preposition words are underlined (L = "to," "like," "at," and so on). The prepositions' objects are compound nouns, with a few articles added here and there. The nouns are also underlined (N_o = "the maintenance technician," "dampened resonance," "6 AM," and so on). As you review the examples, note the italicized journalist's prompt on the right of each listing and how it characterizes the type of specificity associated with the modifier phrase on the left.

<u>to</u> <u>the maintenance technician</u>	*whom?*
<u>like</u> <u>dampened resonance</u>	*what?*
<u>at</u> <u>6 AM</u>	*when?*
<u>under</u> <u>the chrome plating</u>	*where?*
<u>for</u> <u>customer satisfaction</u>	*why?*
<u>with</u> <u>a statistical process control tool</u>	*how?*

The above six samples are all soundly constructed prepositional phrases. Yet most of these phrases also allow a range of possible variations and expansions. For instance, the constructions could feature pronouns rather than nouns as objects. Instead of "to *the maintenance technician*" ($L + N_o$), the first example could be "to *him or her*" ($L + X_o$). Or, we could add adjectives. Instead of "under *the chrome plating*" ($L + N_o$), the fourth example could be "under *the 0.1-millimeter-thick chrome plating*" ($L + (M_n * N_o)$ with the compound adjective M_n = "0.1-millimeter-thick" modifying the compound noun N = "chrome plating"). There are numerous possibilities.

The next example—excerpted from a qualification test document for a new V-8 engine—examines several prepositional phrases in action in a single sentence and shows how they are coded in a corresponding sentence equation. The example also illustrates the flexible positioning that prepositional phrases display when they are inserted into sentence equations and attach to other sentence variables.

In the following sentence, there are two prepositional phrases modifying two compound nouns, with both phrases positioned immediately after the nouns they modify. A third prepositional phrase at the end of the sentence modifies the verb in the middle.

The water pump subassembly at 65 ± 5 rpm rotor speed
must produce a flow rate of 18 ± 0.5 gpm at ≥125 psi steady-state pressure.

To unpack functional layers, let's first look at how the sentence appears stripped down to core nouns and verbs:

The water pump subassembly must produce a flow rate. [core text]

$$N_s + V + N_o \qquad \text{[core equation]}$$

where (excluding articles):

N_s = *water pump subassembly (compound noun)*
V = *must produce (verb)*
N_o = *flow rate (compound noun)*

Next, we add back the additional details that ensure the sentence's technical accuracy. These details are contained in prepositional phrases working as specificity modifiers. Below is the complete, original text along with a full-blown sentence algebra equation that accounts for and defines each element's functional role. Since the prepositional phrases function as modifiers, this equation represents them as Ms (i.e., either M_n or M_v variables). Why one of the terms is coded " . . . $[M_v]^E$" will be explained after the example.

The water pump subassembly *at 65 ± 5 rpm rotor speed* must produce a flow rate *of 18 ± 0.5 gpm* *at ≥ 125 psi steady-state pressure.* [full text]

$$(N_s * M_{n\,1}) + (V) + (N_o * M_{n\,2}) \;\ldots\; [M_v]^E \qquad \text{[full equation]}$$

where the prepositional phrases operating as Ms are

$M_{n\,1}$ = *at 65 ± 5 rpm rotor speed*	[adjectival phrase]
$M_{n\,2}$ = *of 18 ± 0.5 gpm*	[adjectival phrase]
$[M_v]^E$ = *at ≥ 125 psi steady-state pressure*	[adverbial phrase]

and the core variables (excluding articles) are

N_s = water pump subassembly	[compound noun]
V = must produce	[verb]
N_o = flow rate	[compound noun]

As shown above, the complete sentence algebra equation begins with a subject noun (N_s) modified by an adjectival prepositional phrase ($M_{n\,1}$) located immediately to the right of the subject noun ($N_s * M_{n\,1}$). Next, after the subject elements, the formula adds a verb (V) with no modifier next to it. Thereafter, the formula adds an object noun (N_o) modified by an adjectival prepositional phrase ($M_{n\,2}$) positioned right ($N_o * M_{v\,2}$), same as the subject noun and its prepositional phrase modifier. Last, we encounter what sentence algebra codes as a *free-floating, end-loaded modifier*, or "… $[M_v]^E$." We will study free-floating modifiers more in the next chapter on basic sentence equations.

Here, the free-floating modifier represents the third prepositional phrase, "at ≥125 psi steady-state pressure," which modifies the verb (V) in the middle, "must produce," and specifies "*what is* … the pressure output criterion." Prepositional phrases (and adjectives and adverbs) that function as modifiers can move around in sentence equations. Sometimes they attach directly to other sentence variables positioned immediately on their right, and sometimes the order switches, and they multiply "*" specificity on the left. For instance, another version of our example could begin "*At 65 ± 5 rpm rotor speed,* the water pump subassembly…" ($M_{n\,1} * N_s$) and still be a correct sentence. Other times, as illustrated, prepositional phrases tack onto the end of sentences as free-floating modifiers. They can be attached to the beginnings of sentences too.

C = Conjunction

The next part of speech discussed in this chapter is the *conjunction*, denoted by the sentence algebra variable "C." A word or word group playing the conjunction role in a strand of text functions as a connector between items of equal or unequal importance or ranking. The equal-to-equal Cs are the *coordinating conjunctions*—such as "and," "but," and "or." The Cs that join unequally weighted pairings are the *subordinating conjunctions*. Some examples of these are "if," "although," "though," "when," "while," "as," "since," and "because."

Coordinating and subordinating are the two major conjunction types, and they are the focus of this segment. We will look at other types of connector words in subsequent chapters.

When conjunctions connect words and word groups in a sentence equation, the words being connected can play a variety of roles, including N, V, M_n, and M_v. Conjunctions also connect multivariable constructions—for example, a prepositional phrase L + (N_o or X_o) linked together with another prepositional phrase or a clause (N_s or X_s) + V joined with another clause.

Punctuation Conventions for Coordinating and Subordinating Conjunctions

When conjunctions connect clauses, as well as phrases and words, there are four punctuation conventions that apply:

1. *Comma Before Coordinating Conjunction Joining Two Independent Clauses*: Always insert a comma before a coordinating conjunction joining two *independent* clauses (i.e., clauses that can function as stand-alone sentences)—for example, "the gear rotated, *and* the shaft rotated too." Note that, not coordinated, we could also correctly write the two clauses as two stand-alone units: "The gear rotated. The shaft rotated too."
2. *Comma After Subordinate Clause That Precedes an Independent Clause*: Always insert a comma after a subordinate clause (i.e., a clause beginning with a subordinating conjunction that cannot function as a stand-alone sentence) when that clause *precedes* an independent clause, for example, "*although* he applied the brake, the wheels continued turning." In this case, we could alternatively write the independent clause as a complete sentence, "The wheels continued turning." On the other hand, the subordinate clause "Although he applied the brake" could not function as a stand-alone structure.
3. *No Comma Before Subordinate Clause Following Independent*: Generally, do *not* place a comma before a subordinate clause when it *follows* an independent clause—for example, "he rebuilt the brake system *because* it wasn't working." Once again, you could write the independent clause as a complete sentence, "He rebuilt the brake system." However, the subordinate clause "Because it wasn't working" cannot stand alone.
4. *No Comma for Two Words or Two Phrases Joined with a Coordinating Conjunction*: Do not place a comma before a coordinating conjunction when it connects two words—for example, "gear *and* shaft." Also, do *not* use commas when two phrases pair together with a coordinating conjunction—for example, "on land *or* on the sea." (Note: Subordinating conjunctions do not regularly connect word and phrase pairs, and when they do, punctuation is determined case by case—for example, "large, though lightweight.")

The above punctuation conventions do *not* necessarily apply to series of three or more clauses, phrases, or words. Punctuating series of three or more items (Oxford commas) will be covered in Part II, Sentence Optimization, Chap. 10.

Examples of Coordinating and Subordinating Conjunctions in Action

Coordinating Conjunctions

Let's examine a series of examples showcasing *coordinating conjunctions* in action—first joining *variable and variable* (N and N, V and V, etc.), then *phrase and phrase*, and finally *clause and clause*. All items in the series below are displayed two different ways—sentence algebra version on the left and text on the right.

The initial entries of each series (separated by "...") establish an equation pattern that morphs into shorthand symbols: first, "$\mid$" replaces "$+ C_{(coordinating)} +$", the variable for a connection made with a coordinating conjunction; next, "LP" replaces "$(L + (M_n * N_o)$", the equation for a *prepositional phrase*; and then finally "B" replaces "$(N_s + (V * M_v))$", the equation for a *basic clause*.

Sentence algebra	**Text**
...	...
$N + C_{(coordinating)} + N$	statics *and* dynamics
$V \mid V$	shimmies *but* advances
$M_n \mid M_n$	efficient *or* inefficient
$M_v \mid M_v$	fast *and* furiously
$X \mid X$	we *or* they
...	...
$(L + (M_n * N_o) \mid (L + (M_n * N_o)$	in the 3-sigma zone *and* near the curve's tail
$LP \mid LP$	under the stealth radar *but* at supersonic speed
...	...
$(N_s + (V * M_v)) \mid (N_s + (M_v * V))$	the rocket flared brightly, *and* the booster quickly detached
$B \mid B$	paint was applied immediately, *but* the flange still corroded
...	...

In reading through the examples above, after the first five routine variable-and-variable examples, you most likely observed that for the phrase-and-phrase examples (examples 6 and 7), as well as for the clause-and-clause examples (examples 8 and 9), each standard equation is expanded to include a specificity multiplier—an M_n (adjective) or M_v (adverb).

Subordinating Conjunctions

Next let's examine a series of *subordinating conjunctions* in action, joining units that are unequal in importance. Similar to the previous series showcasing coordinating conjunctions in action, these examples display both the sentence algebra and text versions for each listing.

Once again, you will notice some sentence algebra shorthand notation as you read through the listings. For a connection placing the item that is *less than* in importance on the right (">"), the coding is "$> C_{(subordinating)} +$", and the shorthand notation is the forward slash "/". Conversely, for a connection placing the *less than* item on the left ("<"), the coding is "$+ C_{(subordinating)} <$", and the shorthand notation is backward slash "\". Put another way, whether "/" or "\", the slash leans in the direction of the subordinate clause.

The two columns below display two sets of examples, with each example joining together two basic clauses (two Bs) in various ways. The first two examples place the main clause on the left and the subordinate clause on the right. The second two examples place the subordinate clause on the left and main clause on the right.

Sentence algebra	**Text**
...	...
$(N_s + V) > C_{(subordinating)} + (N_s + V)$	the actuator chatters *when* the oil orifice clogs
B / B	the press-fit bearing will fail *because* it will crack
...	...
$(N_s + V) + C_{(subordinating)} < (N_s + V)$	*when* the test station works, the output improves
B \ B	*if* pneumatic pressure peaks, the relief poppet opens
...	

The above text examples of subordinating conjunctions in action reveal a nuance worth mentioning. Unlike coordinating conjunctions, which are always placed *between* the clauses they connect, a subordinating conjunction is always placed *at the beginning of the subordinate clause*, regardless of that clause's placement (before or after) relative to the independent (principal) clause with which it pairs.

Coordinating conjunction

pneumatic pressure peaks *and* the relief poppet opens	*equal-equal*
the relief poppet opens *and* pneumatic pressure peaks	*equal-equal (flipped)*

Subordinating conjunction

the relief poppet opens *if* pneumatic pressure peaks	*principal-subordinate*
if pneumatic pressure peaks, the relief poppet opens	*subordinate-principal*

Review of Full Code versus Shorthand Symbols for Conjunctions

coordinating conjunction — “$+ C_{(coordinating)} +$” = “|”

subordinating conjunction—main clause on left and subordinate clause on right — “$> C_{(subordinating)} +$” = “/”

subordinating conjunction—subordinate clause on left and main clause on right — “$+ C_{(subordinating)} <$” = “\”

I = Interjection

In a sentence equation, an *interjection* is a word or group of words that expresses a *short burst of emotion*. Sentence algebra denotes the eighth variable with an “I.” Although this book’s system recognizes all eight parts of speech, this discussion of the eighth variable is very brief because interjections are rarely used in engineering writing.

What are examples of words that function as interjections? Here, just for the record, are some of my own personal favorites, which I use freely (and for fun) when I am *not* doing engineering writing:

hey	*yeah*	*oh my*
cool	*wow*	*bummer*
whatever	*yep*	*right-on*
ah-hum	*yo*	*alrighty then*

Omitting interjections in engineering writing has nothing to do with the engineering profession being somehow devoid of “emotional flare.” It is a matter of style and convention. In engineering writing, we strive to avoid subjective, qualitative words and instead try to express our technical hurrahs and thrills using objective and quantitative phrasing.

Action Items

α	Consider the vocabulary words you use regularly as an engineering writer. Within this vocabulary, what are the three adjectives you most commonly use or find most useful? What are your three (most commonly used, ostensibly most useful) adverbs? Which of your favorite adverbs end in “-ly”? The next few times you use your favorite modifier words in sentences, consider their value, and delete them (especially “-ly” adverbs) if their purpose is more extraneous than useful.

(continued)

α	After you write your next email, before you press send, search through the text for a prepositional phrase, and then determine what function it plays in the sentence. Also consider whether you could use a single-word adjective or adverb in its place to achieve a similar or improved meaning.
α	In another future email, before you press send, search through the text for coordinating and subordinating conjunctions. Determine where they connect words, where they connect prepositional phrases, and where they connect basic clauses. In places where coordinating conjunctions connect two or more basic clauses, consider what would happen if you eliminated the connection and revised long sentences into two or more short sentences? Would this maneuver make the text choppy, or would it disentangle a convoluted log jam of words and make the text quicker and easier to read?

Recap

Key Learnings/Takeaways

- The sentence algebra variable for **adjective** is $\mathbf{M_n}$, and an adjective's function is to modify a noun. In sentence algebra equations, adjectives are considered "specificity multipliers" and are coded as such: $M_n * N$, $M_n * M_n * N$, etc.
- The four key features impacting appropriate and correct usage of adjectives in text equations are the following:
  - **Hyphenation** can be used to distinguish **compound adjectives** from compound nouns when adjectives and nouns pair together; thus, engineering writers must be decisive about the intended meaning of word clusters representing adjectives vs. word clusters representing nouns and punctuate accordingly.
  - **Serial commas** distinguish a string of **serial adjectives** from a multi-adjective cluster; thus, engineering writers should be decisive about their intended meaning of one vs. the other and punctuate accordingly.
  - **Degree of comparison** signifies an adjective's magnitude of impact; these variations in degree are signaled by variations in the adjective's form.
  - The sentence algebra system considers the **articles** "the," "a," and "an" to be a special type of adjective; however, sentence algebra equations generally treat articles as elliptical (i.e., invisible) components, and thus do not code them as M_n variables. Occasionally, when an article is coded in a sentence algebra equation, it's assigned the special variable "$\mathbf{m_n}$."
- The three special conventions associated with using articles in text and sentence equations are the following:
  - An article placed before another adjective(s) never takes a comma, no matter whether the following adjective(s) is singular, compound, collective, or serial.
  - The indefinite article "a" partners with nouns that begin with a consonant sound, and the indefinite article "an" partners with nouns that begin with a vowel sound—"a," "e," "i," "o," or "u."

  - Some nouns are not compatible with articles at all—for example, “sentence algebra is a new writing system” not “a sentence algebra is a new writing system.”

- The sentence algebra variable for **adverb** is **M_v**, and its usual function is to modify a verb. In sentence algebra equations, adverbs are considered “specificity multipliers” and are frequently (though not always) coded as $V * M_v$ or, sometimes, when modifying adjectives, as $M_v * M_n * N$.
- Often, the increase in specificity associated with an adverb is driven in purpose by the journalist’s prompts—“who?” “what?” “when?” “where?” “why?” and “how?” (5W + H). Although adverbs serve a useful purpose in engineering writing, it is best to use them judiciously and sparingly.
- The sentence algebra variable for **preposition** is **L**. Prepositions are construction words (rather than content words) that “**lead**” or head-up **prepositional phrases**. A basic phrase consists of a lead L followed by core noun object or pronoun object (N_o or X_o). The resultant sentence algebra equation is **$L + (N_o$ or $X_o)$**, which can be expanded a variety of ways with specificity modifiers. The sentence algebra shorthand for a prepositional phrase is **LP**.
- The sentence algebra variable for a **conjunction** is **C**. A conjunction’s function in sentence equations is to form a variety of connections between words, phrases, and clauses. The two major and most frequently used conjunction types are **coordinating** and **subordinating**. The former connects items of equal importance or ranking in sentence equations, and the latter connect items unequal in importance or ranking.
- When conjunctions connect clauses (and words or phrases), several punctuation conventions apply:

  - Place a comma before a coordinating conjunction when it joins two independent clauses.
  - Place a comma after a subordinate clause that precedes an independent clause.
  - Do **not** place a comma before a subordinate clause that follows an independent clause.
  - Do **not** place a comma before a coordinating conjunction joining two words or two phrases.

- In sentence algebra equations, coordinating conjunctions code in long form as “$+ C_{(coordinating)} +$” and in short form as “|”. Subordinating conjunctions code in long form as “$+ C_{(subordinating)} >$” (for independent-subordinate pairs) or “$< C_{(subordinating)} +$” (for subordinate-independent pairs) and in short form as “/” or “\”, respectively.
- The sentence algebra variable for **interjection** is **I**. An interjection expresses a short burst of emotion in a word(s). Though sometimes useful in other forms of communication, interjections are not generally considered to be useful or stylistically appropriate in engineering writing. Subsequent chapters do not include them.

Chapter 5
Basic Sentence Equations: B_1, B_2, and B_3

Learning Objectives

- *Know how to expand the basic sentence template ($S + P_n$) into B_1, B_2, and B_3 sentence equations.*
- *Know what an intransitive verb is, what a transitive verb is, and what a linking verb is.*
- *Know how to scan and code B_1, B_2, and B_3 sentence equations using the sentence algebra coding conventions.*
- *Understand the concept of "flow" and how it operates in a B_2 sentence.*
- *Understand the concept of "two-way flow" and how it operates in a B_3 sentence.*
- *Know what a subject complement is and what its various types are.*

Basic Sentence Equation One (B_1)

You already have some familiarity with basic sentence equation one (B_1). At its core is the most basic of basic clauses, the spark-producing noun and verb pair, "$N_s + V$" (see Fig. 5.1).

The difference between the B_1 core version and the full-blown B_1 sentence formula (shown below) is that the latter adds variation and specificity options. For instance, the expanded version allows the B_1 sentence subject to be either a subject noun (N_s) or pronoun (X_s), both having the capacity to be made more specific with words and word groups functioning as adjectives (M_ns).

These adjectives can number zero (no adjectives) to many adjectives. In typical engineering writing, the number of adjectives modifying a noun usually range from zero to three. The expanded equation also allows the B_1 verb (V_i) to be made more specific with words or word groups functioning as adverbs (M_vs). These adverbs can number zero to many adverbs. Although, once again, the formula provides the option to include numerous modifiers, real-life B_1 sentences typically multiply specificity of the main verb with just one or two adverbs, or none at all.

B. Henderson, *A Math-Based Writing System for Engineers*,
https://doi.org/10.1007/978-3-030-10756-7_5

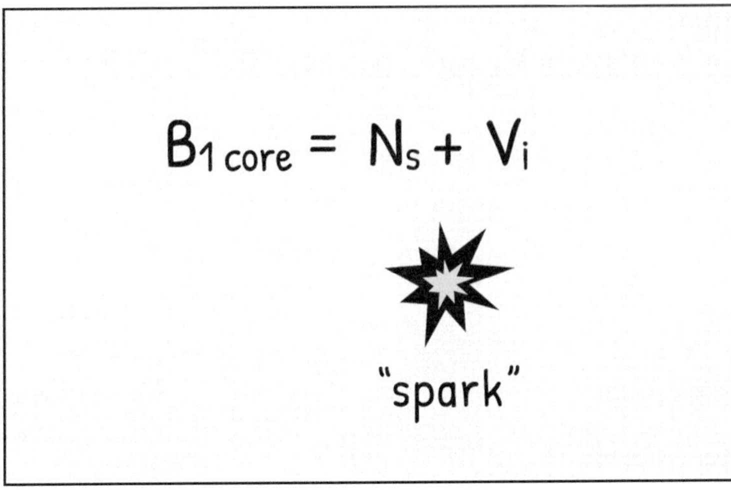

Fig. 5.1 Basic sentence equation one (B_1) core formula for sentence algebra

Note that the B_1 verb is tagged with the subscript "i" which stands for "intransitive." All verbs express action. *Intransitive verbs* are a subcategory of verbs that *express actions in-and-by-themselves* and that *take no objects*. An example of an intransitive verb (V_i) would be "glow," as in "The fluorescent light bulbs glow." This sentence expresses a complete unit of thought. The agent of the action is "the fluorescent light bulbs," and the autonomous, intransitive action is "glow."

Take a moment now to review the fully accessorized B_1 equation and its key components.

$$\mathbf{B_1} = \mathbf{S} + \mathbf{P_1}$$

$$= (\mathbf{M_{n\,a}} * (\mathbf{N_s} \textbf{ or } \mathbf{X_s})) + (\mathbf{M_{v\,b}} * \mathbf{V_i})$$

where: $S = M_{n\,a} * (N_s \text{ or } X_s)$

$P_1 = M_{v\,b} * V_i$

V_i = an *intransitive verb* that expresses action in-and-by-itself and *does **not** transmit* action onto an object

N_s or X_s = a noun or pronoun *subject* of the verb

$M_{n\,a} = \{(M_{n\,0}) * (M_{n\,1}) * (M_{n\,2}) * (M_{n\,3}) \ldots\}$

a series of *adjectives* modifying a noun, numbering anywhere from 0, no adjectives, to multiple adjectives, 1, 2, 3, and so on

$M_{v\,b} = \{(M_{v\,0}) * (M_{v\,1}) * (M_{v\,2}) * (M_{v\,3}) \ldots\}$

a series of *adverbs* modifying a verb, numbering anywhere from 0, no adverbs, to multiple adverbs, 1, 2, 3, and so on

Now that you are acquainted with B_1's components and structure, let's take the equation out for a test drive. The next segment examines five applied demonstrations (demos), 5.1 through 5.5. Each presents a B_1 text sentence accessorized in various ways and then shows how it scans (left-to-right) and codes into a full-blown B_1 equation. The examples represent first-order disassemblies, where the text sentences are divided into their main functional components. Thus, we code word groups that operate cohesively as wholes—such as prepositional phrases (LPs) that modify—as singular items.

Relative to the complete B_1 equation's standard arrangement of variables, which places all specificity multipliers immediately before the terms they modify—that is, $(M_{n\,a} * (N_s \text{ or } X_s)) + (M_{v\,b} * V_i)$—the five applied demonstrations of B_1 display some rearrangements of modifiers. Most are sideways shifts, from multiplication on the left to multiplication on the right, and combinations thereof. Additionally, you will see that demo 5.5 introduces a new sentence algebra term, "$[M]^F$," the *front-loaded, free-floating modifier*. The discussion following demo 5.5 further explains how this variation of an adjective "$[M_n]^F$" or adverb "$[M_v]^F$" works.

Five Applied Demonstrations of B_1

Coding Demonstration 5.1

Original: The lightweight robotic arm swivels quickly.

	M_n	N_s	V_i	M_v
Scanned:	The lightweight	robotic arm	swivels	quickly.

Coded: $(M_n * N_s) + (V_i * M_v)$

Discussion: The core variables in sentence 5.1 are the noun subject, N_s = "robotic arm," and intransitive verb, V_i = "swivels." One adjective, "[The] lightweight," is located in standard position immediately to the left of the noun subject it modifies and specifies *how much* the "robotic arm" weighs $(M_n * N_s)$. One adverb, "quickly," located in an alternate position immediately to the right of the verb it modifies, specifies *how fast* the verb's action "swivels" will occur $(V_i * M_v)$.

Coding Demonstration 5.2

Original: The causeway over the flood zone might collapse from the traffic load.

Scanned:

N_s	$LP(M_n)$	V_i	$LP(M_v)$
The causeway	over the flood zone	might collapse	from the traffic load.

Coded: $(N_s * LP(M_n)) + (V_i * LP(M_v))$

Discussion: Sentence 5.2 also features two modifiers, one for each of the two core variables. An adjectival prepositional phrase, $LP(M_n)$ = "over the flood zone," is positioned immediately after the noun subject, "[The] causeway," and specifies *where* the causeway is located $(N_s * LP(M_n))$. The adverbial prepositional phrase $LP(M_v)$ = "from the traffic load," positioned immediately after the noun subject, "[The] causeway," specifies *why* the action will occur $(V_i * LP(M_v))$.

Coding Demonstration 5.3

Original: Without proper setup and calibration, the new pick-and-place station will drift out of tolerance during steady-state production.

Scanned:

$LP(M_n)$	M_n	N_s
Without proper setup and calibration,	the new	pick-and-place station

V_i	M_v	$LP(M_v)$
will drift	out-of-tolerance	during steady-state production.

Coded: $(LP(M_{n\,1}) * M_{n\,2} * N_s) + (V_i * M_{v\,1} * LP(M_{v\,2}))$

Discussion: Sentence 5.3 is a B_1-type construction with multiple terms modifying each core variable. Two noun modifiers, "Without proper setup and calibration" and "[the] new" (both adding specific details about condition/age) modify the noun subject, N_s = "pick-and-place station," in a standard, modifiers-on-the-left arrangement $(LP(M_{n\,1}) * M_{n\,2} * N_s)$. Two verb modifiers, "out-of-tolerance" and "during steady-state production," modify the verb, V_i = "will drift", on the right. One adverb specifies *how* and the other *when* the station will drift $(V_i * M_{n\,1} * LP(M_{v\,2}))$. Here, sentence algebra notation adds numbers to the modifiers' subscripts to identify relative position of the two adjectives and two adverbs.

Before moving on to sentence 5.4, I should note that for sentence 5.3, there are other functional interpretations of the leading phrase, "Without proper setup and calibration." The above example codes it as an $LP(M_n)$ adjective characterizing the condition of the "pick-and-place station." However, the phrase could also be seen as an $LP(M_v)$ adverb characterizing the *cause* of the core verb's action "will drift" or even the whole message expressed by the clause "the new pick-and-place station will drift out-of-tolerance during steady-state production." All three of these interpretations are reasonable. I chose the one that is the most straightforward to code.

Coding Demonstration 5.4

Original: Nervous at the P.E. Exam, she finally finished anyway.

	$M_{n\,1}$	$LP(M_{n\,2})$	X_s	$M_{v\,1}$	V_i	$M_{v\,2}$
Scanned:	Nervous	at the P.E. Exam,	she	finally	finished	anyway.

Coded: $(M_{n\,1} * LP(M_{n\,2}) * X_s) + (M_{v\,1} * V_i * M_{v\,2})$

Discussion: Two adjectives, "Nervous" (specifies condition of the subject, *what is ...?*) and "at the P.E. Exam" (specifies *where is ...?*), modify sentence 5.4's subject, the pronoun X_s = "she". The adjectives are arranged in uniform, multiply-on-the-left sequence (codes: "$M_{n\,1} * LP(M_{n\,2}) * X_s$"). The predicate's equation introduces another acceptable arrangement of modifiers, where one (or more) modifier multiplies a core element (N, X, or V) on the left and another (or more) multiplies that same core element on the right. In this case, one adverb, "finally" (characterizes *when* the action occurs), is located before the main verb (core element) "finished," and another adverb "anyway" (characterizes *why* and *how so* the action occurs) is located after $(M_{v\,1} * V_i * M_{v\,2})$.

Thus far, in all four of our B_1 sentence demos, we have applied the system's basic coding conventions. Along with not coding articles ("the," "a," "an"), we also have not coded the original text sentence's initial cap letter and terminal punctuation mark (i.e., the period "." at the end). You also may have noticed that we have not coded any other internal punctuation marks—in particular, the comma (",") setting off introductory words in sentence 5.4 above ("Nervous at the P.E. Exam, ...") as well as in sentence 5.3 before ("Without proper setup and calibration, ..."). The next demo, sentence 5.5, also contains an introductory comma omitted in its sentence equation.

For all three of these sentences (5.3, 5.4, and 5.5), if instead we were going from equation to text (sentence cooking), or if we were composing these sentences in an engineering document, the sentences' text versions would require introductory commas to be considered correct, according to punctuation conventions. If you are interested in learning more about text-sentence punctuation, you will find this topic covered in Part II—Sentence Optimization's Chap. 11.

Coding Demonstration 5.5

Original: Once-a-year, we traveled to the offshore facility.

	$[M_v]^F$	(X_s)	V_i	$LP(M_v)$
Scanned:	Once a year,	we	traveled	to the offshore facility.

Coded: $[M_{v\,1}]^F \ldots (X_s) + (V_i * LP(M_{v\,2}))$

Discussion: The fifth demo sentence does not code as neatly as the previous four. The text-to-equation coding does follow the basic convention of same variable order as word order. But as you can see, 5.5's sentence equation displays a bracketed variable in front, $[M_{v\,1}]^F$, and this variable is *not connected* to the rest of the equation with an arithmetic operator, "+" or "*". Instead, the equation incorporates the symbol "...", which in sentence algebra notation stands for *link to free-floating modifier*. Here, the bracketed variable is a free-floating adverb tagged with the superscript "F" (for "front").

Placed where it is, the function of the bracketed adverb, "Once a year," does not offer a logical interpretation that maps back to the standard B_1 equation template (presented at the beginning of this segment), if we apply our math analog and the Commutative Law ($a * b = b * a$). If we were to code "Once a year" as a standard position adjective, the result, $M_n * X_s$, does not make sense. There is no such thing as a "Once a year [,] we." If we consider "Once a year" to be an adverb (which it is), there is no verb, other adjective, or other adverb butted up against for it to *multiply*, that is, to code as $M_v * (V, M_n,$ or $M_v)$. Instead, it links remotely to the verb in the middle, $V_i =$ "traveled," and specifies *when.*

This is one of those situations where *natural syntax*—that is, the natural order of words in English sentences—and our system's math analog and sentence equations diverge. As our sentence algebra lessons progress, you will see that free-floating, *front-loaded* modifiers, especially adverbs, appear frequently in English sentences. Free-floating, *end-loaded* modifiers appear regularly, too. The sentence algebra system accounts for this variation by offering a specialized protocol for recognizing and coding this type of variation. Some examples of "$[M]^E$" terms ("E" as in *end*) appear in the next segment's examples.

Sentence Engineering Tip

Up to this point, we have regularly referred to the central spark-producing equation, $N_s + V$, as a basic clause. We have also now pointed out that it is the basic core (if we distinguish V as "V_i") of basic sentence equation one (B_1).

(continued)

So what's the difference between a basic clause coded "$N_s + V_i$" and a basic equation coded "$N_s + V_i$"? At the sentence algebra level, a stand-alone (independent) clause equation and stand-alone basic sentence equation are the same. However, a basic text sentence framed with the equation $N_s + V_i$— for example, "The fluorescent light bulbs glow."—must be typed as a sentence to be a sentence, with an initial cap letter and terminal punctuation mark. When typed without that initial cap and final period, it is still a clause.

Basic Sentence Equation Two (B_2)

At its core, *basic sentence equation two* contains a "spark"-producing $N_s + V$ pair—just like basic sentence equation one. What's different about the B_2 structure is that its main verb is transitive rather than intransitive. Instead of drawing on resultant "spark" to express action in-and-by-itself, like B_1's intransitive "V_i" verb, a B_2 *transitive verb*—or "V_t"—does more. In addition to forming a dynamic partnership with the subject noun, a V_t also engages in a secondary pairing that *transfers* verb action onto a *noun object*, or N_o. Figure 5.2's illustration of the B_2 core shows this transfer with an arrow and labels it as "flow." In sentence algebra, *flow* refers to a dynamic relationship that *extends beyond basic $N_s + V$ "spark" to and between other variables in a sentence*, usually nouns and pronouns.

A specific example of a transitive verb transferring action to an object would be V_t = "*built*" (the action/doing) and N_o = "*a prototype*" (the result/receiving). If we add an action doer (N_s = "[The] technician"), a B_2 sentence would be "The technician *built a prototype*."

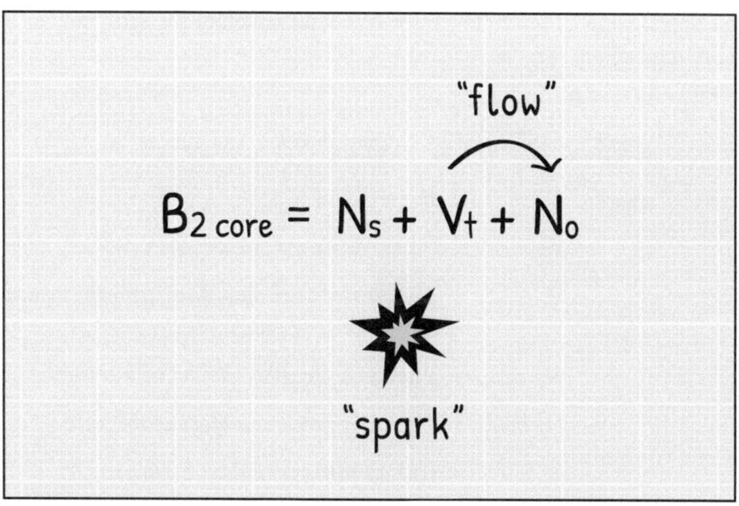

Fig. 5.2 Basic sentence equation two (B_2) core formula for sentence algebra

Another feature to point out about B_2's "power plant" and "frame" is that the subject noun and object noun represent two different grammatical cases. Chapter 3 introduced the *property of case* when it discussed the functional roles of nouns and pronouns. To review, when a noun or pronoun presides over a verb as the agent/doer of the verb's action, it is considered to be in the *subjective case*; conversely, when a noun or pronoun receives a verb's action, it is in the *objective case*. Case does *not* inflect (or change) a noun's spelling (except for adding an *'s* for the possessive case), but case *does* affect pronoun inflection (for instance, "she," "her," and "hers"), and we will consider this when we examine the five B_2 demonstrations.

Sentence Engineering Tip

Some verbs can function both as a transitive verb that transfers action onto an object *and* as an intransitive verb that expresses action in-and-by-itself. In the sentence "The test driver *shifted* gears," the verb *shifted* is transitive because the driver *did something* to the gear shift lever that causes a gear ratio change. However, in the sentence "Tectonic plates *shift*," the verb "shift" is intransitive because the plates themselves are shifting around; they are not transferring shifting action onto something else. A dictionary, in addition to recording the other already-mentioned features and qualities of the words it lists, *also* indicates whether a verb is transitive (often noted as "t") and/or another type of verb (such as an intransitive verb, often noted as "i").

Below is the complete B_2 sentence formula, with provisions for added modifiers and pronoun replacements. Take a moment to review the B_2 equation and all of its parts. As compared to B_1's full coding in the previous segment, the recipe for B_2's S = subject is the same. The main difference is that B_2's predicate adds one more term, an object noun or pronoun (N_o or X_o), with potential for zero, one, or several specificity multipliers ($M_{n\,a}$)s.

$$\mathbf{B_2} = \mathbf{S} + \mathbf{P_2}$$

$$= (\mathbf{M_{n\,a}} * (\mathbf{N_s} \textbf{ or } \mathbf{X_s})) + ((\mathbf{M_{v\,b}} * \mathbf{V_t}) + (\mathbf{M_{n\,a}} * (\mathbf{N_o} \textbf{ or } \mathbf{X_o})))$$

where: S $= M_{n\,a} * (N_s \text{ or } X_s)$

P_2 $= (M_{v\,b} * V_t) + (M_{n\,a} * (N_o \text{ or } X_o))$

V_t = a *transitive verb* that ***does*** *transmit* action onto an object

N_s or X_s = a noun or pronoun *subject* of the verb

N_o or X_o = a noun or pronoun *object* of the verb

$M_{n\,a}$ $= \{(M_{n\,0}) * (M_{n\,1}) * (M_{n\,2}) * (M_{n\,3}) \ldots\}$

a series of adjectives modifying a noun, numbering anywhere from 0, no adjectives, to multiple adjectives, 1, 2, 3, and so on

$$M_{v\,b} = \{(M_{v\,0}) * (M_{v\,1}) * (M_{v\,2}) * (M_{v\,3}) \ldots\}$$

a series of adverbs modifying a verb, numbering anywhere from 0, no adverbs, to multiple adverbs, 1, 2, 3, and so on

The next five demonstrations showcase B_2 sentences with transitive verbs in action. Whereas the B_1 examples featured a front-loaded adverb in demo 5.5, these demos feature a couple of sentences with free-floating modifiers appearing at the sentences' *ends*. The examples that include the end-loaded modifiers (5.9 and 5.10), in addition to displaying the original text-into-equation versions, also display alternate versions (tagged "§") that reconfigure the sentence equations, so that the free-floating modifiers are repositioned directly next to the term they modify. As you will see, the reconfigured sentences are not improvements upon the originals. In fact, they serve as excellent examples of the English language being steered tidily by *natural syntax*, rather than standardized equations.

Five Applied Demonstrations of B_2

Coding Demonstration 5.6

Original: The arc welder mended a broken truss.

Scanned:

N_s	V_t	M_n	N_o
The arc welder	mended	a broken	truss.

Coded: $(N_s) + (V_t) + (M_n * N_o)$

Discussion: Demo sentence 5.6 conforms exactly with the standard B_2 equation template. The core is N_s ("[The] arc welder") + V_t ("mended") + N_o ("[a] truss"), and it is accessorized with one adjective, "broken," which is situated immediately before the noun object it modifies, that is, "[a] broken truss" (coded: $M_n * N_o$).

Coding Demonstration 5.7

Original: An out-of-calibration tool milled out-of-spec flats on the aluminum blocks.

Scanned:

$M_{n\,1}$	N_s	V_t	$M_{n\,2}$	N_o
An out-of-calibration	tool	milled	out-of-spec	flats

$LP(M_{n\,3})$
on the aluminum blocks.

Coded: $(M_{n\,1} * N_s) + (V_t) + (M_{n\,2} * N_o * LP(M_{n\,3}))$

Discussion: Demo sentence 5.7 incorporates several adjective-type modifiers. All are positioned directly next to the nouns they modify (on the left or right) and connected with multiplication symbols. $M_{n\,1}$ = "[An] out-of-calibration" modifies N_s = "tool." The term that receives the action or *flow* from the sentence's transitive verb, "milled" (V_t), is the noun object, "flats" (N_o). The noun object is modified on the left as well as on the right: $M_{n\,2}$ ("out-of-spec") * N_o ("flats") * $LP(M_{n\,3})$ ("on the aluminum blocks").

Coding Demonstration 5.8

Original: Ultra-precise nano-scale robots someday will remove cholesterol from human arteries.

Scanned:

$M_{n\,1}$	$M_{n\,2}$	N_s	$M_{v\,1}$	V_t	N_o
Ultra-precise	nano-scale	robots	someday	will remove	cholesterol

$LP(M_{n\,3})$
from human arteries.

Coded: $(M_{n\,1} * M_{n\,2} * N_s) + (M_{v\,1} * V_t) + (N_o * LP(M_{n\,3}))$

Discussion: For demo sentence 5.8, once again, there are multiple modifiers positioned next to the terms that they modify. The sentence's subject noun, "robots" (N_s), is modified by a pair of adjectives forming a modification chain: $M_{n\,1}$ ("ultra-precise") * $M_{n\,2}$ ("nano-scale") * N_s ("robots"). In this case, adverb $M_{v\,1}$ ("someday" *when?*) modifies the verb V_t ("will remove"), which transfers action onto the noun object N_o ("cholesterol"), modified by $LP(M_{n\,3})$ = "from humans arteries."

Coding Demonstration 5.9

Original: They successfully perfected the artificial knee cartilage after thirty-one trials.

Scanned:

X_s	$M_{v\,1}$	V_t	N_o
They	successfully	perfected	the artificial knee cartilage

$[LP(M_{v\,2})]^E$
after thirty-one trials.

Coded: $(X_s) + (M_{v\,1} * V_t) + (N_o) \ldots [LP(M_{v\,2})]^E$

or rearranged to more closely conform with the standard B_2 equation:

$(X_s) + (M_{v\,1} * V_t * [LP(M_{v\,2})]^E) + (N_o)$

They successfully perfected *after thirty-one trials* the artificial knee cartilage. §

Discussion: Demo sentence 5.9's core equation is subject pronoun X_s ("They") + transitive verb V_t ("perfected") + compound noun object N_o ("[the] artificial knee cartilage"). Two adverb-type modifiers further accessorize the core equation. The first modifier ("successfully") multiplies the verb on the left and specifies *how so?* (codes: $M_{v\,1} * V_t$), and the second, a prepositional phrase ("after thirty-one trials"), free-floats at the end of the sentence and specifies *when?* $(\ldots [LP(M_{v\,2})]^E)$. Underneath, the example also displays alternate equation and text versions (§) that reposition the free-floating modifier directly next to the verb it modifies.

Coding Demonstration 5.10

Original: The manager fired him *for his grossly unethical behavior without any notice at 5 PM.*

Scanned:

N_s	V_t	X_o	$LP[M_{v\,1}]^E$
The manager	fired	him	*for his grossly unethical behavior*
$LP[M_{v\,2}]^E$	$LP[M_{v\,3}]^E$		
without any notice	*at 5 PM*.		

Coded: $(N_s) + (V_t) + (X_o) \ldots LP[M_{v\,1}]^E\ LP[M_{v\,2}]^E\ LP[M_{v\,3}]^E$

or rearranged to more closely conform with the standard B_2 equation:

$(N_s) + ((V_t) * LP[M_{v\,1}]^E * LP[M_{v\,2}]^E * LP[M_{v\,3}]^E) + (X_o)$

The manager fired *for his grossly unethical behavior without any notice at 5 PM* him. §

Discussion: Demo sentence 5.10 is the stripped-down B_2 core equation subject noun N_s ("[The] manager") + transitive verb V_t ("fired") + pronoun object X_o ("him"). No modifiers directly attach to and multiply any of the core variables. However, three short prepositional phrases, each functioning as an adverb, float on the end of the sentence and modify the sentence's main verb—"for his grossly unethical behavior" (specifies *why?*), "without any notice" (specifies *how so?*), and "at 5 PM" (specifies *when?*). The three end-loaded modifiers code "... $LP[M_{v\,1}]^E$ $LP[M_{v\,2}]^E$ $LP[M_{v\,3}]^E$."

Like example 5.9, underneath the coding, example 5.10 displays an alternate equation and text version (§) that reposition the free-floating modifiers directly next to the verb they modify.

In the above series, the two demo sentences that deviate from the standard B_2 equation template are 5.9 and 5.10. Both have end-loaded, free-floating modifiers at their ends. To examine how each of these sentences maps back to the standard B_2 equation as "home base," underneath each demo's original text and standard coding, the demo displays how, alternatively, the sentence's components can be rearranged into standard B_2 equation sequencing, with core elements and their modifiers, positioned side by side (rearrangements tagged "§").

In both cases, the rearranged equations do resolve in standard B_2 formulae. On the other hand, the two alternate text sentences that result (repeated below) sound unnatural when read aloud.

5.9 Alternate Phrasing

They successfully perfected *after thirty-one trials* the artificial knee cartilage.

(awkward)

5.10 Alternative Phrasing

The manager fired *for his grossly unethical behavior without any notice at 5 PM* him.

(absurd)

One additional thing to note about the two alternate phrasings (shown above) is that the syntactical rough spot in each occurs in the same place—that is, where an adverb or adverbs have been positioned *between* a transitive verb and its object. As a general rule, English sentences designed this way violate natural syntax and do not work. They result in constructions that sound unusual—"unlike how a native speaker would naturally say it." As viewed through the lens of sentence algebra, we can alternatively characterize and express this rule another way: for smoothly operating B_2 equations, do not position adverbs modifying a transitive verb so that they stand directly in the way of a B_2 equation's verb-to-object "flow."

Basic Sentence Equation Three (B_3)

The *basic sentence equation three* (B_3) features a third type of verb. It is neither transitive (V_t) nor intransitive (V_i). It does not transfer action to an object nor does it express action in-and-by-itself. Instead, after first partnering with a subject noun to

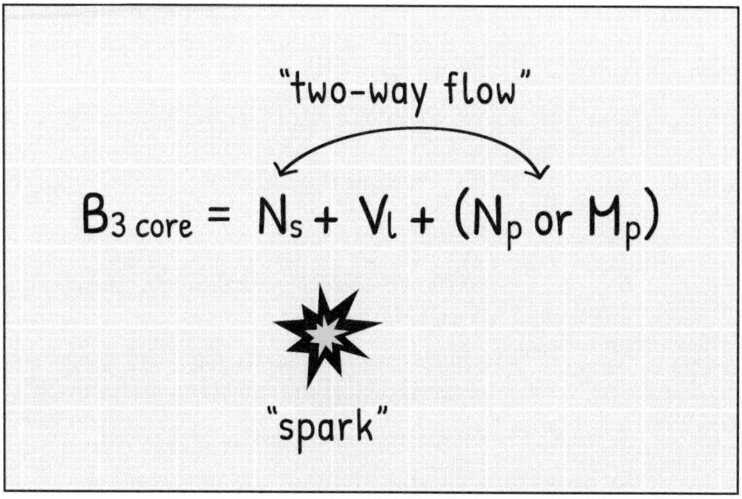

Fig. 5.3 Basic sentence equation three (B_3) core formula for sentence algebra

produce "spark," it *links* a variable in the sentence's subject with a variable in the sentence's predicate. The B_3 core version shown in Fig. 5.3 illustrates this linkage with a two-way flow line. Here, the variables being linked are the subject noun (N_s), positioned to the left of the verb, and either a *predicate noun* or a *predicate adjective* (N_p or M_p), located to the right of the main verb in the predicate. The symbol for the third type of verb, known as a linking verb, is "V_l": "V" subscript little "l" for "linking."

In applied engineering writing, the most frequently used linking verb is "be"—which conjugates into a variety forms: *is*, *was*, *were*, *will be*, *has been*, and so forth. Sentence engineers should also note that words like *seem*, *look*, *become*, *feel*, and *appear* can also operate as linking verbs—although, again, the verb "be," more often than not, occupies the main verb slot in a B_3 sentence in engineering writing.

Beyond its core essentials, the expanded version of the B_3 structure is the most lengthy sentence equation encountered thus far (see below). The length is not so much because basic sentence equation three is significantly more complex in function than a B_1 or B_2 structure, but because of the variety of options it allows for its *subject complement (SC)* component. Located to the right of the linking verb in the predicate, the subject complement can be a *predicate noun* (N_p) or a *predicate adjective* (M_p), as displayed in the B_3 core illustration. It can also be a *predicate pronoun* (X_p). All can be made more specific with an appropriate modifier—an adjective term for a predicate noun or pronoun, or an adverb term for a predicate adjective. On the other hand, the B_3 recipe uses the same sequencing and accessorizing protocols for its subject noun or pronoun and verb as do the complete equations for B_1 and B_2.

$$\mathbf{B_3} = \mathbf{S} + \mathbf{P_3}$$
$$= (\mathbf{M_{n\,a}} * (\mathbf{N_s} \textbf{ or } \mathbf{X_s})) + ((\mathbf{M_{v\,b}} * \mathbf{V_l}) + \mathbf{SC})$$
$$= (\mathbf{M_{n\,a}} * (\mathbf{N_s} \textbf{ or } \mathbf{X_s})) + \left((\mathbf{M_{v\,b}} * \mathbf{V_l}) + \left((\mathbf{M_{n\,a}} * \left(\mathbf{N_p} \textbf{ or } \mathbf{X_p}\right)\right) \textbf{ or } \left(\mathbf{M_{v\,b}} * \mathbf{M_p}\right)\right)\right)$$

where:

S = $(M_{n\,a} * (N_s \text{ or } X_s))$

P_3 = $((M_{v\,b} * V_l) + SC)$

SC = *subject complement* = $((M_{n\,a} * (N_p \text{ or } X_p))$ or $(M_{v\,b} * M_p))$

V_l = a *linking verb* that forms a two-way connection between subject and subject complement (SC)

N_s or X_s = a noun or pronoun *subject* of the verb

N_p or X_p = a *predicate noun* or *predicate pronoun* acting as a subject complement in the sentence's predicate and *linking* back to a *subject noun* or *subject pronoun* in the sentence's subject

M_p = a *predicate adjective* acting as a subject complement in the sentence's predicate and *linking* back to a *subject noun* or *subject pronoun* in the sentence's subject

$M_{n\,a}$ = $\{(M_{n\,0}) * (M_{n\,1}) * (M_{n\,2}) * (M_{n\,3}) \ldots\}$

a series of adjectives modifying a noun, numbering anywhere from 0, no adjectives, to multiple adjectives, 1, 2, 3, and so on

$M_{v\,b}$ = $\{(M_{v\,0}) * (M_{v\,1}) * (M_{v\,2}) * (M_{v\,3}) \ldots\}$

a series of adverbs modifying a verb, numbering anywhere from 0, no adverbs, to multiple adverbs, 1, 2, 3, and so on

Prior to moving on to our more detailed demonstrations of B_3 sentences in action, let's glance at some brief, preliminary examples of each of the three kinds of B_3 subject complements (SCs) at the core equation level.

First, we will look at the *predicate noun*. The B_3 formula makes it possible for a basic sentence to be layered, to simultaneously capture and express two different labels for the same person, place, or thing. Here is a sentence about a man whose *birth name* is "Jeff" whose *job title* is "Wafer Fab Manager":

Jeff is the Wafer Fab Manager.

Next is an example of SC as a *predicate pronoun*. This B_3 variation links flow between the sentence's subject (in this case, the man, "Jeff") and a pronoun in the predicate—in this case, the personal pronoun, "he," a *placeholder* word for "Jeff."

Jeff is he.

The above illustrates "the somewhat mysterious *case*" of the predicate pronoun. Why is the pronoun inflected "he" rather than "him"—that is, written "Jeff is *he*" rather than "Jeff is *him*"? Functional logic provides the answer. If "he" were the object of a transitive verb—for example, "Management promoted *him*"—then "he" would be in the objective case and inflected as "him" and not "he." However, we are dealing with the linking verb "is," and "he" is *not* the object of "is." In this construction, "he" links back and provides another label for the subject, "Jeff." Thus, "he" takes the *subjective case*.

A third B_3 variation to preview is the SC as a *predicate adjective*. Continuing with our sentence about Jeff—also known as the "Wafer Fab Manager" (as well as "he")—let's examine what the sentence looks like when the subject links to a predicate adjective:

Jeff is ambitious.

In the above sentence, the predicate adjective helps readers to characterize "Jeff" not just as someone with a specific job title and name but rather as an individual person with a specific human trait. Jeff, the Wafer Fab Manager, is also a man whose motivational level is "ambitious"—"ambitious Jeff." We could also use a prepositional phrase to serve as the predicate adjective by writing a sentence about "on-the-fast-track Jeff"—that is, "*Jeff* is *on the fast track*." There are myriad B_3 sentences that could possibly characterize "Jeff." Knowing the B_3 equation options allows sentence engineers to actively and knowingly apply any of them, at the writer's discretion.

Next, use your new knowledge of the B_3 equation to trace through the structures of the following example sentences' text and equation versions.

Five Applied Demonstrations of B_3

Coding Demonstration 5.11

Original: The aerospace engineer was an expert on finite element analysis of aluminum castings.

	N_s	V_l	N_p	$LP(M_n)$
Scanned:	The aerospace engineer	was	an expert	on finite element analysis of aluminum castings.

Coded: $(N_s) + (V_l) + (N_p * LP(M_n))$

Discussion: Sentence 5.11's verb (V_l = "was") links the subject noun (N_s = "[The] aerospace engineer") with another label (N_p = "[an] expert") for the same person. Additionally, there is a long phrase (actually, two prepositional phrases chained together as one adjective) that modifies the predicate noun (specifies *what* type of expert . . .). The long phrase with both prepositions highlighted, for review, is $LP(M_n)$ = "*on* finite element analysis *of* aluminum castings."

Coding Demonstration 5.12

Original: Our new-hire chemical engineer is the leading expert on optimal bubblegum elasticity.

Scanned:

X'	$M_{n\,1}$	N_s	V_l	N_p
Our	new-hire	chemical engineer	is	the leading expert

$LP(M_{n\,2})$
on optimal bubblegum elasticity.

Coded: $X'\text{-}(M_{n\,1} * N_s) + (V_l) + (N_p * LP(M_{n\,2}))$

Discussion: Sentence 5.12 is similar in design to 5.11. The sentence's verb (V_l = "is") links the subject noun (N_s = "chemical engineer") with another name/label (N_p = "[the] leading expert") for the same person. There is also a prepositional phrase ($LP(M_{n\,2})$ = "on optimal bubblegum elasticity") that modifies the predicate noun (specifies *what* type of expert . . .). What's different here is that both a possessive pronoun (X' = "Our") and an adjective ($M_{n\,1}$ = "new-hire") attach to the subject noun, increasing its specificity two different ways.

Coding Demonstration 5.13

Original: After the convincing presentation, they felt ecstatic about MEMS technology.

Scanned:

$[LP(M_{v\,1})]^F$	X_s	V_l	M_p	$[LP(M_{v\,2})]^E$
After the convincing presentation,	they	felt	ecstatic	about MEMS technology.

Coded: $[LP(M_{v\,1})]^F \ldots (X_s) + (V_l) + (M_p) \ldots [LP(M_{v\,2})]^E$

Discussion: The coding of 5.13 reveals that underneath the original word (text) sentence, there is a B_3 equation featuring a subject pronoun (X_s = "they"), linking verb (V_l = "felt"), and predicate adjective (M_p = "ecstatic") linking back to and modifying the subject (*ecstatic* they). There are also two free-floating modifiers—a front-loaded prepositional phrase functioning as an adverb, $[LP(M_{v\,1})]^F$ = "After the convincing presentation" (specifies *when* they "felt"), and an end-loaded prepositional phrase, $[LP(M_{v\,2})]^E$ = "about MEMS technology," also functioning as an adverb (specifies *what* they "felt").

Coding Demonstration 5.14

Original: The leaky cryogenic valve is very unsafe.

Scanned:

$M_{n\,1}$	N_s	V_l	$M_{v\,1}$	M_p
The leaky	cryogenic valve	is	very	unsafe.

Coded: $(M_{n\,1} * N_s) + (V_l) + (M_{v\,1} * M_p)$

Discussion: Sentence 5.14 showcases another B_3 construction with a predicate adjective (M_p = "unsafe") as its subjective complement. This time, however, the adjective has a specify multiplier, an adverb ($M_{v\,1}$ = "very"). In turn, the predicate adjective links back (V_l = "is") to the subject noun (N_s = "cryogenic valve"), which is directly modified by $M_{n\,1}$ = "[The] leaky." Thus, the sentence is about a cryogenic valve that is characteristically "leaky" as well as "very unsafe."

Coding Demonstration 5.15

Original: After working on seven major projects all at once, Bob seemed incredibly stressed out.

Scanned:

$[LP(M_{v\,1})]^F$	N_s	V_l
After working on seven major projects all at once,	Bob	seemed

$M_{v\,2}$	M_p
incredibly	stressed out.

Coded: $[LP(M_{v\,1})]^F \ldots (N_s) + (V_l) + (M_{v\,2} * M_p)$

Discussion: Aside from the front-loaded, free-floating adverb "After working on seven major projects all at once," sentence 5.15 echoes the design of sentence 5.14. The subject "Bob" is linked = via the verb "seem" to a predicate adjective "stressed out" that is further amplified in degree by the adverb "incredibly."

Action Items

α	Select a text sample from a recent email written by you, a colleague, or both. The sample(s) should be a paragraph or two long, about 100 to 200 words. Now, scan the text, and, as best as you can, identify all core B_1 equations, that is, $B_1 = ((N_s \text{ or } X_s) + V_i)$. Ignore words and word groups that appear to serve as modifiers as well as connectors between basic sentences, such as conjunctions ("and," "but," "although," "because," and so on). Also ignore connector punctuation marks like semicolons ";" and colons ":". You may identify B_2 and B_3 core equations, too, if you wish, but keep principal focus on identifying the B_1 cores. How many B_1 cores did you find? What types of actions did each sentence express—*You doing/thinking/feeling something? Someone else doing/thinking/feeling something?* or *A thing doing something?* Note: There are no right or wrong answers here. Just speculate and go with your hunches. Carry any insights you discover in this action item forward to the next one.
α	In anticipation of your next performance review (or, if you're a student, the next evaluation of a major graded assignment), *identify* three job-related accomplishments that you feel are noteworthy and indicative of your "value-add" as an engineer (or engineering student) in your work group (or class). Some generic examples of accomplishments would be (1) "I developed a solution for problem X," (2) "I created a new design for customer Y," and (3) "I finished project Z ahead of schedule." Now, briefly summarize each of your three specific accomplishments as B_2-type sentences. Can you do this? Most likely, yes. Note: The core formula for your three sentences should be $X_s + V_t + N_o$, where X_s = "I" (that is, you, the reader). Considering your experience doing action item one, do you think that you could describe the three accomplishments that you expressed in action item two using B_1 core formulas rather than B_2 formulas? Why or why not? Briefly explain.
α	This action item has to do with B_3-type sentences said aloud—specifically, as part of a phone conversation. The next time you receive a phone call and the person on the other end says hello and then asks "Is this <insert your name>?", consider, if you don't already do so, replying with the grammatically correct "Yes, it is I." A theme emphasized in this book is that sentence algebra allows you to understand the functional inner workings of sentences and make good, solid, professional, and conscious choices when you build messages out of sentences. In the preceding how-to-answer-a-phone-call scenario, as a professional engineer, whether to go with "It is I" or "It's me" is a choice you should consciously make. If you know the person on the other end to be a nit-picky, fussy grammarian, and if you strategically want to impress him or her, go with the former.

Recap

Key Learnings/Takeaways

- The basic sentence template $(S + P_n)$ expands into the *basic sentence equation one* formula as follows: $B_1 = S + P_1 = (M_{n\ a} * (N_s \text{ or } X_s)) + (M_{v\ b} * V_i)$, where $M_{n\ a}$ stands for zero, one, or multiple adjectives and $M_{v\ b}$ stands for zero, one, or multiple adverbs.

- A key feature of the B_1 equation is that it creates "spark" with an *intransitive verb*. This type of verb expresses action in-and-by-itself (and does not transfer action to an object). Sentence algebra represents an intransitive verb as "V_i."
- A new coding variation and notation appears in this chapter for sentence engineers to add to their sentence algebra tool kits. It is the modifier positioned at the front or end of a sentence, rather than directly next to the core term it modifies within the sentence. We code this element either "$[M]^F$ . . .", where the subscript "F" stands for "front" and ". . ." stands for "free-floating" modifier, or ". . . $[M]^E$" at the "end" of a sentence. Free-floating modifiers can be adjectives ($[M_n]^F$ or $[M_n]^E$) or adverbs ($[M_v]^F$ or $[M_v]^E$).
- The basic sentence template (S + P_n) expands into the *basic sentence equation two* formula as follows: $B_2 = S + P_2 = (M_{n\,a} * (N_s \text{ or } X_s)) + (M_{v\,b} * V_t) + (M_{n\,a} * (N_o \text{ or } X_o))$, where $M_{n\,a}$ stands for zero, one, or multiple adjectives and $M_{v\,b}$ stands for zero, one, or multiple adverbs.
- The B_2 equation incorporates a *transitive verb*, which—instead of expressing action in-and-by-itself like an intransitive verb—transfers verb action onto an object. Sentence algebra represents a transitive verb as "V_t" and its noun or pronoun object as "N_o" or "X_o."
- Some verbs can function as *transitive* verbs that transfer action onto an object *and also* as *intransitive* verbs that express action in-and-by-themselves, for example, "Engineers calculate shear stress." versus "Engineers calculate." Dictionaries indicate whether a verb is typed as a transitive verb, intransitive verb, both, or something else.
- In sentence algebra, "flow" refers to a dynamic relationship that extends beyond basic N_s + V "spark"—to and between other variables in a sentence. In a B_2 sentence, the "flow" line goes from transitive verb (V_t) to noun or pronoun object (N_o or X_o).
- The basic sentence template (S + P_n) expands into the *basic sentence equation three* formula as follows: $B_3 = S + P_3 = (M_{n\,a} * (N_s \text{ or } X_s)) + (M_{v\,b} * V_l) + ((M_{n\,a} * (N_p \text{ or } X_p)) \text{ or } (M_{v\,b} * M_p))$, where $M_{n\,a}$ stands for zero, one, or multiple adjectives and $M_{v\,b}$ stands for zero, one, or multiple adverbs.
- The B_3 equation incorporates a *linking verb*, which—instead of expressing action in-and-by-itself like an intransitive verb *or* transferring verb action onto an object like a transitive verb—*links* a noun subject or equivalent in the subject to a *subject complement* in the predicate. The subject complement associates the subject noun or equivalent with an alternative name, label, or reference *or* with an adjective. The sentence algebra symbol for a linking verb is "V_l"—a "V" with a subscript "l" for "linking."
- In sentence algebra, "two-way flow" refers to the dynamic flow relationship that goes back-and-forth between the two linked items in a B_3 sentence, a subject to the left of V_l with a subject complement to the right of V_l. A B_3 construction is versatile. There are two choices for subject—subject noun (N_s) and subject pronoun (X_s)—and there are three choices for the subject complement: predicate noun (N_p), predicate pronoun (X_p), and predicate adjective (M_p).

- As demonstrated throughout Chap. 5, the standard B_1, B_2, or B_3 sentence equations and basic sentence algebra conventions apply when coding B_1, B_2, or B_3 sentences from text to equation. However, in the chapter's coding demonstrations, we observed that *natural syntax* sometimes also guides the positioning of modifiers in a sentence to ensure that the sentence's text version reads "like a native speaker would naturally say it."

Chapter 6
More Basic Sentence Equations: B_4 and B_5

Learning Objectives

- *Know how to expand the basic sentence template ($S + P_n$) into B_4 and B_5 sentence equations.*
- *Know how to scan and code B_4 and B_5 sentences using equations and the sentence algebra coding conventions.*
- *Understand how "flow" operates in B_4 and B_5 sentences.*
- *Know what an indirect object is and what an object complement is and the various types.*
- *Know how to scan and code a longer text sample containing a stream of basic sentences scaffolded by a variety of basic sentence equations.*

Basic Sentence Equation Four (B_4)

The 4th basic sentence formula (B_4) embodies a standard subject-noun-plus-verb "spark" plus two instances of flow. Like basic sentence B_2, the main verb transfers action directly onto a noun object (N_o) in the predicate (stage one flow). Yet with B_4, there is a second line of flow that extends "spark" further (stage two flow). This transfer goes from the noun object (N_o) to a second noun in the predicate, known as the indirect object noun "N_i." Figure 6.1 displays B_4 in stripped-down, core version form.

With options for modifiers and pronouns, the B_4 equation shapes into the expanded version shown below.

B. Henderson, *A Math-Based Writing System for Engineers*,
https://doi.org/10.1007/978-3-030-10756-7_6

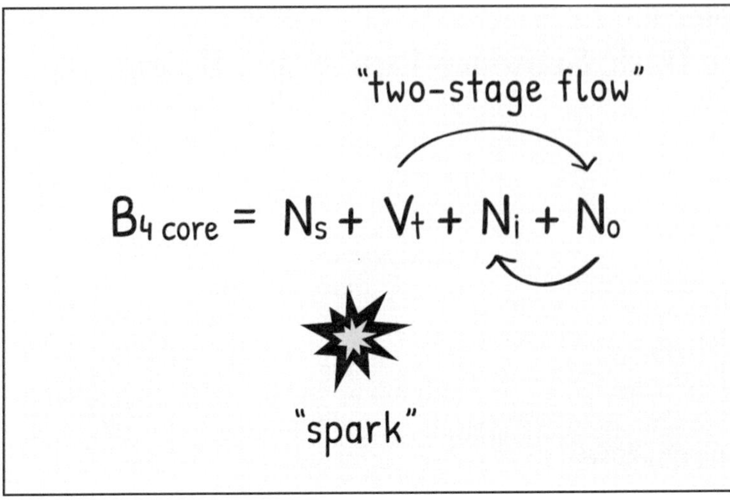

Fig. 6.1 Basic sentence equation four (B_4) core formula for sentence algebra

$$\mathbf{B_4 = S + P_4}$$
$$\mathbf{= S + ((M_{v\,b} * V_t) + (M_{n\,a} * (N_i \text{ or } X_i)) + (M_{n\,a} * (N_o \text{ or } X_o)))}$$

where: S $= M_{n\,a} * (N_s \text{ or } X_s)$

V_t = a *transitive verb* that *transmits* action onto an object

N_s or X_s = a noun or pronoun *subject* of the verb

N_o or X_o = a noun or pronoun *direct object* of the verb

N_i or X_i = a noun or pronoun *indirect object* of the verb

$M_{n\,a}$ $= \{(M_{n\,0}) * (M_{n\,1}) * (M_{n\,2}) * (M_{n\,3}) \ldots\}$

a series of *adjectives* modifying a noun, numbering anywhere from 0, no adjectives, to multiple adjectives, 1, 2, 3, and so on

$M_{v\,b}$ $= \{(M_{v\,0}) * (M_{v\,1}) * (M_{v\,2}) * (M_{v\,3}) \ldots\}$

a series of *adverbs* modifying a verb, numbering anywhere from 0, no adverbs, to multiple adverbs, 1, 2, 3, and so on

The next segment's coding demonstration features five sentences with B_4 equations at their cores. To better examine how the action flows from subject to objects, an alternate word ordering, or syntax, of the sentence predicate's core elements is shown below each of the standard, left-to-right codings. In the alternate phrasings, arrows (⇨) show how the action is transferred first to the sentence's direct object and next to its indirect object. The alternate phrasings also reveal how adding the preposition "to" helps complete the logic of a B_4 sentence's predicate. In practice, text sentences developed using the B_4 equation can be written using either the standard structure or rearranged structure that adds "to." Here, we focus on the standard structure.

Five Applied Demonstrations of B_4

Coding Demonstration 6.1

Original: The airframe manufacturer offered the bidder a subcontract.

	N_s	V_t	N_i	N_o
Scanned:	The airframe manufacturer	offered	the bidder	a subcontract.

Coded: $N_s + V_t + N_i + N_o$

Alternate phrasing of predicate: *... offered [⇨] a subcontract [to] [⇨] the bidder.*

Discussion: Demo sentence 6.1 is an example of a B_4 sentence that communicates its message using the core equation and no accessory modifiers.

Coding Demonstration 6.2

Original: Someone anonymously relayed the offshore facility an urgent message about the hurricane.

	X_s	M_v	V_t	$M_{n\,1}$	N_i
Scanned:	Someone	anonymously	relayed	the offshore	facility

$M_{n\,2}$	N_o	$LP(M_{n\,3})$
an urgent	message	about the hurricane.

Coded: $X_s + (M_v * V_t) + (M_{n\,1} * N_i) + (M_{n\,2} * N_o * LP(M_{n\,3}))$

Alternate phrasing of predicate: *... relayed [⇨] an urgent message [to] [⇨] the offshore facility.*

Discussion: Instead of a noun subject, demo sentence 6.2 begins with a pronoun subject, X_s = "Someone." Specifically, this sentence's subject is an indefinite pronoun (for instance, "someone," "something," "anywhere," and so on). Here, four modifiers (M_v, $M_{n\,1}$, $M_{n\,2}$, and LP ($M_{n\,3}$)) accessorize the B_4 core formula, multiplying the core variables V_t, N_i, and N_o on their left, as well as in one case (N_o) also on its right.

Coding Demonstration 6.3

Original: After reviewing the schematic, she gave them a stern, brutally-to-the-point warning about patent infringement.

Scanned:

$[LP(M_v)]^F$	X_s	V_t	X_i	$M_{n\,1}$
After reviewing the schematic,	she	gave	them	a stern,

$M_{n\,2}$	N_o	$LP(M_{n\,3})$
brutally-to-the-point	warning	about patent infringement.

Coded: $[LP(M_v)]^F \ldots X_s + V_t + X_i + (M_{n\,1} * M_{n\,2} * N_o * LP(M_{n\,3}))$

Alternate phrasing of predicate: *... issued [⇨] a warning [to] [⇨] them.*

Discussion: A free-floating, front-end adverbial phrase, $[LP(M_v)]^F$ = "After reviewing the schematic," heads up sentence 6.3. This prepositional phrase modifies the verb V_t = "gave" and specifies *when* the action occurred. At this point in the sentence algebra chapters, you have encountered a number of sentences containing free-floating modifiers, in both front-loaded and end-loaded positions, and have most likely observed that applied writing regularly features these constructions. Two of the sentence's core variables are personal pronouns, X_s = "she" and X_i = "them." The last core variable N_o = "a warning" is modified by two adjectives on the left ($M_{n\,1}$ = "a stern" and $M_{n\,2}$ = "brutally-to-the-point") and another term on its right, LP ($M_{n\,3}$) = "about patent infringement."

Coding Demonstration 6.4

Original: The distillation column *will economically supply* the seaside community purified water at about 1 ¢/gal.

Scanned:

N_s	V_t	N_i
The distillation column	*will economically supply*	the seaside

	N_o	$LP(M_n)$
community	purified water	at about 1 ¢/gal.

Coded: $N_s + V_t + N_i + (N_o * LP(M_n))$

Alternate phrasing of predicate: *... will economically supply [⇨] purified water [to] [⇨] the seaside community.*

Discussion: Sentence 6.4's coding maps back to the standard B_4 equation in a fairly routine fashion. One variation worth pointing out, however, is the verb construction "will economically apply," which is highlighted in *italics*. The example codes it as a multi-word verb (V_t). It is actually a two-word verb form—the auxiliary verb V_{aux} = "will" and root verb V_{root} = "supply"—with the adverb M_v = "economically" positioned in between. This type of "split verb" construction codes "$V_{aux} * M_v * V_{root}$." In English sentences, in addition to adverbs modifying verbs either on the left or right, it is also common to locate them in between main verb words. As already established, along with a set of standard formulas, *natural syntax* also guides the final structure of sentence equations scaffolding text. In the case of sentence 6.4, although legitimate specificity multiplication at an equation level is $(M_v * V_{aux} * V_{root}) = (V_{aux} * V_{root} * M_v) = (V_{aux} * M_v * V_{root})$, placing the adverb before or after the verb words, as in "economically will apply" or "will apply economically," results in an awkward-sounding construction for this text sentence.

Coding Demonstration 6.5

Original: The Chief Engineer assigned Bob McSupercalc the tedious, yet also critical task of investigating phase I's failed prototype with a root-cause analysis.

Scanned:

N_s	V_t	N_i	$M_{n\,1}$
The Chief Engineer	assigned	Bob McSupercalc	the tedious,

C	M_v	$M_{n\,2}$	N_o	$LP(M_{n\,3})$
yet	also	critical	task	of investigating phase I's failed prototype

with a root-cause analysis.

Coded: $(N_s + V_t + N_i + ((M_{n\,1} \mid (M_v * M_{n\,2})) * N_o * LP(M_{n\,3}))$

Alternate phrasing of predicate: *... assigned [⇨] the task [to] [⇨] Bob McSuperCalc.*

Discussion: The final example in this sequence, sentence 6.5, features three core variables with no modifiers: N_s = "The Chief Engineer," V_t = "assigned," and N_i = "Bob McSupercalc." The fourth core variable, N_o = "task," on the other hand, is modified by four terms. There is a three-modifier cluster consisting of two adjectives, $M_{n\,1}$ = "the tedious" and $M_{n\,2}$ = "critical," and one adverb, M_v = "also," on the left. The coordinating conjunction C = "yet" pairs together the two adjectives, and the one adverb, M_v = "also," modifies the adjective "critical." On the right of the noun object "task" is the adjectival phrase, $LP(M_{n\,3})$ = "of investigating phase I's failed prototype with a root-cause analysis." Thus, altogether the four modifiers and the fourth core variable code "$((M_{n\,1} \mid (M_v * M_{n\,2})) * N_o * LP(M_{n\,3}))$."

Basic Sentence Equation Five (B_5)

The final basic sentence formula, B_5, when stripped down to charge and flow is also a cousin to the B_2 formula. As Fig. 6.2 reveals, the equation generates noun-plus-verb spark and then transfers charge from main verb to an object (stage one flow). Like B_4, the B_5 equation features an additional instance of flow.

In the case of B_5, the second flow line (stage two) goes from the main verb's object to another term in the predicate, represented in Fig. 6.2 as (N_c or M_c). This term is known as the *object complement (OC)* and can take shape as either an *object complement noun* (N_c) or *object complement modifier* (M_c).

Note also in the illustration of B_5's simplified core equation that the second-stage flow line displays with two arrows, one aiming forward and one aiming back. This is because the link between B_5's main (direct) object and its object complement is similar to the linking of subject noun (or pronoun) with subject complement in the B_3 equation. Yet, in the B_5 sentence partnering, there is no linking verb in between. However, in the B_5 coding demos, our alternate phrasing of each sentence's predicate shows that adding the invisible (elliptical) linking verb form "to be" clarifies the logic of the equation underneath.

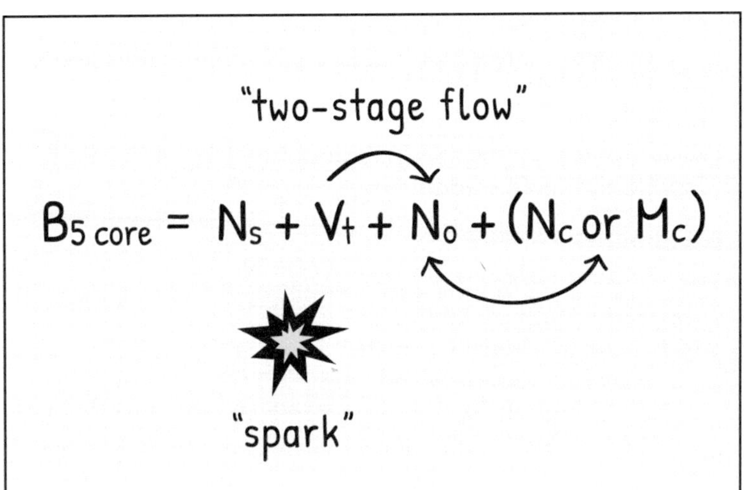

Fig. 6.2 Basic sentence equation five (B_5) core formula for sentence algebra

The complete B_5 equation is shown below. It identifies a third variable option for the objective complement—an object complement pronoun swapped in for object noun. Thus, the complete sentence algebra notation for objective complement is OC = N_c, X_c, or M_c. Whether noun, pronoun, or adjective, the purpose of an object complement is consistently the same. It adds uniqueness and/or specificity to the object with which it is partnered (linked) in one of two ways—through *noun-type renaming* (in the case of N_c or X_c) or through *adjectival modifying* (in the case of M_c).

$$\mathbf{B_5 = S + P_5}$$
$$\mathbf{= S + ((M_{v\,b} * V_t) + (M_{n\,a} * (N_o \text{ or } X_o)) + ((M_{n\,a} * (N_c \text{ or } X_c)) \text{ or } (M_{v\,b} * M_c)))}$$

where:

S	$= M_{n\,a} * (N_s \text{ or } X_s)$
V_t	= a *transitive verb* that *transmits* action onto an object
N_s or X_s	= a noun or pronoun *subject* of the verb
N_o or X_o	= a noun or pronoun *object* of the verb
N_c or X_c	= a noun or pronoun *object complement*
M_c	= an adjective *object complement*
$M_{n\,a}$	$= \{(M_{n\,0}) * (M_{n\,1}) * (M_{n\,2}) * (M_{n\,3}) \ldots\}$ a series of *adjectives* modifying a noun, numbering anywhere from 0, no adjectives, to multiple adjectives, 1, 2, 3, and so on
$M_{v\,b}$	$= \{(M_{v\,0}) * (M_{v\,1}) * (M_{v\,2}) * (M_{v\,3}) \ldots\}$ a series of *adverbs* modifying a verb, numbering anywhere from 0, no adverbs, to multiple adverbs, 1, 2, 3, and so on

As usual, next we will look at five demonstration sentences to advance your understanding of how B_5 sentences work "in action." Along with the scanning and coding, there is a strand of text showing the core variables and two-stage flow pattern in each example sentence's predicate. This time, there is no rearrangement of the syntax. Like the B_4 examples, however, we can note that an invisible word—in this case two words, "to be"—exists underneath the text. This linking verb form "to be" stands between the objects (object of the verb and object complement) and clarifies their partnership logic.

Five Applied Demonstrations of B_5

Coding Demonstration 6.6

Original: The corporate board named a mechanical engineer CEO.

Scanned:

N_s	V_t	N_o	N_c
The corporate board	named	a mechanical engineer	CEO.

Coded: $N_s + V_t + N_o + N_c$

Alternate phrasing of predicate: *. . . named [⇨] a mechanical engineer [to be] [⇨] CEO.*

Discussion: The B_5 core equation scaffolds sentence 6.6 without additional modifiers. Two multi-word compound nouns fill the slots of subject and direct object. The object complement is the well-known corporate initialism for "chief executive officer."

Coding Demonstration 6.7

Original: The fastidious quality control inspector reluctantly rated the shipment of etched silicon wafers within tolerance

Scanned:

$M_{n\,1}$	N_s	M_v	V_t
The fastidious	quality control inspector	reluctantly	rated

N_o	$LP(M_{n\,2})$	$LP(M_c)$
the shipment	of etched silicon wafers	within tolerance

Coded: $(M_{n\,1} * N_s) + (M_v * V_t) + (N_o * LP(M_{n\,2})) + (LP(M_c))$

Alternate phrasing of predicate: *. . . rated [⇨] the shipment [to be] [⇨] within tolerance.*

Discussion: Demo sentence 6.7 maps to an accessorized B_5 equation with an object complement that functions as an adjective, $LP(M_c)$ = "within tolerance," modifying the direct object, N_o = "shipment," rather than renaming it. The subject noun and main verb are made more specific with modifier words—first the adjective, $M_{n\,1}$ = "fastidious," modifies the subject N_s = "quality control inspector," and second, the adverb M_v = "reluctantly" modifies V_t = "rated." The direct object N_o = "the shipment" is also modified by an adjectival prepositional phrase, $LP(M_{n\,2})$ = "of etched silicon wafers," working as a specificity multiplier.

Coding Demonstration 6.8

Original: After learning that both the lift force and stability data exceeded the design parameters, the quality assurance engineer judged the new wing design's wind tunnel test very successful.

Scanned:

$[LP(M_{v\,1})]^F$				
After learning that both the lift force and stability data exceeded				
	N_s	V_t		
the design parameters,	the quality assurance engineer	judged		
N'	N_o	$M_{v\,2}$	M_c	
the new wing design's	wind tunnel test	very	successful.	

Coded: $[LP(M_{v\,1})]^F \ldots (N_s) + (V_t) + (N'\text{-}N_o) + (M_{v\,2} * M_c)$

Alternate phrasing of predicate: *. . . judged [⇨] the wind tunnel test [to be] [⇨] successful.*

Discussion: Demo sentencc 6.8 begins with a free-floating, front-loaded modifier, a long prepositional phrase led off with "After learning that . . . ." The phrase functions as an adverb that places two additional conditions on the main verb, "judged,"—specifies *when* the action occurred and also *what* critical information informed the judgment. A possessive noun term, "the new wing design's," attaches to the direct object and clarifies *what* is being tested in the wind tunnel. Like 6.7, the object complement is an adjective—in this case, an adjective word, "successful," which itself is modified by an adverb, "very."

Coding Demonstration 6.9

Original: The Legal Department classified the new formula for the resin designed by my engineering team "internal-access-only."

Scanned:

N_s	V_t	$M_{n\,1}$	N_o	$LP(M_{n\,2})$
The Legal Department	classified	the new	formula	for the resin
	M_c			
designed by my engineering team	"internal-access-only."			

Coded: $N_s + V_t + (M_{n\,1} * N_o * LP(M_{n\,2})) + M_c$

Alternate phrasing of predicate: *classified [⇨] the formula [to be] [⇨] "internal-access-only"*

Discussion: Sentence 6.9 features an adjective object complement, specifically the compound adjective, "internal-access-only," adding further specificity to the noun object, "formula." The noun object is also modified directly by the adjective "the new" on its left and the adjectival phrase "for the resin designed by my engineering team" on the right.

Coding Demonstration 6.10

Original: After the biomedical engineer relinquished intellectual property rights to her prosthesis design to investors, the business transaction made future profits theirs but left the credit for the altruistic invention hers.

Scanned:

$[LP(M_{v\,1})]^F$		
After the biomedical engineer relinquished intellectual property rights		

	N_s	$V_{t\,1}$
to her prosthesis design to investors,	the business transaction	made

$M_{n\,1}$	$N_{o\,1}$	$X_{c\,1}$	C	$V_{t\,2}$	$N_{o\,2}$	$LP(M_{n\,2})$
future	profits	theirs	but	left	the credit	for the altruistic

	$X_{c\,2}$
invention	hers.

Coded: $[LP(M_{v\,1})]^F \ldots N_s + [(V_{t\,1} + (M_{n\,1} * N_{o\,1}) + X_{c\,1}) \mid (V_{t\,2} + (N_{o\,2} * LP(M_{n\,2})) + X_{c\,2})]$

Alternate phrasing of predicate: *... made [⇨] profits [to be] [⇨] theirs <u>but</u> left [⇨] the credit [to be] [⇨] hers.*

Discussion: Finally, demo sentence 6.10 illustrates the use of pronouns as objective complements. The sentence begins with a long front-loaded, adverbial phrase $[LP(M_{v\,1})]^F$ containing content that not only modifies the sentence's verbs and establishes *when* the action occurs but also provides information about *who* and *what* that informs the sentence's message on the whole. The sentence's subject is "the business transaction." After this, however, there are two P_5-type predicates joined together with the coordinating conjunction C = "but." Accessory elements aside, for one of the two predicates, the verb is "made," the object is "profits," and the objective complement is the possessive personal pronoun "theirs." The second part of this "compound predicate" is the verb "left," the object "the credit," and the objective complement (also a possessive personal pronoun) "hers."

Just like the adjective pair in sentence 6.5 in the earlier B_4 examples, sentence 6.10 above illustrates another type of pairing of basic sentence elements using a connector word. In this case, the coordinating conjunction C = "but" joins a pair of main verbs and what follows after each of those verbs. The result is two P_5 predicates that link back to a singular subject. In general, when forming text sentences in applied writing, you will discover many options for expanding the standard basic sentence equations. You can group together core elements in a variety of ways using conjunctions. You can also incorporate connector words to join two (or more) subassemblies, such as predicates, within a basic sentence framework.

A Review of Basic Sentence Algebra Coding and Structures

Before concluding our coverage of basic sentences and moving on to explore more advanced sentences in Chaps. 7 and 8, here is a list of all of the coding conventions that we have discussed and demonstrated thus far. Some were established in Chap. 2, and others were added along the way. You will also find a review box of the five basic sentence formulas and their specific components included with the list. After the review, the final part of Chap. 6 applies all that we have learned thus far about coding basic sentences and basic sentence elements to a longer sample of engineering text.

The Basic Coding Conventions for Sentence Algebra Equations:

- *Code left-to-right, the same as you read and write:* When coding a text sentence as a sentence algebra equation, the left-to-right sequencing of variables in the equation should match the original left-to-right arrangement of words in the sentence.
- *Use the plus sign (+) and asterisk (*) to signal basic operations.* To connect sequences of core variables (N_s, X_s, and V_s) to form basic equations, use the plus sign; to add specificity to core variables with modifiers (such as M_n or M_v), use the asterisk as a specificity multiplier.
- *Use parentheses and brackets to organize terms.* For example, by adding parentheses, we can organize "$M_{n1} * N_s + M_v * V_t + M_{n2} * N_i + M_{n3} * X_o$," into a form that better reveals the subject and predicate of the B_4 equation underneath, "$[(M_{n1} * N_s)] + [(M_v * V_t) + (M_{n2} * N_i) + (M_{n3} * X_o)]$" or $[S] + [P_4]$.
- *Identify and code top-level functional components in equations.* When you are scanning and coding a sentence, strive first to identify and code its top-level functional components, even though further disassembly is often possible. Sentence algebra calls this *first-order coding*.
  - For instance, disassembling a prepositional phrase, an "LP," into its sub-parts can introduce another layer of interesting analysis. But if the phrase's chief role in a sentence is to modify a noun or verb, coding it as a prepositional phrase that plays the role of modifier, $LP(M_n)$ adjective or $LP(M_v)$ adverb, is usually sufficient for applied sentence engineering. For instance, code the prepositional phrase "beyond the stratosphere" as "$LP(M_n)$", rather than "$(L + (m_n * N_o))$".

- Additionally, with longer prepositional phrases composed of multiple sub-phrases (one phrase nested into the other), if the lead phrase and subsequent phrases all work together as a modifier of a noun or verb, it is also generally best to code the phrase as one collective unit in top-level coding. For example, "*In an industrial park near the Mississippi River on the outskirts of Memphis, Tennessee*, a startup company thrives," codes as "$[LP(M_v)]^F\ldots N_s + V_i$". The long phrase, all of it, specifies *where* the "startup company" is thriving.

- *Sentence equations, usually consider articles ("the," "a," and "an") elliptical (invisible) and do not code them with variables.* If necessary, you can represent articles with the special variable "m_n."
- *Sentence equations omit their text versions' initial capital letters and internal and terminal punctuation marks.* Internal punctuation marks include commas setting off introductory words and word groups, items in lists, or sentence elements inserted between two commas or two dashes. Chapter 11 covers proper use of internal punctuation marks in text sentences.
- *The Basic Math Laws apply selectively to sentence algebra equations.* The sentence algebra system is a math analog, not a pure math model. Natural syntax, what the system calls the natural order of words in English sentences, and native speaker conventions also guide sentence equations. More specially,
  - The *Commutative Law* **does** sometimes apply to *specificity multiplication* within sentences, especially when it comes to M_vs (adverbs). For example, the basic sentence containing the subject "engineers," main verb "succeed," and the adverb "frequently" can be written correctly two different ways: "Engineers *frequently* succeed." (coded as $N_s + (M_v * V_i)$ or "Engineers succeed *frequently*." [coded as $N_s + (V_i * M_v)$]. On the other hand, the sentence "The pilot *successfully* landed the plane." with the adverb "successfully" placed before the verb "landed" becomes an awkward construction if we multiply specificity on the right, as in "The pilot landed *successfully* the plane."
  - The *Commutative Law* generally does *not* apply to sentence algebra *addition of core elements* (with or without accessorizing) in equations. For example, the spark-producing pair "$N_s + V$" is not functionally equivalent to "$V + N_s$" nor is "The pilot successfully landed the plane" equivalent to "The plane successfully landed the pilot."
  - The *Associative Law* often *does* apply, since it involves organizing terms with parentheses (and brackets) and not operations and order of terms.
  - The *Distributive Law* generally does *not* apply.

- *Free-floating, front-loaded, or end-loaded adverbs and adjectives* present another coding situation where natural syntax trumps the rudimentary math principles—often with regard to adverbs modifying a basic sentence's main verb. When this type of sentence element modifies another element in a sentence, it does so from a "free-floating" position at the sentence's beginning or end, not as specificity "multiplier" directly next to the element on its left or right side.
  - We code these elements as either "$[M_n \text{ or } M_v]^F$" with superscript "F" for front of the sentence or "$[M_n \text{ or } M_v]^E$" with superscript "E" for end. Three ellipses

"..." set off free-floating terms. For example, the sentence "*Periodically*, we traveled to the offshore facility" codes as "$[M_{v\,1}]^F \ldots (X_s) + V_i * LP(M_{v\,2})$."

– A prepositional phrase can also function as a free-floating modifier. In this case, we code the phrase as "$[LP(M_n \text{ or } M_v)]^F$" or "$[LP(M_n \text{ or } M_v)]^E$." For example, the sentence "*After the storm*, we traveled to the offshore facility" codes as "$[LP(M_{v\,1})]^F \ldots (X_s) + V_i * LP(M_{v\,2})$."

- *Split verbs* are multi-word verb forms that position an adverb (M_v) *between* an *auxiliary verb* and *principal verb* (intransitive, transitive, or linking) using the following construction, "$V_{aux} * M_v * (V_i$ or V_t but seldom $V_l)$." For example, the sentence "The sentence engineer *had already finished* the coding process" codes as "$N_s + (V_{aux} * M_v * V_t) + N_o$."

Review Box: The Five Basic Sentence Formulae

The *Basic Formula* is as follows:

$$B_n = S + P_n \qquad \text{where } n = 1, 2, 3, 4, \text{ or } 5$$

The *Subject (S)* expands into one sub-formula:

$$S = M_{n\,a} * (N_s \text{ or } X_s)$$

where: N_s or X_s = a noun or pronoun subject of the main verb

$M_{n\,a} = \{(M_{n\,0}) * (M_{n\,1}) * (M_{n\,2}) * (M_{n\,3}) \ldots\}$
a series of *adjectives* modifying the subject noun or pronoun, numbering anywhere from "0," no adjectives, to multiple adjectives, "1," "2," "3," and so on

The *Predicate (P_n)* expands into five sub-formulas:

$$P_1 = (M_{v\,b} * V_i)$$
$$P_2 = (M_{v\,b} * V_t) + (M_{n\,a} * (N_o \text{ or } X_o))$$
$$P_3 = (M_{v\,b} * V_l) + \left((M_{n\,a} * (N_p \text{ or } X_p)) \text{ or } (M_{v\,b} * M_p)\right)$$
$$P_4 = (M_{v\,b} * V_t) + (M_{n\,a} * (N_i \text{ or } X_i)) + (M_{n\,a} * (N_o \text{ or } X_o))$$
$$P_5 = (M_{v\,b} * V_t) + (M_{n\,a} * (N_o \text{ or } X_o)) + ((M_{n\,a} * (N_c \text{ or } X_c)) \text{ or } (M_{v\,b} * M_c))$$

where: V_i = an *intransitive verb* that does **not** transmit action onto an object

V_t = a *transitive verb* that **does** transmit action onto an object

V_l = a *linking verb* that links either a predicate noun or pronoun or a predicate adjective back to the subject

(continued)

$M_{vb} = \{(M_{v0}) * (M_{v1}) * (M_{v2}) * (M_{v3}) \ldots\}$
a series of *adverbs* modifying a verb, numbering anywhere from "0," no adverbs, to multiple adverbs "1," "2," "3," and so on

N_o or X_o = a noun or pronoun *object of the verb*

N_i or Xi = a noun or pronoun *indirect object of the verb*

N_p or X_p = a noun or pronoun acting as a *predicate noun*

M_p = an adjective acting as a *predicate adjective*

N_c or X_c = a noun or pronoun *object complement*

M_c = an adjective *object complement*

Sentence Coding Demo of a Paragraph

Chapter 6's final segment leverages all that precedes it. The demonstration scans and codes a sentence stream forming a continuous block of text or paragraph. The stream consists of 13 sentences and 212 words total. All 13 of the sentences are scaffolded by basic sentence structures—B_1 through B_5—and include a variety of the sentence elements that we have studied thus far. The text sample comes from an early draft of a textbook on agricultural engineering processes by engineer S. Milton Henderson (my grandfather) and one of his colleagues. This draft is from S. Milton Henderson's personal archives and is not a direct excerpt from the final manuscript that eventually went into print [1].

I have made a few modifications to the text to ensure that it contains only basic sentence structures and basic sentence elements that we have studied thus far in Part I. Otherwise, the sample's original writing style and content have been preserved. Unlike the single-sentence scan and code examples that I have used to demonstrate sentence algebra applied to each type of basic sentence, aside from several short notes, this example is not comprehensively annotated, sentence by sentence, with a discussion about each equation's coding rationale. Here, the book challenges you, the reader, to follow along the scan and code process and make sense of it as you move along. Immediately after the coding, you will find that the example tallies up frequency of occurrence of each basic sentence type. This tally illustrates one way that sentence algebra can be used as an indicator of an engineering writer's individual writing style.

All in all, I believe that you will see in this demonstration that good engineering writing can be accomplished with only basic sentence equations and components, if properly applied. Simple rather than fancy works. This is not to say that advanced sentence equations and components have no useful purpose. They certainly do, and you will learn more about properly applying these tools in Chaps. 7 and 8.

Sample Text

[1]Some designate engineering "both an art and a science". [2]The artistic side of engineering involves creative ideas, trial and error, extrapolations, and conclusions from patterns. [3]It does not so much require abstract theories. [4]For instance, centuries ago in North America, Native Americans developed effective systematic practices for fertilizing corn with dead fish. [5]They developed these methods with logical thinking, holistic beliefs, and empirical evidence, and not with molecular-scale models from plant biologists and soil scientists. [6]On the other hand, the science part of engineering is math-centered, theoretical, and research-based. [7]For a given system and range of possible inputs and operational conditions, engineering science can predict, monitor, and measure the system's performance, with rigor and precision, sometimes in atomic increments. [8]For example, for a specific set of parameters, engineers can calculate the required air for burning a pound of ethyl alcohol and the resultant heat. [9]However, in practice, art and science combine. [10]A sample of commercial ethyl alcohol might actually produce 12,950 Btu per lb instead of 13,170 Btu as predicted by calculations. [11]In a real-world applied engineering scenario, the preceding 1.67% variation could be acceptable. [12]Regardless, in the end, most would agree on one thing. [13]Engineering innovation has provided humankind, other species, and the earth an array of useful products and processes. [212 words]

Scan and Code

	X_s	V_t	N_o		$N_{c\,1}$	C	$N_{c\,2}$
1.	Some	designate	engineering	"*both*	an art	*and*	a science."

$$B_5 = X_s + V_t + N_o + (N_{c\,1}|N_{c\,2})$$

Note: In this sentence "both/and" work together as a correlative conjunction, which codes as a coordinating conjunction "|".

	$M_{n\,1}$	N_s	$LP(M_{n\,2})$	V_t	$M_{n\,3}$	$N_{o\,1}$	$N_{o\,2}$	C
2.	The artistic	side	of engineering	involves	creative	ideas,	trial	and

	$N_{o\,3}$	$N_{o\,4}$	C	$N_{o\,5}$	$LP(M_{n\,4})$
	error,	extrapolations,	and	conclusions	from patterns.

$$B_2 = (M_{n\,1} * N_s * LP(M_{n\,2})) + V_t + \left[(M_{n\,3} * N_{o\,1})\,|\,(N_{o\,2}|N_{o\,3})\,|\,(N_{o\,4})\,|\,(N_{o\,5} * LP(M_{n\,4})\right]$$

	X_s	V_{aux}	$M_{v\,1}$	$M_{v\,2}$	V_t	$M_{n\,1}$	N_o
3.	It	does	not	so much	require	abstract	theories.

$$B_2 = X_s + (V_{aux} * M_{v\,1} * M_{v\,2} * V_t) + (M_{n\,1} * N_o)$$

4.

$[LP(M_{v1})]^F$	$[M_{v2}]^F$	$[LP(M_{v3})]^F$	N_s	V_t
For instance,	centuries ago	in North America,	Native Americans	developed

M_{n1}	N_o	$LP(M_{n2})$
effective	systematic practices	for fertilizing corn with dead fish.

$$B_2 = [LP(M_{v1})]^F [M_{v2}]^F [LP(M_{v3})]^F \ \ldots \ N_s + V_t + (M_{n1} * N_o * LP(M_{n2}))$$

5.

X_s	V_t	M_n	N_o	$[LP(M_{v1})]^E$
They	developed	these	methods	with logical thinking, holistic beliefs,

	C	M_{v2}	$[LP(M_{v3})]^E$
and empirical evidence,	and	not	with molecular-scale models from

plant biologists and soil scientists.

$$B_2 = X_s + V_t + (M_n * N_o) \ \ldots \ [LP(M_{v1})]^E \mid M_{v2} * [LP(M_{v3})]^E$$

Note: In sentence (5), two end-loaded adverbial phrases formed using the preposition "with" are joined together as an opposing pair with the conjunction "and" (coded "|") and negating adverb "not" (coded "M_{v2}"). Also, as used here, "these" (the plural of "this") functions as a specialized modifier known as a "determiner." We simplify and code it as a regular adjective "M_n."

6.

$LP(M_v)$	N_s	$LP(M_n)$	V_l	M_{p1}
On the other hand,	the science part	of engineering	is	math-centered,

M_{p2}	C	M_{p3}
theoretical,	and	equation-based.

$$B_3 = [LP(M_v)]^F \ \ldots \ (N_s * LP(M_n)) + V_l + \left(M_{p1} | M_{p2} | M_{p3}\right)$$

Note: This B_3 sentence features predicate adjectives joined together by the conjunction "and."

7.

$[LP(M_{v1})]^F$
For a given system and range of possible inputs and operational conditions,

N_s	V_{aux}	V_{t1}	V_{t2}	C	V_{t3}	N'
engineering science	can	predict,	monitor,	and	measure	the system's

N_o	$[LP(M_{v2})]^E$
performance,	with rigor and precision, sometimes in atomic increments.

$$B_2 = [LP(M_{v1})]^F \ \ldots \ N_s + V_{aux} * (V_{t1} | V_{t2} | V_{t3}) + \left(N'\text{-}N_o\right) \ \ldots \ [LP(M_{v2})]^E$$

8.

$[LP(M_{v\,1})]^F$	$[LP(M_{v\,2})]^F$	N_s	V_t	$N_{o\,1}$
For example,	for a specific set of parameters,	engineers	can calculate	the amount

$LP(M_n)$	C	$N_{o\,2}$
of required air for burning a pound of ethyl alcohol	and	the resultant heat.

$$B_2 = [LP(M_{v\,1})]^F[LP(M_{v\,2})]^F \ldots N_s + V_t + ((N_{o\,1} * LP(M_n))|N_{o\,2})$$

9.

$[M_{v\,1}]^F$	$[LP(M_{v\,2})]^F$	$N_{s\,1}$	C	$N_{s\,2}$	V_i
However,	in practice,	art	and	science	combine.

$$B_1 = [M_{v\,1}]^F[LP(M_{v\,2})]^F \ldots (N_{s\,1}|N_{s\,2}) + V_i$$

10.

N_s	$LP(M_{n\,1})$	M_v	V_t
A sample	of commercial ethyl alcohol	actually	might produce

N_o	$LP(M_{n\,2})$
12,950 Btu per lb	instead of 13,170 Btu as predicted by calculations.

$$B_2 = (N_s * LP(M_{n\,1})) + (M_v * V_t) + (N_o * LP(M_{n\,2}))$$

11.

$[LP(M_v)]^F$	$M_{n\,1}$	$M_{n\,2}$	N_s
In a real-world applied engineering scenario,	the preceding	1.67%	variation

V_l	M_p
could be	acceptable.

$$B_3 = [LP(M_v)]^F \ldots (M_{n\,1} * M_{n\,2} * N_s) + V_l + M_p$$

12.

$[M_{v\,1}]^F$	$[LP(M_{v\,2})]^F$	X_s	V_{aux}	$M_{v\,3}$	V_{root}	$LP(M_{v\,4})$
Regardless,	in the end,	many	can	probably	agree	on one thing.

$$B_1 = [M_{v\,1}]^F\ [LP(M_{v\,2})]^F \ldots X_s + (V_{aux} * M_{v\,3} * V_t) * LP(M_{v\,4})$$

13.

N_s	V_t	$N_{i\,1}$	$M_{n\,1}$	$N_{i\,2}$	C
Engineering innovation	has provided	humankind,	other	species,	and

$N_{i\,3}$	N_o	$LP(M_{n\,2})$
the earth	an array	of products and processes.

$$B_4 = N_s + V_t + (N_{i\,1}|(M_{n\,1} * N_{i\,2})|N_{i\,3}) + (N_o * LP(M_{n\,2}))$$

Sentence Stream Underneath the Text

(1) B_5—(2) B_2—(3) B_2—(4) B_2—(5) B_2—(6) B_3—(7) B_2—(8) B_2—(9) B_1—(10) B_2—(11) B_3—(12) B_1—(13) B_4

Frequency

2× B_1
7× B_2
2× B_3
1× B_4
1× B_5

Action Items

α	As you go through your next work week (or school week if you are an engineering student), be on the lookout for B_4-type sentences in everything you read and write. Keep track of how many you discover and copy and paste or write each one into an electronic holding file or just jot each one down on a piece of paper. At the end of a week, how many did you find? Can you scan and underline the core variables in each of these B_4 sample sentences? For reference: B_4 core = (N_s or X_s) + V_t + (N_i or X_i) + (N_o or X_o).
α	Repeat the above action item for B_5 sentences. For reference: B_5 core = (N_s or X_s) + V_t + (N_o or X_o) + (N_c or X_c or M_c). Note: B_4 and B_5 equations occur less frequently in English sentences than B_1, B_2, and B_3 equations. So if you did not find an occurrence of a B_4 and/or B_5 sentence this past week, do not fret. Using the core formulas for each, instead, can you build an example B_4 and B_5 text sentence yourself?
α	If you have read all of Chaps. 2–6, take a one-paragraph (or so) sample of your own writing (or someone else's) and do a scan and code similar to the one illustrated in the final demonstration in Chap. 6. However, to make the process more straightforward, for your scan and code, focus **only** on identifying the core variables in each sentence rather than all sentence elements. If you come across sentences that contain multiple basic structures, that's okay. Just code all of the B_ns that you find. If you come across a sentence that you have trouble coding, skip it—no problem. It may contain advanced sentence structures and elements, which you will learn about soon. Use this activity to become better acquainted with the sentence equation types you (or someone else) tend to favor in your writing. Is your "go-to" sentence a B_1 (subject + intransitive verb), B_2 (subject + transitive verb + object), B_3 (subject + linking verb + subject complement), or another?

Recap

Key Learnings/Takeaways

- The basic sentence template (S + P_n) expands into the *basic sentence equation four* formula as follows: $B_4 = S + P_4 = S + ((M_{v\,b} * V_t) + (M_{n\,a} * (N_i \text{ or } X_i)) + (M_{n\,a} * (N_o \text{ or } X_o)))$.

- The key features of the B_4 equation are that it embodies a standard subject-noun-plus-verb *plus a direct and indirect object.* Like basic sentence B_2, the main verb transfers action directly to a noun or pronoun object (N_o or X_o) in the predicate. Yet with B_4, the action is further transferred from the noun object (N_o) to a second noun in the predicate, known as the indirect object noun or pronoun coded "N_i or X_i."
- The basic sentence template ($S + P_n$) expands into the *basic sentence equation five* formula as follows: $B_5 = S + P_5 = S + ((M_{v\,b} * V_t) + (M_{n\,a} * (N_o \text{ or } X_o)) + ((M_{n\,a} * (N_c \text{ or } X_c)) \text{ or } (M_{v\,b} * M_c)))$.
- The key features of the B_5 equation are that, like a B_4 structure, it transfers action from the main verb's object to another term in the predicate. In this case, the term is known as an *object complement*, which can be either an object complement noun or a pronoun coded "N_c or X_c) or an object complement modifier coded "M_c."
- The relationship between B_5's main (direct) object and its object complement is similar to the relationship linking subject noun (or pronoun) and subject complement in a B_3 equation. However, in the B_5 partnering, there is no linking verb in between. In a B_5 text sentence, you can observe that there is an invisible (elliptical) linking verb form "to be" between the object noun (or pronoun) and the object complement, which clarifies the sentence's logic.
- The sentence algebra coding conventions that have been covered, thus far, in Chaps. 2–6 are as follows:

  – *Code left-to-right, the same as you read and write.*
  – *Use the plus sign (+) and asterisk (*) to signal basic operations.*
  – *Use parentheses and brackets to organize terms.*
  – *Strive to identify and code top-level functional components in sentence equations.*
  – *In sentence equations, the articles ("the," "a," and "an") are usually considered elliptical (invisible) and are not coded with variables.*
  – *In basic and advanced sentence equations, sentence algebra omits the text version's typographic start and stop signals.*
  – *Sentence algebra omits internal punctuation marks inside individual basic sentences.*
  – *The Basic Math Laws apply selectively to sentence algebra equations.*
  – *Free-floating, front-loaded or end-loaded adverbs and adjectives can appear at the beginning or end of a sentence.*
  – *Split verbs are multi-word verb forms that position an adverb (M_v) between an auxiliary verb and principal verb.*

Reference

1. Henderson, S.M., Perry, R.L.: Agricultural Process Engineering, 3rd edn. The AVI Publishing Company, Inc., Westport, CN (1976)

Chapter 7
Advanced Sentence Structures: Compound and Complex Sentences and Relative Clauses

Learning Objectives

- *Understand what an advanced sentence is.*
- *Understand how to combine basic sentences together to form compound and complex sentence structures and the various connector mechanisms used to join one basic sentence to another.*
- *Understand what both a relative pronoun and a relative clause are, how to use them correctly in a text sentence, and how to code them correctly in sentence algebra equations.*
- *Know the difference between standard relative pronouns and compound relative pronouns.*
- *Understand what is meant by "scrambled" and "unscrambled" relative clause construction and (when applicable) how to determine relative pronoun number and case.*

What Are Advanced Sentences?

Chapters 3–6 have examined the sentence algebra fundamentals. Chapters 3 and 4 presented the eight functional roles that words and word groups can play in a sentence, that is, the eight parts of speech and the algebraic variables associated with them. In Chaps. 5 and 6, we studied the five basic sentence structures and the corresponding five basic equations. Chapters 5 and 6 also covered how to vary and accessorize basic equations by incorporating modifiers, pronouns, and prepositional phrases. Chapters 7 and 8 build upon the preceding fundamentals and examine a variety of advanced sentence structures and components.

The basic framework underneath an advanced sentence (A) is always one or more basic sentences or basic sentence variations, and because of this framework, advanced sentences always contain at least one subject noun and main verb pair (N_s + V) creating at least one instance of spark. Furthermore, within advanced

B. Henderson, *A Math-Based Writing System for Engineers*,
https://doi.org/10.1007/978-3-030-10756-7_7

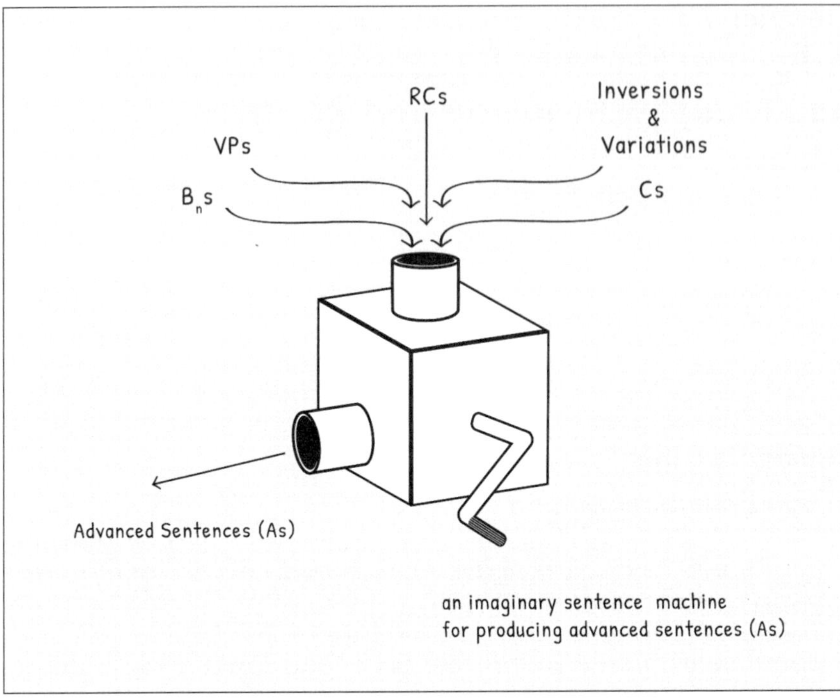

Fig. 7.1 An imaginary sentence engineering machine for producing advanced sentences (As)

sentences, the core sentence structures, whether one or two or more, have additional freedom to display variations and rearrangements in sentence element word order, which grammarians call syntax, and also to incorporate a greater variety of accessory elements.

Here is an overview of the advanced constructions and components to be covered in detail in Chap. 7 as well as the following Chap. 8:

- Compound and complex: basic sentences joined together with conjunctions (C) to create compound and complex combinations
- Bonded sentences: basic sentences (usually two) joined together with a semicolon (";"), colon (":"), or dash ("—")
- Relative pronouns and relative clauses (RCs)
- Verbals and verbal phrases (VPs)
- Inversions and variations (permutations): basic sentences rearranged and rephrased into passive voice, as questions and commands, and with expletive introductions

Figure 7.1 is a creative illustration of a sentence engineering machine. This imaginary machine builds advanced sentences (As). The ingredients that go into the machine's intake manifold include the basic equations (B_ns) already covered in Chaps. 5 and 6, as well as the advanced elements listed above that will be covered in Chaps. 7 and 8.

Compound and Complex Sentence Combinations

Advanced sentence structures can be formed by joining together two or more basic sentences to create longer, more intricate constructions containing two or more core N_s + V *spark*-producing pairs. Like any sentence, these combinations encode a human thought, can be characterized by a sentence algebra equation, and, when written in text, begin with a capital letter and end with a terminal punctuation mark. Because they contain more elements and multiple subject-verb combinations, advanced sentences have the ability to encapsulate more content per unit as compared to a basic sentence.

This segment presents several types of constructions that combine two (or more) basic sentence equations—*compound*, *complex*, *compound-complex*, and *bonded* sentences. Compound, complex, and compound-complex sentence connections are formed using connector parts of speech—specifically, three different types of conjunctions—*coordinating*, *subordinating*, and *correlative conjunctions* (all discussed in Chap. 4). Bonded sentences are formed using connector punctuation: *semicolons*, *colons*, and *dashes*.

Compound Sentences

A *compound sentence* joins together two basic sentences (B_ns) with a coordinating conjunction. In this construction, the two basic sentences that form the advanced sentence are independent subunits (also known as independent clauses). Each subunit is of an equivalent level of importance relative to the other and to the overall idea being expressed by the sentence. Each subunit could function by itself as a stand-alone basic sentence. To review, the main coordinating conjunctions are "and," "but," "or," and "nor"—and also include "for," "yet," and "so." In sentence equations, coordinating conjunctions are coded "|"—which is algebraic shorthand for "+ $C_{(\text{coordinating})}$ +". Coordinating conjunctions are always placed between the items they connect and are not necessarily considered attached to either item. For example:

Engineers design products. (basic sentence, B_2 structure)
and (coordinating conjunction, "|")
Engineers write reports. (basic sentence, B_2 structure)
Engineers design products, ***and*** engineers write reports. (compound sentence, $B_2 | B_2$)

Complex Sentences

A *complex sentence* combines two basic sentences (B_ns) together with a subordinating conjunction. In this construction, within the advanced sentence, one basic sentence functions as an independent subunit. If extracted from the complex sentence, this subunit could stand alone as a grammatically correct sentence. The other basic sentence functions as a subordinate subunit or dependent clause. The subordinate subunit supports the meaning of the independent subunit, and its relative level of importance is lower than or subordinate to its independent partner.

We think of the subordinating conjunction as being attached to the subordinate subunit. If extracted from the parent sentence, the subordinate subunit cannot stand alone as a grammatically correct sentence. To review, the subordinating conjunctions include "if," "although," "though," "when," "while," "as," "since," and "because," among others. In sentence equations, they are coded either "\" (subordinate, or dependent, clause on the left side of the independent clause) or "/" (subordinate, dependent, clause on the right side of the independent). "\" and "/" are sentence algebra shorthand for "$+ C_{(\text{subordinating})} <$" and "$> C_{(\text{subordinating})} +$", respectively. For example:

> Engineers write reports. (basic sentence, B_2 structure)
> ***because*** their jobs require this task (subordinate basic clause, / B_2 structure)
> Engineers write excellent reports ***because*** their jobs require this task. (complex sentence, B_2 / B_2)

Compound-Complex Sentences

Compound-complex structures combine both of the preceding types of advanced sentences. The result is an advanced sentence composed of two or more independent subunits and one or more dependent (subordinate) subunits. Note that sentence algebra considers a compound sentence that is composed of more than two independent clauses (three or more) still to be a compound sentence, not compound-complex, and, likewise, a complex sentence composed of one independent clause and two or more dependent clauses still to be a complex sentence, rather than compound-complex. The recipes discussed here for combining basic sentences to form compound, complex, and compound-complex constructions also apply when combining the basic sentence inversions and variations to be discussed later in Chap. 8. For example:

> Engineers design products ***and*** engineers write reports ***because*** their jobs require these tasks. (compound-complex sentence, all three previous example structures combined, B_2 | B_2 / B_2)

One variation of the compound sentence deserves discussion prior to moving on to this segment's series of examples. This is when two basic sentences, each of relatively equivalent import, are joined together with a *correlative conjunction*, a special coordinating conjunction consisting of paired instead of single words—for example, "not only-but also," "either-or," or "neither-nor." For this variation, a subtle difference appears in the text version: one part of the two-word conjunction attaches to the front of one basic sentence, and the second word attaches to the front of the second basic sentence. At the equation level, however, the algebraic coding stays the same. To illustrate, here are two B_1 sentences joined together to form a compound advanced sentence, first with a coordinating conjunction and then with a correlative conjunction (the same formula guiding both):

$B_1 \mid B_1$	(compound sentence equation)
I win, ***and*** you lose.	(text with coordinating conjunction)
Either I win, ***or*** you lose.	(text with correlative conjunction)

Compound and Complex Sentences In-action

Below are five more detailed examples showcasing conjunctions joining B_n subunits (basic clauses with modifiers). The conjunctions are shown in bold ***italic***, the core variables are underlined, and each resultant advanced sentence coding is displayed underneath the text.

Our team designed a top-notch, quad-redundant hydraulic actuator for controlling dual ailerons on large-scale commercial airframes, ***and*** we sold it to the highest bidder. 7.1

$(B_2 \mid B_2)$ *compound sentence*

Our team designed a top-notch, quad-redundant hydraulic actuator for controlling dual ailerons on large-scale commercial airframes, ***although*** unfortunately, our competitors designed an even better one. 7.2

(B_2 / B_2) *complex sentence*

While the linear coefficient of expansion, α_L, of water is 0.00012, the coefficient for liquid ammonia calculates at 0.00136, ***and*** this 11X increase makes the liquid ammonia a better choice for working fluid inside the temperature-controlled expansion valve. 7.3

$(B_3 \backslash B_1 \mid B_5)$ *compound-complex*

After she reviewed the manufacturing cost estimates for both options, the project manager made the corrugated cylinder design the top candidate over the piston-in-cylinder design, ***even though*** the failure mode and effects analysis rated the piston design slightly more robust. 7.4

$(B_2 \backslash B_5 / B_5)$ *compound-complex*

After testing prototypes and achieving poor results, the project manager gave the team more time, ***and*** then the team went back to the proverbial drawing board, ***and*** they redesigned the hydraulic valve, ***so*** the version 2.0 valve incorporated the original corrugated cylinder mechanism. 7.5

$(B_4 \mid B_1 \mid B_2 \mid B_2)$ *compound sentence*

Generally, the examples above demonstrate that using the variable "C" and the corresponding connection-type symbols "|," "\," and "/" to combine and organize B_n subunits into advanced sentences is a relatively straightforward task. However, just because it is easy to use conjunctions to chain together basic sentences to form advanced sentences does not mean that this practice is routinely preferable. Examples 7.1 through 7.4 above are effective compound/complex structures. On the other hand, sentence 7.5 probably would be easier for readers to understand if the sentence were revised into two (or more) shorter segments. For instance:

After testing prototypes and achieving poor results, the project manager gave the team more time. Then the team went back to the proverbial drawing board, and they redesigned the hydraulic valve, so the version 2.0 valve incorporated the original corrugated cylinder mechanism.

Sentence Engineering Tip

The sentence algebra formula for a complex sentence will either look like this "$B_n \setminus B_n$" or look like this "B_n / B_n." Think of it this way:

1. When a subordinate (dependent) B_n clause joins with an independent B_n clause, the subordinate clause goes *under* the slanted line.
2. When an independent B_n clause joins with a subordinate (dependent) B_n clause, the independent clause goes *above* the slanted line.

For example, consider the following pair of B_2 structures, with "because" attached to and subordinating the first: "*Because* visual perspective can often assist the engineering writer, a smart writer applies it." Here, subordinate comes first, and thus is located *under* the slanted line in the complementary equation, $B_2 \setminus B_2$.

Bonded Sentences

Instead of using a C = conjunction, another way to connect two (or more) basic sentences to form advanced sentences is with an appropriate *connector punctuation mark*, such as one of the following:

- Semicolon ";"
- Colon ":"
- Dash "—"

Later, in the Part II chapters on sentence optimization, you will learn more about how to correctly punctuate text sentences—both how to organize elements within a sentence using commas and how to join stand-alone basic sentences (independent clauses) together with connector punctuation marks.

At the equation level in sentence algebra, basic sentences (independent clauses) joined together with connector punctuation marks like a semicolon (";") are referred to as *bonded sentences*. For all connector punctuation marks, the sentence algebra notion for a B_n to B_n bond is a tilde "~" as in "$B_n \sim B_n$." Here is an example:

Two basic sentences
B_2 = The managers observed the X2 Drone's excellent performance at the prototype trials.
B_5 = They selected it best choice for commercial development.

Example bonded sentence
$B_2 \sim B_5$ = The managers observed the X2 Drone's excellent performance at the prototype trials; they selected it best choice for commercial development.

Relative Pronouns (X_rs) and Relative Clauses (RCs)

Many of the sentences that you have encountered thus far in Chaps. 2–7 have contained prepositional phrases (LPs)—multi-word subassemblies that function as adjectives or adverbs. *Relative clauses (RCs)* are another type of multi-word subassembly that perform a part-of-speech function. Like a prepositional phrase (LP), a relative clause is built around a lead element. However, whereas a preposition (L) heads up an LP, a relative pronoun (X_r) heads up an RC.

Like the other pronouns, a relative pronoun can function as a placeholder word that "stands in" for a noun located elsewhere—its antecedent. Sometimes, the antecedent is nearby in the same sentence and precedes the relative clause that the relative pronoun heads up. Sometimes, relative pronouns serve as placeholders for persons, places, and things outside of the sentence in which their companion RC resides. And sometimes, relative pronouns have no antecedent at all and head up RCs as a construction device. In this case, the relative pronoun defines a word cluster (more specifically, the relative pronoun and the clause of which it is a part) as a relative clause and something that grammar logic allows as a legitimate accessory element in English sentences.

Regardless of whether a relative pronoun functions as a placeholder word or construction device and signal word, once you use a relative pronoun to form a relative clause, the clause becomes an accessory element. Also, like subordinate clauses that begin with subordinate conjunctions, RCs are dependent. They play a subordinate role relative to the sentence in which they reside. RCs cannot stand alone as grammatically complete sentences. Always situated within a parent sentence, relative clauses regularly serve as adjectives. In some cases, relative clauses function as nouns.

The standard relative pronouns include:

who, whose, whom, which, that

Relative pronouns that serve as placeholder words (and that link to an antecedent) also embody the properties of *number* and *case*. Regarding number, a relative pronoun adopts the same number as its antecedent *outside the relative clause.* Regarding case, it's the opposite. A relative pronoun adopts case (subjective, objective, or possessive) based on the role it plays *within (inside) the relative clause* it leads. The only standard relative pronoun that changes its spelling based on case, however, is "who"—as in *who*, *whom*, and *whose*, respectively. When a relative pronoun serves as a construction device and signal, both number and case are irrelevant. It's always used and spelled the same way.

There are also compound relative pronouns. These include

whatever, whoever, whomever, whosever, whichever

Compound relative pronouns behave the same way that standard relative pronouns do regarding function, number, and case. The compound relative pronouns also include a variety of "-soever" versions in addition to the "-ever" versions—for example, "whatsoever" and "whosoever." The more straightforward, concise "-ever" spellings are preferred in engineering writing.

Here are some key fundamentals to know and keep in mind when you are creating and using relative clauses (RCs) in parent sentences:

- When a relative pronoun functions as a placeholder word, remember that the *number* of that relative pronoun—standard or compound—is determined by the pronoun's *antecedent outside the relative clause*, and not its functional role within the clause.
- Conversely, remember that the *case* of relative pronoun is determined by its *functional role inside the relative clause*, and not its antecedent outside of the clause.
- No matter what role a relative pronoun plays within a relative clause, its word order *always* places the relative pronoun *first*, even if doing so requires "scrambling" the standard word order (syntax) of the clause.
- Relative pronouns—especially *that*—are sometimes "written in invisible ink" (that is, used as elliptical elements).

  For example, unless we factor in an elliptical X_r, the sentence "This is an assignment I want" might appear to be two sentences fused together, not one, (i.e., "This is an assignment" fused with "I want"). Actually, what the writer means here is "This is an assignment [that] I want." Although leaving a relative pronoun in or out of the sentence often makes no difference to most native speakers of English, sometimes leaving them out can cause confusion for multilingual speakers. To make your engineering sentences readily understandable to engineering readers in the global workplace, I recommend minimizing your use of elliptical relative pronouns.

Below are five examples of relative clauses nested inside parent sentences, with each parent sentence's text scanned and then coded. In the original text sentences, the RCs are italicized to make them easy to see. In the equations, all relative clauses are labeled "RC" with each RC's specific part-of-speech function defined in parentheses, for example, $RC(M_n)$ = "relative clause functioning as an adjective."

In addition, after doing a first-order coding of each example's parent sentence, the relative clause(s) in the sentence is extracted, unscrambled (if applicable), and then further decoded at the clause level. Here, the disassembly of the RCs illustrates the process of going beyond first-order coding of a sentence and doing *second-order coding* of specific sentence elements.

Coding Demonstration 7.6

Original: The reason *that he went to Cleveland* was the annual ASME conference on fluid dynamics in commercial applications.

Scanned:

N_s	$RC(M_{n\,1})$	V_l	N_p
The reason	*that he went to Cleveland*	was	the annual ASME

N_p (cont.)	$LP(M_{n\,2})$
conference	on fluid dynamics in commercial applications.

Coded: $B_3 = (N_s * RC(M_{n\,1})) + V_l + (N_p * LP(M_{n\,2}))$

$RC(M_{n\,1}) =$ *that he went to Cleveland* $\quad X_r =$ *that*

$= X_r + (X_s + V_i * LP(M_v))$ (not scrambled)

$= X_r + (B_1)$

RC function = *adjective* modifying subject noun

X_r function = relative pronoun used as a *construction device*

Discussion: Example 7.6 illustrates a scenario where the grammatical role of the relative pronoun "that" is to head up an RC construction, with the pronoun itself acting as a construction device. In the parent sentence (a B_3 structure), the relative clause $RC(M_{n\,1})$ = "that he went to Cleveland" functions as an adjective modifying the noun subject N_s = "The reason." Disassembly of the relative clause reveals the clause has a standard (not scrambled) B_1 structure with X_r = "that" added to the front.

Coding Demonstration 7.7

Original: Because of her extensive technical experience gained from 15 years of prior employment at Boeing, we predict *that she will make a significant contribution at Bui Aerospace*

Scanned:

$[LP(M_v)]^F$: Because of her extensive technical experience gained from 15 years of prior employment at Boeing,

X_s: we

V_t: predict

$RC(N_o)$: *that she will make a significant contribution at Bui Aerospace.*

Coded: $B_2 = [LP(M_v)]^F \ldots (X_s + V_t + RC(N_o))$

$RC(N_o)$ = *that she will make a significant contribution at Bui Aerospace* $X_r =$ *that*

$= X_r + (X_s + V_t + (M_{n1} * N_o) \ldots [M_v]^E)$ (not scrambled)

$= X_r + (B_2)$

RC function = *noun object* of "predict"

X_r function = relative pronoun used as a *construction device*

Discussion: Example 7.7 also illustrates a scenario where the relative pronoun "that" heads up an RC construction and functions as a construction device. This time, the parent sentence is a B_2 structure that begins with a free-floating, frontloaded, prepositional phrase, which is led off with the compound preposition "because of." The phrase modifies the main verb V_t = "predict" (specifies why the action occurs). Inside the parent sentence, the clause $RC(N_o)$ = "that she will make a significant contribution at Bui Aerospace" functions as the noun object of the main verb. Here, the second-order coding of the relative clause shows that its frame is a standard (not scrambled) B_2 structure with X_r = "that" added onto the front.

Coding Demonstration 7.8

Original: The electrical engineer calculated the proper deposition layer thickness *that the IC chip required.*

Scanned:

N_s	V_t	$M_{n\,1}$	N_o
The electrical engineer	calculated	the proper	deposition layer

	$(RC(M_{n\,2}))$
thickness	*that the IC chip required*.

Coded: $B_2 = N_s + V_t + (M_{n\,1} * N_o * (RC(M_{n\,2}))$

$RC(M_{n\,2})$ = *that* *the IC chip* *required* $(X_o)_r$ = *that*

$= (X_o)_r + N_s + V_t$ (scrambled)

$= N_s + V_t + (X_o)_r$ (unscrambled)

$= B_2$

= *the IC chip* *required* *that*

RC function = *adjective* modifying object noun

$(X_o)_r$ function = object of the verb "required" (case *objective*)

$(X_o)_r$ antecedent = "deposition layer thickness" (*singular*)

Discussion: Example 7.8 displays a relative pronoun going beyond the function of construction device. It both heads up a relative clause and plays a grammatical role within the clause itself. The parent sentence is a B_2 structure. Inside the parent sentence, the clause $(RC(M_{n\,2}))$ = "that the IC chip required" functions as an adjective modifying the compound noun object N_o = "deposition layer thickness." A further scan and code of the relative clause itself unveils a "scrambled" B_2 structure underneath—the relative clause and also verb object "that" is positioned first, the noun subject "IC chip" comes next, and the transitive verb "required" comes last. The coding also shows the RC's "unscrambled" version. Here, the antecedent of "that" is the singular compound noun, "deposition layer thickness," located outside of the relative clause in the parent sentence.

Coding Demonstration 7.9

Original: I assigned the technician *whom I had just hired* and *who scored 100% on the application test* the lead role on my qualification test team for the butterfly valve project.

Scanned:

X_s	V_t	N_i	$RC(M_{n\,1})$	C
I	assigned	the technician	*whom I had just hired*	and

$RC(M_{n\,2})$	N_o
who scored 100% on the application test	the lead role

$LP(M_{n\,3})$
for my qualification test team for the butterfly valve project.

Coded:

$$B_4 = X_s + V_t + (N_i * (RC(M_{n\,1}) \mid RC(M_{n\,2})) + (N_o * LP(M_{n\,3}))$$

$RC(M_{n\,1})$ = *whom I had just hired* $\quad (X_o)_r$ = *whom*

$= (X_o)_r + X_s + (V_{aux} * M_v * V_t)$ (scrambled)

$= X_s + (V_{aux} * M_v * V_t) + (X_o)_r$ (unscrambled)

$= B_2$

= *I had just hired whom*

RC function = *adjective* modifying indirect object noun

X_r function = object of the verb "had just hired" (case *objective*)

X_r antecedent = "technician" (*singular*)

$RC(M_{n\,2})$ = *who scored 100% on the application test* $\quad (X_s)_r$ = *who*

$= (X_s)_r + V_t + (N_o * LP(M_v))$ (not scrambled)

$= B_2$

RC function = *adjective* modifying indirect object noun

X_r function = subject of the verb "scored" (case *subjective*)

X_r antecedent = "technician" (*singular*)

Discussion: In example 7.9, we get a chance to examine a parent sentence (a B_4 structure) that contains two relative clauses within it—one "scrambled" and one "unscrambled." Example 7.9 also validates that RCs, like most other sentence elements, can be compounded (two, three, or so on) using a conjunction, in this case, "and." Both relative clauses are led by a relative pronoun that stands in for the noun "technician," which refers to a singular person. And both relative clauses function as adjectives modifying "technician." However, the first RC (a scrambled B_2 structure) leads off with an objective case pronoun "whom," and the second RC (a standard B_2 structure) leads off with a subjective case pronoun "who." The second-order coding and analysis of the RCs trace through why one spelling is "whom" and the other is "who."

Coding Demonstration 7.10

Original: The idea of using Teflon, instead of stainless steel, to make the poppet valve came from a summer intern *whose ideas often rivaled those of manufacturing engineers already on our payroll.*

Scanned:

N_s — The idea
$LP(M_n)$ — of using Teflon, instead of stainless steel, to make the poppet valve
V_i — came
$LP(M_v)$ — from a summer intern *whose ideas often rivaled those of manufacturing engineers already on our payroll.*

Coded: $B_1 = (N_s * LP(M_n)) + (V_i * LP(M_v))$

where

$LP(M_v)$ = from a summer intern *whose ideas often rivaled those of manufacturing engineers already on our payroll*

$= L + (M_{n1} * N_o * RC(M_{n2}))$

and

$RC(M_{n2})$ = *whose ideas often rivaled those of manufacturing engineers already on our payroll* $(X_r)' = whose$

$= (X_r)'\text{-}N_s + (M_v * V_t) + (X_o * LP(M_n))$ (not scrambled)

$= B_2$

RC function = *adjective* modifying "summer intern" (noun object of preposition "from")

X_r antecedent = "intern" (*singular*)

X_r case = *possessive* (indicates possession of "ideas")

Discussion: Finally, example 7.10 illustrates that, just as a relative clause can be nested in a parent sentence, a relative clause can also be nested inside another phrase or clause—in this scenario, a prepositional phrase led by "from." The first four examples in the series show relative pronouns taking either the subjective or objective cases. In 7.10, the relative pronoun, "whose," operates in the possessive case. Like the other RCs examined in examples 7.6 through 7.9, the antecedent of "whose," which is "summer intern" (singular), determines the relative pronoun's number, and its function within the relative clause that it leads determines its case.

As mentioned already, sometimes relative pronouns can function elliptically in sentences—that is, they can be grammatically present but written in "invisible ink." In the above series, you may have noticed, for instance, that the original version of sentence 7.6, "The reason *that he went to Cleveland* was the annual ASME conference on fluid dynamics. . ." reads smoothly with "that" eliminated, "The reason *he went to Cleveland* was the annual ASME conference on fluid dynamics. . ." In fact, the relative pronouns "that," "who," and "whom" can drop out of the text of sentences 7.7, 7.8, and 7.9 as well without catastrophic impact. On the other hand, in sentence 7.10, the last example of the series, eliminating "whose" from the text and making the relative pronoun elliptical causes the remaining sentence to be awkward and confusing, "The idea of using Teflon … to make the poppet valve came from a summer intern *ideas often rivaled those of manufacturing engineers …?*" Even when dropping out a relative pronoun might work for some readers, to ensure accessibility to all readers, especially multilingual professionals in the global workplace, most of the time, it's best to keep relative pronouns "visible." Doing so also helps ensure rigor and precision.

Action Items

α	Take a 5-minute time-out during your work day, and write three basic sentences that describe three things on your "to do list" for tomorrow. For example: "I need to call several parts supply vendors to determine best pricing for servo motors." "Before I begin work, I plan to go to the gym." "I will finish my CAD layout of the redesign." Then assemble three advanced sentences, using coordinating and subordinating conjunctions and following the following three sentence recipes $(B_n \mid B_n \mid B_n)$, $(B_n \setminus B_n \mid B_n)$, and $(B_n / B_n \mid B_n)$
α	Select a short document (one page or less) written by a peer or colleague, and play the role of editor. Print out a hard copy of this sample and read through it, underlining each relative clause you come across (and hopefully you will come across some). Also be on the lookout for any invisible (elliptical) "that"s, "who"s, and "whom"s. Consider adding them back in. Also, determine which of the RCs that you identify are "scrambled," and write an "unscrambled" version of each

(continued)

α	The next time *you* write a short document (one page or less), play the role of editor again. Print out a hard copy of this sample and read through it, underlining each relative clause you come across (and hopefully you will come across some). Also be on the lookout for any invisible (elliptical) "that"s, "who"s, and "whom"s. Consider adding them back in. Also, determine which of the RCs that you identify are "scrambled," and write an "unscrambled" version of each. Now consider your findings about relative clauses from your examination of a peer or colleague's writing sample versus yours. What's different and what's the same about your uses of relative clauses versus theirs?

Recap

Key Learnings/Takeaways

- An advanced sentence (A) is one or more basic sentences (B_ns) or basic sentence variations. Like a basic sentence, an advanced sentence's communication purpose is to encode and express a thought—although an advanced sentence's level of specificity and encoded content can, by design, be more intricate and multifaceted than that expressed by a basic sentence.
- To form a compound sentence, a writer joins two or more basic sentences using coordinating conjunctions (and sometimes correlative conjunctions).
- To form a complex sentence, a writer joins two or more basic sentences using subordinating conjunctions.
- The connector mechanisms for joining basic sentences to form longer advanced sentences include the following:
  - Coordinating conjunctions (e.g., "and," "but," "or"), which are coded in sentence equations using the symbol "|."
  - Subordinating conjunctions (e.g., "because," "although," "since"), which are coded in sentence equations using the symbols "\" (dependent left, independent right) or "/" (independent left, dependent right).
  - Connector punctuation marks, namely, semicolons (";"), colons (":"), and dashes ("—"). Sentence algebra calls these constructions "bonded sentences," and in equations, these connections are represented by a tilde "~" symbol.
- Like a prepositional phrase (LP), a relative clause is a subassembly built around a lead element. Whereas a preposition (L) heads up an LP, a relative pronoun (X_r) heads up an RC. In relative clauses, the relative pronoun can function as a construction word and signal of clause type. Or the relative pronoun can function as a placeholder word linked to an antecedent outside of the relative clause, with the relative pronoun also functioning as a core variable within the clause.
- The standard relative pronouns include *who*, *whose*, *whom*, *which*, and *that.*
- There are also compound relative pronouns. These include *whatever*, *whoever*, *whomever*, *whosever*, and *whichever.*

- Relative pronouns are impacted by the properties of number and case.
  - Regarding *number*, relative pronouns adopt the same number as their antecedent outside the relative clause.
  - Regarding *case*, however, relative pronouns adopt case (subjective, objective, or possessive) based on the role that they play within the relative clause they lead.
  - The only standard relative pronoun that changes its spelling based on case, however, is "who"—as in who, whom, and whose, respectively.
- Coding, decoding, and incorporating relative clauses into engineering text require sentence engineers to be able to discern and interpret "scrambled" relative clauses (i.e., clauses with a word order that does not map back to a standard basic sentence structure) and also to be able to determine a relative pronoun's type, number, and case.
- Finally, sentence engineers must also be aware that sometimes relative pronouns leading relative clauses are written in "invisible ink." In general, however, this book does not recommend this practice and advises engineering writers to make relative pronouns "visible."

Chapter 8
More Advanced Sentence Structures: Verbal Phrases, Inversions, and Variations

Learning Objectives

- *Understand what both a verbal and verbal phrase are, how to use them correctly in text sentences, and how to code them correctly in sentence algebra equations.*
- *Understand how to phrase a sentence in the passive voice; as a question, command, or request; and with an expletive introduction.*
- *Understand conceptually what "inversion" and "variation" mean when applied to sentence algebra's five basic sentences.*
- *Know how to scan and code a longer text sample containing a stream of sentences containing both basic and advanced sentence structures.*

Verbals (V_vs) and Verbal Phrases (VPs)

In sentences, another multi-word element that can function as a part of speech (adjective, adverb, or noun) is the *verbal phrase* (*VP*). Verbal phrases are constructed around a verb form known as a *verbal* (V_v), and just as prepositions head up prepositional phrases and relative pronouns head up relative clauses, verbals head up verbal phrases. Verbals do not partner with subject nouns or pronouns, and do not assume the main verb slot in sentences. Instead, verbals and the verbal phrases that they lead are located in a variety of other places in sentences. A verbal can also appear alone.

When a verb is used as a verbal rather than as a main verb, it can appear in several forms. A verbal's *infinitive form* takes the shape of the verb's root form or first dictionary listing. When used in a sentence, however, an infinitive verbal is typically distinguished with *the sign of the infinitive* "*to*" placed immediately before it. As we learned in Chap. 3, the second verb form listed in a dictionary is simple past tense,

B. Henderson, *A Math-Based Writing System for Engineers*,
https://doi.org/10.1007/978-3-030-10756-7_8

and then beyond this (depending on the dictionary and whether the verb is regular or irregular), there exist two other forms—*past participle* and *"-ing" form*. These can both function as verbals.

We learned in Chap. 3's segment on V = verb that the *past participle* for a regular verb is spelled identically to its simple past tense form—that is, add the suffix "-ed." Past participles for irregular verbs have unique spellings. On the other hand, the "-*ing*" form, whether regular or irregular verb, is *always* spelled the same way; that is, it adds the suffix "-ing." This form represents not one but rather *two different verbal types*—the *present participle* and also something known as the *gerund*. Thus, with respect to dictionary listing, a verb's first form (root), third form, and fourth form can function as verbals. The second form, simple past tense, technically, cannot—although as just mentioned, for regular verbs both the second and third forms are spelled the same way (by adding "-ed").

The ***four types of verbals*** (V_v***s)*** and the ***four types of verbal phrases (VPs)*** that they lead—infinitive, past participle, present participle, and gerund—are examined in more detail below and illustrated with brief examples:

- An ***infinitive*** is a root verb with the word "to" placed in front of it—for example, "to incorporate," "to calculate," and "to listen." It is important to note that in this formation, "to" acts as "the sign of the infinitive" *and not* as a preposition. Within a sentence, an infinitive verb or verbal phrase can function as a *noun*, *adjective*, or *adverb*. Below are examples of an infinitive verb phrase functioning each of these three ways in a sentence. In the first example sentence, the infinitive phrase is the noun subject of the main verb "would be." In the second sentence, the phrase works as an adjective adding additional specificity to the sentence's subject "The best way." The third example shows an infinitive phrase modifying the main verb "must push" (push *why?*) along with another adverb "firmly" (push *how?*):

  To incorporate the company would be advantageous for employees. (noun)

  The best way to repair the collapsed vessel is with a sonic weld. (adjective)

  You must push firmly to engage the machine's off switch. (adverb)

- A ***past participle*** is a regular root verb with an "-ed" attached to the end. For irregular verbs, though, the past participle is spelled irregularly. For instance, "modeled" and "designed" are regular past participles. "Drawn" and "built" are irregular past participle forms. (Note: how to spell an irregular verb to form its past participle is not a closely guarded secret. For all irregular verbs, this spelling is listed as the third dictionary entry, right after the verb's past tense spelling—for example, [to] draw, drew, *drawn*.) In a sentence, a past participle verbal or verbal phrase usually functions as an *adjective*—modifying either a noun or pronoun element. An example of a past participial phrase functioning as an adjective (modifying the sentence's subject noun, "The conceptual sketch") would be:

  The conceptual sketch drawn by Ethan showcased a breakthrough design. (adjective)

- A ***present participle*** is a root verb with an "-ing" attached to the end. This spelling is consistent for both regular verbs and irregular verbs—for instance, "calculating," "making," and "lying." Present participles alone, and present participle phrases, like past participles, are often used as *adjectives* modifying nouns and pronouns. An illustration of a present participle verb phrase follows. It functions as an adjective modifying "the engineer":

  Calculating stress thresholds silently in her head, the engineer said little while she ate dinner with her husband. (adjective)

- A ***gerund*** takes the same form as a present participle—a root verb with an "-ing" attached to the end. The difference is that the *purpose of gerunds and gerund phrases is to function as nouns* rather than modifiers. For example, another version of the above sentence with the verbal verb "calculating" this time functioning as a gerund in a gerundial phrase would be the following. Here, the verb phrase "calculating stress thresholds" functions as the object of the verb "continued" (as in continued *a specific type of task*).

  After dinner, she continued calculating stress thresholds until she completed the task. (noun)

Table 8.1 is a checklist snapshot showing each verbal's basic recipe. The table also displays how each verbal (and resultant verbal phrase) typically functions in a sentence.

When verbals head up phrases, these phrases are constructed like the predicate of a basic sentence or some variation of a basic predicate. Depending on specific verb choice and phrase construction, verbals can be intransitive $(V_i)_V$, transitive $(V_t)_V$, or linking $(V_l)_V$. Thus, some verbals express action in and by themselves; others transfer action onto and between objects and complements located after (to the right of) a transitive verbal; and infrequently, in verb phrases, they behave as linking verbs (when situated in constructions such as those discussed next).

Below are five examples of verbal phrases nested inside parent sentences, with each parent sentence's text scanned and then coded. In the original text sentences, the VPs are italicized to make them easy to see. In the equations, all verbal phrases

Table 8.1 Conversions and capabilities for root verbs into verbals

		Functions		
Verbals ("V_v"s)	Basic recipe	Noun	Adjective	Adverb
Infinitive	"to" + V_{root}[a]	✓	✓	✓
Past participle	V_{root} + "-ed"[b]		✓	
Present participle	V_{root} + "-ing"		✓	
Gerund	V_{root} + "-ing"	✓		

[a]Note: V_{root} is a verb's first dictionary entry (the "look it up" word) for both regular and irregular verbs

[b]Note: The past participle for irregular verbs is a respelling displayed as the third dictionary entry

are labeled "VP" with each VP's specific part-of-speech function defined in parentheses, for example, $VP(M_n)$ = "verbal phrase functioning as an adjective."

Similar to the Chap. 7 coding demonstrations of RCs in sentences, here, after doing a first-order coding of each example sentence, the verbal phrase(s) in the sentence is extracted and then further decoded at the phrase level. The second-order coding process notes whether the verbal is transitive, intransitive, or linking and also reveals which basic predicate type (P_1, P_2, P_3, P_4, or P_5) scaffolds the phrase's core structure. In places, the second-order coding also further investigates verbal phrases (VPs) nested inside prepositional phrases (LPs).

Coding Demonstration 8.1

Original: As the twenty-first century progresses, the petroleum industry is continuously striving *to develop new methods for safe extraction of oil.*

Scanned:

C_s	N_s	V_i	N_s
As	the 21st first century	progresses,	the petroleum industry

V_{aux}	$M_{v\,1}$	V_i	$VP(M_{v\,2})$
is	continuously	striving	*to develop new methods for safe extraction of oil.*

Coded: $B_1 \backslash B_1 = (N_s + V_i)\backslash(N_s + ((V_{aux} * M_{v\,1} * V_i) * VP(M_{v\,2}))$

$VP(M_{v\,2})$ = *to develop* new methods for safe extraction of oil $\qquad (V_v)_t$ = *to develop*

$= (V_t)_v + (M_{n\,1} * N_o * LP(M_{n\,2}))$

VP function = adverb modifying main verb "is striving"

VP core = *to develop* methods (infinitive phrase)

$= (V_t)_v + N_o$ (scaffolded with a P_2-type predicate)

Discussion: Example 8.1 is a $B_1 \backslash B_1$ complex sentence. The first B_1 is a short, subordinate clause joined (with the C_s = subordinating conjunction "as") to a second, longer B_1-type independent clause. In the second clause, the infinitive verbal phrase, "*to develop* new methods for safe extraction of oil," led by the infinitive verbal "to develop," adverbially modifies the main verb "is striving" (*how so?*). This verb is split and also modified by "continuously" (*when?*), which is situated in between the auxiliary verb "is" and principal verb "striving." An additional level of scanning and coding reveals that the infinitive verbal phrase is framed around a basic P_2 predicate.

Coding Demonstration 8.2

Original: In response to market demand, the petroleum industry has developed an effective yet controversial method *to supply consumers oil from underground shale*.

Scanned:

$[LP(M_v)]^F$	N_s	V_t
In response to market demand,	the petroleum industry	has developed

$M_{n\,1}$	C_c	$M_{n\,2}$	N_o	$VP(M_{n\,3})$
an effective	yet	controversial	method	*to supply consumers oil from underground shale*.

Coded: $B_2 = [LP(M_v)]^F \ldots N_s + V_t + (M_{n\,1} | M_{n\,2}) * N_o * VP(M_{n\,3})$

$VP(M_{n\,3})$ = *to supply* consumers oil from underground shale $\quad (V_t)_v$ = *to supply*

$= (V_t)_v + N_i + (N_o * LP(M_n))$

VP function = adjective modifying noun object "method"

VP core = *to supply* consumers oil (infinitive phrase)

$= (V_t)_v + N_i + N_o$ (scaffolded with a P_4-type predicate)

Discussion: In example 8.2, we see a B_2 sentence accessorized with basic and advanced components. The noun subject is "the petroleum industry," the main verb is "has developed," and its object is "method." A free-floating, front-loaded prepositional phrase functioning as an adverb (*why?*) modifies "has developed." Several adjectives modify "method," two joined together with the coordinating conjunction "yet" and another by itself, the infinitive verbal phrase, "to supply consumers oil from underground shale." In the verbal phrase, the direct object of "to supply" (supply *what?*) is "oil" and the indirect object (supply *to whom?*) is "consumers." Thus, this VP uses a basic sentence four predicate (P_4) as its frame.

Coding Demonstration 8.3

Original: "Fracking technology" offers a quick-fix solution for *dwindling supplies of domestic oil* and *increasing reliance on foreign sources.*

Scanned:

N_s	V_t	M_n	$N_{o\,1}$	L
"Fracking technology"	offers	a quick-fix	solution	for

$VP(N_{o\,2})$	C_c	$VP(N_{o\,3})$
dwindling supplies of domestic oil	and	*increasing reliance on foreign sources.*

Coded: $B_2 = N_s + V_t + (M_n * N_{o\,1}) * (L + (VP(N_{o\,2}) \mid VP(N_{o\,3})))$

$VP(N_{o\,2})$ = *dwindling* supplies of domestic oil $(V_t)_v$ = *dwindling*

$= (V_t)_v + (N_o * LP(M_n))$

VP function = noun object of preposition "for"

VP core = *dwindling* supplies (gerundial phrase)

$= (V_t)_v + N_o$ (scaffolded with a P_2-type predicate)

$VP(N_{o\,3})$ = *increasing* reliance on foreign sources $(V_t)_v$ = *increasing*

$= (V_t)_v + (N_o * LP(M_n))$

VP function = noun object of preposition "for"

VP core = *increasing* reliance (gerundial phrase)

$= (V_t)_v + N_o$ (scaffolded with a P_2-type predicate)

Discussion: Sentence 8.3 is a B_2 with the core elements, *subject* "fracking technology," *transitive verb* "offers," and *object* "solution" ($N_{o\,1}$). What is interesting about this sentence is that it features a prepositional phrase (LP) built around L = "for," which has a compound noun phrase for its object. Specifically, inside the LP, this compound object is built out of two verbal phrases, each headed by gerunds and joined together with the coordinating conjunction C_c = "and"—and coded as "$VP(N_{o\,2}) \mid VP(N_{o\,3})$." In text form, the VPs are "*dwindling* supplies of domestic oil" and "*increasing* reliance on foreign sources." Both VPs are structured as P_2-type predicates. This sentence is an excellent example of phrases nesting inside other phrases.

Coding Demonstration 8.4

Original: However, the process pollutes underground water tables *supplying water for human consumption and agricultural irrigation*, and fracking is not a viable long-term solution.

Scanned:

$[M_{v\,1}]^F$	N_s	V_t	N_o
However,	the process	pollutes	underground water tables

$VP(M_{n\,1})$
supplying water for human consumption and agricultural irrigation,

C_c	N_s	V_l	$M_{v\,2}$	$M_{n\,2}$	$M_{n\,3}$	N_p
and	fracking	is	not	a viable	long-term	solution.

Equation: $B_2 \mid B_3 = [M_{v\,1}]^F \ldots N_s + V_t + (N_o * VP(M_{n\,1})) \mid N_s + (V_l * M_{v\,2}) + (M_{n\,2} * M_{n\,3} * N_p)$

$VP(M_{n\,1})$ = *supplying* water for human consumption and agricultural irrigation $(V_t)_v$ = supplying

$= (V_t)_v + (N_o * LP(M_n))$

VP function = adjective modifying noun object "underground water tables"

VP core = *supplying* water (participial phrase)

$= (V_t)_v + N_o$ (scaffolded with a P_2-type predicate)

Discussion: Example 8.4 is a compound sentence (B_2 joined with a B_3). The sentence begins with free-floating adverb, "However," which functions as a transition word from the sentence before to this one. Sentence 8.4 includes a verbal phrase, "*supplying* water for human consumption and agricultural irrigation," led by a present participle. This phrase is built around a P_2 predicate. Here, the VP functions as an adjective modifying the plural compound noun "underground water tables."

Coding Demonstration 8.5

Original: Net profit is, of course, a key business driver, yet *developing safe, clean, energy-supply processes* is the paramount goal *for* our current generation of energy systems engineers, *committed to professional standards for safety and excellence.*

Scanned:

N_s	V_1	M_v	$M_{n\,1}$	N_p	C_c
Net profit	is,	of course,	a key	business driver,	yet

$VP(N_s)$	V_1	$M_{n\,2}$
developing safe, clean, energy-supply processes	is	the paramount

N_p	$LP(M_{n\,3})$
goal	*for* our current generation of energy systems engineers,

committed to professional standards for safety and excellence.

Equation: $B_3 \mid B_3 = (N_s + (V_1 * M_v) + (M_{n\,1} * N_p)) \mid (VP(N_s) + V_1 + (M_{n\,2} * N_p * LP(M_{n\,3})))$

$VP(N_s)$ = *developing* safe, clean, energy-supply processes $(V_t)_v$ = *developing*

$= (V_t)_v + (M_{n\,1} * M_{n\,2} * M_{n\,3} * N_o)$

VP function = noun subject of main verb "is"

VP core = *developing* processes (gerundial phrase)

$= (V_t)_v + N_o$ (scaffolded with a P_2-type predicate)

$LP(M_{n\,3})$ = *for* our current generation of energy systems engineers, *committed to professional standards for safety and excellence.*

$= (L + (X'\text{-}(M_{n\,1} * N_o)) * LP(M_{n\,2}) * VP(M_{n\,3}))$

and then nested within $LP(M_{n\,3})$:

$VP(M_{n\,3})$ = *committed* to professional standards for safety and excellence $(V_i)_v$ = *committed*

$= (V_i)_v * LP(M_v)$

VP function = adjective modifying noun object N_o = "*generation*" of the preposition L = "*for*"

VP core = *committed* (participle)

$= (V_i)_v$ (P_1-type predicate)

Discussion: The final example in this set, demonstration 8.5, is another compound sentence, a B_3 connected to a B_3. The first B_3 is short and contains no verbal phrases. The second B_3 contains two. The first one, "*developing safe, clean, energy-supply processes*," is a gerundial phrase acting as basic sentence subject. The other VP in the second B_3 is nested within a prepositional phrase, LP = "for our current generation . . . ." The prepositional phrase modifies the predicate noun, N_p = "goal."

The nested VP, "*committed to professional standards for safety and excellence*," is constructed around the past ("-ed" type) participle "committed." This participle expresses action in and by itself and thus leads a phrase that models a P_1-type predicate. The nested VP's function is to adjectivally modify the noun, "generation," which is the object of the preposition inside the "parent" prepositional phrase, LP = "*for* our current *generation*. . . ." Once again, as does example 8.3, example 8.5 reveals the presence of phrases nested into other phrases—something that regularly occurs in longer, more advanced sentence structures.

Inversions and Variations

Passive Voice

The passive voice is one of several syntactical (word order) variations of standard basic sentence (B_n) structures with transitive verbs—namely, B_2, B_4, and B_5. Up to this point, the basic sentence formulas have routinely positioned main verb objects—people, places, and things that receive transferred action—to the right of main verb, in the predicate (P). In passive voice, however, the equation *inverts*, with the object of verb action shifting locations, from original position in the predicate (to the right of main verb) to an alternate position as the sentence's subject (to the left of the main verb).

A passive sentence's subject (the *passive subject*) plays two roles simultaneously. It operates as *subject instigator of "spark,"* and it names the *object* of main verb's action. The official grammar term for this reversal of basic sentence word order is *passive voice*. Though sometimes misunderstood as something that has to do with verb tense, the property of voice has entirely to do with syntax.

For example, here is a standard B_2 sentence in ***active voice*** (with core elements underlined):

HP's John Vaught first proposed thermal excitation for inkjet printing. (8.6a)

subject(doer) + verb(action flows to the right) + object

left → right

Here is the original sentence inverted into ***passive voice*** structure (with core elements underlined):

Thermal excitation for inkjet printing was proposed first by HP's John Vaught. (8.6b)

subject(object) + verb(action flows to left) + doer(nested in phrase)

left ← right

With passive voice, sometimes the doer is placed to the right of the main verb in the predicate, nested in a prepositional phrase usually starting with "by," as in the example above ("by HP's John Vaught"). Other times, the doer of the main verb's action is omitted from the sentence. In these cases, the doer does not disappear entirely. It exists elsewhere, sometimes appearing in an upstream sentence in a block of text, thus remaining within the context of the text's message, and giving *remote agency* to the verb with which it is linked. Other times, the doer is "offstage," implied but not named.

Here is another B_2 sentence in ***active voice***:

The computer automatically calculated pressure inside the vessel using the PV = nRT equation. (8.7a)

subject(doer) + verb(action flows to the right) + object *note: doer included*

left → right

The next version of the sentence changes word order to ***passive voice*** and omits the action doer, "computer":

Pressure inside the vessel was calculated automatically using the PV = nRT equation. (8.7b)

subject(object) + verb(action flows to left) *note: doer is missing*

left ← right

Both of the above sentences' active and passive versions could be effective sentences in an engineering document. Which version is best depends on the document's intended audience and business context. Note: if the writer wanted to include the action doer in sentence (8.7b), the writer need only to tack on "by the computer" to the end of the sentence.

You probably noticed in this segment's first two examples (8.6 and 8.7) that the passive verbs are written using two words—*a form of the verb "be"* (auxiliary verb) combined with another verb ending in "-ed" that expresses the verb's main action (principal or root verb). As we learned in Chap. 3, adding the "-ed" suffix can form simple past tense. Here, however, the "-ed" versions of the main verbs in passive voice sentences represent their *past participle*. Past participles duplicate the past tense "-ed" spellings of regular verbs, but they have unique spellings if a verb is irregular.

The next example shows the active and passive versions of a sentence formed around the irregular verb, "sell," for which there is no "-ed" spelling. It's "sold," not "selled." This example's verb forms (both versions) are presented in three different verb tense types—simple past/perfect past/past progressive, rather than just simple past (like the verbs in examples 8.6 and 8.7):

Active Version

$$\underline{\text{Stephen}}\ \underline{\text{sold/had sold/was selling}}\ \text{an optional}\ \underline{\text{exterior maintenance plan}}\ \text{to the client, resulting in additional \$1.2M net revenue for our Skyscraper project.} \quad (8.8a)$$

$$\textit{subject(doer)} + \textit{verb(action flows to the right)} + \textit{object}$$
$$\textbf{left} \rightarrow \textbf{right}$$

Passive Version

$$\text{An optional}\ \underline{\text{exterior maintenance plan}}\ \underline{\text{was sold/had been sold/was being sold}}\ \text{to the client by}\ \underline{\text{Stephen}},\ \text{resulting in additional \$1.2M net revenue for our Skyscraper project.} \quad (8.8b)$$

$$\textit{subject(object)} + \textit{verb(action flows to the left)} + \textit{doer(nested in phrase)}$$
$$\textbf{left} \leftarrow \textbf{right}$$

Coding Passive Voice Sentences

At the equation level, sentence algebra notation uses the subscript "pass" for "passive." Thus, a basic sentence two in passive voice is "$B_{2\,pass}$." Likewise, to signal a passive (transitive) verb at the equation level, the coding is "V_{pass}." In a general equation (e.g., 8.9b below), to identify an optional prepositional phrase in the predicate containing a nested action doer, the system uses a bracketed prepositional phrase formula tagged with the subscript "opt" for "optional"—that is, "$[L + M_{na} * (N_s \text{ or } X_s)_o]_{opt}$." A multiplication symbol "*" attaches the phrase to the main passive verb, because the phrase functions as a specificity modifier (action done by *what?* or by *who?*).

As established earlier in this segment, the subject of a passive sentence serves a dual purpose. Therefore, when coded, the *passive subject* noun (or pronoun) is tagged within parentheses with the subscript "o" for "object" and noted outside the parentheses with the subscript "s" for "subject"—that is, "$(N_o$ or $X_o)_s$." Similarly, the optional prepositional phrase in the predicate tags its object noun (or pronoun) with a subscript "s" on the inside of its parentheses, and with an "o" subscript outside, to record that this noun (or pronoun) is the main verb's action doer *functioning as* (and taking the case of) an object, nested in a prepositional phrase—that is, we code it as "$(N_s$ or $X_s)_o$."

To illustrate, here is the *general formula* for an *active B_2 sentence* (8.9a) along with its inverted counterpart, a *passive B_2 sentence* (8.9b).

$$B_2 = (M_{n\ a} * (N_s \text{ or } X_s)) + (M_{v\ b} * V_t) + (M_{n\ a} * (N_o \text{ or } X_o)) \tag{8.9a}$$

$$B_{2\ pass} = \left(M_{n\ a} * (N_o \text{ or } X_o)_s\right) + \left(M_{v\ b} * V_{pass}\right) * \left[\left(L + \left(M_{n\ a} * (N_s \text{ or } X_s)_o\right)\right]_{opt} \tag{8.9b}$$

As an end piece for this segment, the gray box demo below illustrates how the sentence algebra system would code three of the specific passive sentences featured earlier.

Advanced Coding Demo: Three Sentence Equations, Three Passive Sentences

(8.6b, 8.7b, 8.8b)

$$(B_2)_{pass} = (N_o)_s * LP(M_n) + V_{pass} * M_v * \left[L + N'\text{-}(N_s)_o\right]$$

Thermal excitation for inkjet printing was proposed[1] *first by HP's John Vaught.*

$$(N_o)_s * LP(M_n) + \left(V_{pass} * M_{v\ 1} * VP(M_{v\ 2})\right)$$

Pressure inside the vessel was calculated automatically using the PV=nRT equation.

$$M_n * (N_o)_s + V_{pass} * LP(M_{v\ 1}) * \left[L + (N_s)_o\right] \ \ldots \ [VP(M_{v\ 2})]^E$$

An optional exterior maintenance plan was sold/had been sold/was being sold to the client by Stephen, resulting in an additional $1.2M net revenue for our Skyscraper project.

[1]Note: Although grammatically correct and simple to code, stylistically, a more natural (and smoother) way to phrase the passive verb and its modifier in sentence 8.6b—that is, "was proposed first"—would be to place the modifier "first" between the passive verb's auxiliary verb "was" and principal verb "proposed." Thus, the text would be "was first proposed" and the coding ($V_{aux} * M_v * V_{pass}$).

Questions

Another variation of the basic sentence formulas that can (but does not always) involve rearrangement of key sentence elements results from phrasing a sentence as a question rather than as a straight declaration of fact, opinion, or feelings. For typical engineering writing, the most frequently used sentence syntax is standard, declarative, active voice phrasing, and the second is usually standard, declarative phrasing recast in passive voice. But along with proposals and outcomes that require reports, engineering activities often also require research and inquiry and the language to record such things. Thus, when called upon to do so, engineering writers must also be able to encode their dynamic thinking in the form of questions as well as statements.

The official name for standard sentence phrasing is *declarative*. The official name for sentences phrased as questions is *interrogative*. Like declarative sentences, interrogative sentences can be presented in either active or passive voice.

The chief signal that a sentence is being phrased as a question is the punctuation that comes at the sentence's end: a question mark (?). Sometimes, more often in conversation than in formal writing, a question mark alone is all it takes to turn the purpose of a sentence from statement to question. For example, below are two versions of a B_5 sentence. While the words and word sequencing of version 8.10a are identical to that of 8.10b, the terminal punctuation in one case is a period and, in the other, a question mark. This change alone causes one sentence to have a different purpose from its counterpart—one states information (8.10a), and the other questions it (8.10b).

Our manager appointed the new hire, Gary Greenhorn, project lead instead of me. (8.10a)

Our manager appointed the new hire, Gary Greenhorn, project lead instead of me? (8.10b)

More often than not, however, basic sentence formulas ***do*** undergo re-wordings when question marks appear at sentences' ends. This rewording usually involves placement of one or more question-creating words at the sentence's beginning. Question-creating words (QWs) come in several grammatical forms: pronoun (X-type), auxiliary verb (V_{aux}-type), and adverb (M_v-type).

The first, the X-type question words, is a subcategory of pronouns known as *interrogative pronouns* (see list below). The sentence algebra notation for this type is "QW_x." Along with "what," you will observe that this subcategory includes words that also function as relative pronouns:

what, *who*, *whose*, *whom*, and *which*

The second type, V_{aux} question words, are "helping verbs" (auxiliary verbs) which, instead of helping to form a part of a verb conjugation as they often do in declarative sentences, in interrogative sentences function as *modal auxiliaries* positioned (up front) as question-creating words (see below). The sentence algebra notation for this type is "$QW_{v\,aux}$." Some examples are as follows:

can, *could,* *will,* *would,* *do,* *did,* *may,* *might,* and so on...

Third are the M_v-type question words, which include a set of terms (see below) known in sentence algebra also as *journalistic adverbs* because they are affiliated with the *journalistic prompts* and specificity equation, 5W+H, previously discussed in Chap. 4. The sentence algebra notation for this type is "QW_{mv}".

when, *where,* *why,* and *how*

To formulate an interrogative sentence, question words can be used as *construction devices* to help phrase a sentence as a question (as shown by 8.11b). Question words *can also* serve the dual purpose of *signaling* the question and *substituting* for a sentence element about which the question is being asked (as shown by 8.11c). Moreover, the re-wording patterns for question-making are sometimes *inversions*, similar to the re-wording pattern for a passive voice sentence—that is, key elements move around and main verb "flow" is re-aimed "right-to-left," rather than "left-to-right."

Here is a set of example sentences (see 8.11a through 8.11g below) that display a standard B_4 sentence rephrased into a question six different ways. All of the rephrasings, except for the last, begin with the addition of a question-creating word or words (QWs). The example sentences are also annotated to show direction of "flow" of the main verb, and you can see in the annotations that each variation (except for the last, which simply adds a question mark) either reconfigures the original sentence's wording, re-aims main verb flow, or both. For each of the re-worded (inverted) examples, an "unscrambled" version is displayed underneath. Since the example sentences are B_4s with both a direct and an indirect object, one twist is a variation where flow "splits" and goes both ways, because one object is in the predicate and one appears before the subject (illustrated below in 8.11c, 8.11d and 8.11e). In all of the example sentences, both the main verb and question word(s), when present, are highlighted in **bold**.

Original Sentence (Declarative)

The biotech firm **gave** the consultants security clearance last week. (8.11a)

subject + ***verb*** + *indirect object* + *direct object* + *adverb*
left ⟶ right

Six Variations (Interrogative)

Did the biotech firm **give** the consultants security clearance last week? (8.11b)

QW(*construction device*) + *subject* + ***verb*** + *indirect object* + *direct object* + *adverb*
left⟶right

What did the biotech firm **give** the consultants last week? (*inversion*) (8.11c)

QWs(*direct object and signal*) + *subject* + ***verb*** + *indirect object* + *adverb*
left ← → right (split flow)

unscrambled: the biotech firm ***did*** **give** the consultants ***what*** last week

Whom did the biotech firm **give** security clearance last week? (*inversion*) (8.11d)

QWs(*indirect object and signal*) + *subject* + ***verb*** + *direct object* + *adverb*
left ← → right (split flow)

unscrambled: the biotech firm ***did*** **give** ***whom*** the security clearance last week

When did the biotech firm **give** the consultants security clearance? (*inversion*) (8.11e)

QWs(*adverb and signal*) + *subject* + ***verb*** + *indirect object* + *direct object*
left ← → right (split flow)

unscrambled: the biotech firm ***did*** **give** the consultants security clearance ***when***

Who **gave** the consultants security clearance last week? (8.11f)

QW(*functions as subject*) + ***verb*** + *indirect object* + *direct object* + *adverb*
left⟶right

The biotech firm **gave** the consultants security clearance last week? (**add** "?") (8.11g)

subject + ***verb*** + *indirect object* + *direct object* + *adverb*
left⟶right

Coding Interrogative Sentences (Questions)

To account for all of the syntactical maneuvers that can turn a declarative sentence into an interrogative sentence in a generalized sentence algebra equation would require a long, cumbersome construction with multiple "or" statements. Nonetheless, for sentence engineers, it is useful to capture the recipe for phrasing a sentence as a question in a manner that is as objective, concrete, and quantitative as possible.

Exhibit 8.12 below consolidates the crux of what has been covered in this segment into a generic template rather than comprehensive equation. To rephrase a basic sentence from standard (declarative) to question (interrogative) format, the template reminds you to consider all of the following:

(a) Optionally, add a question-making word or words (QWs) to the sentence's beginning as a *construction device*.
(b) Optionally, rephrase or invert the sentence so it leads off with a question-making word or words (QWs) that *replace (stand in for) a sentence element* around which the *question is being framed*.
(c) *Always* use a *question mark* at the end of the sentence as the terminal punctuation mark.

Template for $(\mathbf{B_n})_{\mathbf{question}}$ **is . . .**

$$[\textit{\textbf{sometimes}}\text{ add QW(s) @ beginning}] + [\textit{\textbf{sometimes}}\text{ rephrase and/or invert the }B_n\text{ and use QWs to replace words}] + [\textit{\textbf{always}}\text{ add "?" to end}] \quad (8.12)$$

As an end piece for this segment, the gray box demo below illustrates how the sentence algebra system would code three specific sentences from this segment as questions.

Advanced Coding Demo: Three Sentence Equations, Three Questions

(8.11b, 8.11c, 8.11f)

$$QW_{v\ aux} + N_s + V_t + N_i + N_o \ldots [M_v]^E + (?)$$

Did the biotech firm give the consultants security clearance last week?

$$QW_x(N_o) + QW_{v\ aux} + N_s + V_t + N_i \ldots [M_v]^E + (?)$$

What did the biotech firm give the consultants last week?

$$QW_x(N_s) + V_t + N_i + N_o \ldots [M_v]^E + (?)$$

Who gave the consultants security clearance last week?

Commands and Requests

In addition to *declarative* and *interrogative* sentences (also referred to as "statements" and "questions"), there are two more sentence types to note—*imperative* and *exclamatory*. Imperative sentences—sentences phrased as *commands or requests*—are eminently useful in engineering writing, especially in instructional and procedural documents. On the other hand, exclamatory sentences (that end in "!") are almost never used in the domain of engineering writing except in cautions or warnings, and will not be covered in this segment.

To optimize a document's readability at both the paragraph (horizontal) and bulleted/numbered list (vertical) levels, excellent engineering writers use sentences phrased as commands to call out priority status, promote concision, achieve symmetry (especially in vertical lists), guide peers, and direct support personnel (especially in technical procedures/instructions).

Like a declarative sentence, an imperative sentence ends with a period. In fact, the phrasing of an imperative sentence is nearly identical to that of a standard basic sentence. One distinction is that a command routinely begins with the subject word "you." What can be confusing is that the lead word "you" in an imperative sentence is usually stated *implicitly* rather than explicitly and is *invisible*. In a sentence algebra equation, the emphatic "You" beginning a text sentence command still codes as an "X_S." However, to indicate the "You" is invisible, as it usually is, the term is bracketed, "[X_S]."

All of the sentences below are commands or requests. Each is built out of one of the five basic sentence formulas, spawned from and organized around "spark" generated by a subject pronoun and main verb. All five commands/requests are legitimate sentences, not fragments. Yet, in all instances, the leading "You" (the subject pronoun) is implied.

Stop at 5:45 PM. (B_1) (8.13)

Please remove the governor from the air compressor's engine. (B_2) (8.14)

Be yourself. (B_3) (8.15)

Do not assign Jeff aftermarket warranty and repair duties for the beta product. (B_4) (8.16)

Consider Spacely Sprockets our #1 supplier in light of the Cogswell Cogs fiasco. (B_5) (8.17)

Sentences 8.13 through 8.17 illustrate all five basic sentence formulas phrased as commands, with each sentence subject (S) being the invisible word "You." Going from declarative to imperative changes nothing in the sentences' predicates. They are all standard P_1, P_2, P_3, P_4, and P_5 structures, respectively. It is also useful to note how the word "please" adds politeness to the command stated by 8.14 and transforms it into a request and also how in 8.16 the helping verb "do" combined with the adverb “not” work to negate a directive or demand. Thus, beyond the imperative “you,” there are sometimes “accent” words (AWs) that heighten or sharpen the command; these words, when present, are coded “AW” in sentence algebra equations.

Exhibit 8.18 below is a template for constructing basic commands based on standard frameworks. To rephrase any one of the five basic sentences, from standard (declarative) to command/request (imperative) format, the template reminds you to consider all of the following:

(a) Always use the emphatic “You,” implicit or explicit (usually implicit and *invisible*) for the sentence subject.
(b) Optionally, add “accent words” (AWs) such as “please,” “do,” and “not,” to the sentence subject to heighten or sharpen the command’s meaning.
(c) From the main verb onward, for a standard B_n sentence, use a standard P_n predicate.

$$\textbf{Template for } (\mathbf{B_n})_{\mathbf{command}} \textbf{ is} \ldots$$

$$\left[\textbf{\textit{always}} \text{ use emphatic “You” as subject, implicit or explicit}\right] + \left[\textbf{\textit{sometimes}} \text{ add AWs, e.g., “please,” “do,” “not,” etc.}\right] + \left[\text{use a standard } P_n \text{ predicate}\right] \quad (8.18)$$

As an end piece for this segment, once again, a gray box demo illustrates several sentence algebra coding applications—specifically, how the system would code three specific sentences from this segment as commands.

Advanced Coding Demo: Three Sentence Equations, Three Commands/ Requests

(8.13, 8.14, 8.16)

$$B_{1\,\text{command}} = [X_s] + (V_i * LP(M_v))$$

Stop at 5:45 PM.

$$B_{2\,\text{command}} = [X_s] + AW + V_t + (N_o * LP(M_v))$$

Please remove the governor from the air compressor’s engine.

$$B_{4\,\text{command}} = [X_s] + AW + AW + V_t + N_i + (M_v * M_{n\,1} * N_o * LP(M_{n\,2}))$$

Do not assign Jeff aftermarket warranty and repair duties for the beta product.

Expletive Introductions

Another variation on the standard B_n sentence occasionally seen in engineering writing is a construction that leads off with either "It" or "There," which is then partnered with a form of the verb "to be"—usually "is," "was," or "were." In this phrasing, known as an *expletive introduction*, "It" or "There" acts as the sentence's syntactical subject; a form of "to be" comes next and serves as the main verb; and then, in the predicate, the rest of the sentence follows, including the sentence's true subject, the principal doer of the action of the sentence's true main verb. The true main verb is often an additional verb form in the predicate beyond "to be". Two simple examples of expletive introductions are

It is important that you follow all safety regulations.

There are many options presented in our partnership agreement.

In sentence algebra equations, the expletives "it" and "they" are both coded as "X_e"—"X" for pronoun and subscript "e" for expletive. The "to be" verb that the expletive partners with gets coded as a standard linking verb, "V_l." There are parallels between a sentence constructed with an *expletive introduction* and a sentence incorporating passive voice. Both feature "mock subjects" to the left of the main verb and true subjects (the bona fide doers of main verb action) to the right, or in some cases for passive constructions, action doers that are "offstage." The term *expletive* means "filler."

Here are five examples of basic sentences that have been rephrased with expletive introductions. Each features either "it" or "there" in the role of syntactical subject, and each reveals its true subject in *italics* in the predicate. The true subjects range from verb phrase, noun, or pronoun, to noun nested in a prepositional phrase. For each example, the sentence's standard *un-inverted* phrasing and sentence type are shown immediately beneath in parentheses.

It is satisfying *to apply calculus to real-world problems.* (8.19)

(*To apply calculus to real-world problems* is satisfying. B_1)

It was *he* who held the key to the office supplies cabinet. (8.20)

(*He* [was who] held the key to the office supplies cabinet. B_2)

There is *one viable option* for the prototype's final test. (8.21)

(*One viable option* exists for the prototype's final test. B_1)

There are *too many sales persons* at the technical consortium. (8.22)

(*Too many sales persons* are at the technical consortium. B_3)

It has been said by *some* that humans will travel far beyond our galaxy in the 22nd century. (8.23)

(*Some* have said that humans will travel far beyond our galaxy in the 22nd century. B_2)

These examples show that going back and forth between phrasing with an expletive introduction and standard phrasing is fairly straightforward. The only irregularities that arise in 8.19 through 8.23 are that the standard version of 8.20 (below the expletive version) achieves a more concise phrasing with "was who" eliminated (and treated as an elliptical element).

We can also see in the examples above that the standard version of 8.21 reads more smoothly (but retains the same meaning) if "is" is swapped out for "exists." Additionally, you should note the number of the verb in 8.21 versus 8.22. The first is *singular*, and then the next is *plural* because the true subject in the first case (8.21) is "one viable option" (one) and the true subject of the second (8.22) is "many sales persons" (multiple). "One is ...." "Many are ...." Finally, also notice that 8.23 represents a sentence that is phrased both with an expletive introduction *and* the passive voice. The rewrite immediately beneath it removes both and makes this version of the sentence standard, active voice.

Exhibit 8.24 below is template for constructing a basic sentence with an expletive introduction. To rephrase any one of the five basic sentence structures from standard to expletive, the template reminds you to consider all of the following:

(a) Always begin with the expletive "it" or "there" as expletive sentence subject.
(b) Always follow with a form of the verb "to be" by itself or combined with an additional verb.
(c) From the expletive main verb onward, use a standard P_n predicate framework, usually with the sentence's true subject ("ts") nested somewhere within the predicate.

Template for $(\mathbf{B_n})_{\mathbf{expletive}}$ (basic w/expletive intro) is...

[***always*** begin with "it" or "there" as expletive subject]

+ [***always*** follow with "to be" or "to be" + another verb]

+ [nest **true subject** in a **standard $\mathbf{P_n}$ predicate**] (8.24)

The gray box demo below illustrates several sentence algebra coding applications—specifically, how the system might code three specific sentences from this segment with expletive introductions. One characteristic of sentences with expletive introductions that is not optimal is that they lack conciseness (and vigor), since they take the real subject out of the sentence core and replace it with an empty word. Nevertheless, sometimes this construction comes in handy, and it's worthwhile having in your sentence engineering tool kit (but I suggest you use expletive introductions sparingly).

Advanced Coding Demo: Three Sentence Equations, Three Sentences with Expletive Beginnings

(8.19, 8.21, 8.23)

$$B_{1\,\text{expletive}} = X_e + (V_i * VP(M_v)^{ts})$$

It is satisfying to apply calculus to real-world problems.

$$B_{3\,\text{expletive}} = X_e + V_1 + ((M_{n\,1} * M_{n\,2} * (N_p)^{ts}) * LP(M_{n\,3}))$$

There is one viable option for the prototype's final test.

$$B_{2\,\text{expletive}} = X_e + (V_{pass} * (LP(M_v)) + RC(N_o)$$

$$\textit{where: } LP(M_v) = (L + X_o^{ts}).$$

It has been said by some that man will travel far beyond our galaxy in the 22nd century.

Note: "ts" superscript indicates "true subject."

Review

Below is a *review box* that presents a general template for creating either a basic or advanced sentence (text version) using the material covered thus far in the Part I—Sentence Algebra chapters. You may find this template useful as a quick refresher.

Review Box: General Template for Constructing a Basic or Advanced Sentence (Text Version)

$$\text{initial capital letter} + \left\{\begin{array}{l}\text{basic sentence, } B_n\text{, or}\\ \text{advanced sentence, A}\end{array}\right\} + \begin{array}{c}\text{period or question mark}\\ (\text{"." or "?"})\end{array}$$

where: $n = 1, 2, 3, 4,$ or 5

A = an inversion or variation sentence of a basic sentence ($B_{n\,\text{passive}}$, $B_{n\,\text{question}}$, $B_{n\,\text{command/request}}$, or $B_{n\,\text{expletive}}$); compound or complex sentence ($B_n \mid B_n$, B_n / B_n, or $B_n \setminus B_n$); bonded sentence ($B_n \sim B_n$); multiple compound, complex, compound-complex, or bonded sentences; *and similar but more lengthy combinations.*

(continued)

| = coordinating conjunction, for example, "and," "but," "or," etc.

/ = subordinating conjunction (e.g., "although," "since," "because," etc.) with principle clause on left and *subordinate clause on right*

\ = subordinating conjunction with principal clause on right and *subordinate clause on left*

~ = two stand-alone B_n sentences (standard, inverted, or variation) joined together with a semicolon ";", colon ":", or dash "—".

Key Concepts and Coding Conventions for Advanced Sentence Equations

Here also is a list of key sentence algebra concepts and coding conventions covered in the advanced sentence chapters (Chaps. 7 and 8). The list does not include items already presented at the end of Chap. 6 in the basic sentence concepts and coding review segment.

Relative clauses:	Always construct a relative clause (coded "RC") with a relative pronoun "X_r" (e.g., "who," "that," or "whom") heading it up, even when the relative pronoun within the syntax of that clause is not the true subject; also remember that the relative pronoun's antecedent outside of the clause dictates its number, while at the same time, its true grammatical function inside the clause determines its case.
Verbals:	Be aware of and correctly apply the four different types and functions of verbals: (1) the *infinitive* is formed by partnering "to" with the verb's root or first dictionary listing (functions as a *noun*, *adjective*, or *adverb*); (2) the *past participle* is formed with the "-ed" suffix for regular verbs and with a specialized spelling for irregular verbs (functions as an *adjective*); (3) the *present participle* is formed with the "-ing" suffix for all verbs (functions as an *adjective*); and (4) the *gerund* is also formed using the "-ing" suffix for all verbs (functions as a *noun*). Code all verbals as "V_v" and, if you wish, more specifically, "$V_{v\,i}$," "$V_{v\,t}$," or "$V_{v\,l}$" to distinguish *intransitive*, *transitive*, or *linking* verb forms.
Verbal phrases:	Form a verbal phrase (coded VP) with a verbal (V_v) in the lead position as the phrase's main verb. For structure, use any one of the five basic predicate frames (standard syntax or a variation)—that is, VP (from P_1, P_2, P_3, P_4, or P_5); and then treat a VP within a parent sentence as a sentence element that functions either as a modifier or noun, depending on the type of verbal that leads the verbal phrase.

(continued)

Passive sentences:	Use the subscript "pass" to indicate a passive verb (V_{pass}), and use double subscripts, "($(N_o$ or $X_o)_s$)," to indicate a *passive subject*, that is, a noun or pronoun main verb object functioning as subject of a passive verb; likewise, use the term "($(N_s$ or $X_s)_o$)" to indicate a main verb subject functioning as an object (usually the object of a preposition) in the predicate of a passive sentence.
Phrasing questions:	To phrase a sentence as a question, *always* place a *question mark* at the *end* of the sentence; also, as needed, add one or more question-making words (coded "QW") to the sentence's beginning, for example, "What," "When," "Do," and so on.
Phrasing commands:	To phrase a sentence as a command, *always* use the emphatic "You" (usually implied and "invisible") to *begin* the sentence as the sentence's subject, coded in brackets, "[X_s]." You can also add optional accent words such as "please" (coded AW).
Phrasing expletive beginnings:	To phrase a sentence with an expletive introduction, *always* begin the sentence with an expletive "It" or "They" followed with a form of the verb "to be," either by itself or combined with an additional verb (code "It" or "They" as "X_e").

Coding Demo of a Sentence Stream Containing Basic and Advanced Structures

Similar to the capstone example at the end of Chap. 6, Chap. 8 also ends with a demonstration that scans and codes a sentence stream (one paragraph). However, this stream contains both basic and advanced sentence structures. The text sample comes from an early draft of a textbook on engineering graphics and problem-solving methods written by engineer Jerald M. Henderson (my father) and a colleague. This draft comes from his personal archives and is not an excerpt from the final draft that eventually went into print [1]. Though further perfected later, this sample in its earlier form provides an interesting discussion on the origin of engineering, and does so using a variety of sentence components and equations, both basic and advanced.

Aside from short notes, the example coding is not comprehensively annotated. Just like the Chap. 6 example, I challenge you to follow the scan-and-code process and make sense of it as you move along, using all that you have learned thus far. After the coding, you will find a very brief text analysis that tallies up frequency of occurrence of each basic sentence type as it presents in both standard and alternate phrasings (such as passive voice)—as well as the number of advanced sentences present in the sample that are formed using two or more basic sentence clauses.

Sample Text

[1]Who, in human history, were the first engineers? [2]Were they the Chinese architects of the Great Wall, the Aztec temple builders, the Roman aqueduct designers, or the primitive men and women who invented the wheel? [3]Some say that it was [4]Imhotep—a great physician, architect, and statesman in ancient Egypt. Imhotep designed the first known pyramid at Saqqara about 2650 B.C. as a tomb for King Zoser. [5]Built as a series of steps and known as "the Step Pyramid," the project marked the first documented use of hewn stone for a civil structure. [6]Over the next century and a half, many other pyramids were built by the Egyptians. [7]These structures were truly remarkable in both scope and size, and for the engineering discipline, they marked one of the first applications of arithmetic and geometry to guide and improve designs. [8]One innovation was making the sides of the pyramids smooth instead of stepped. [9]Nearby Saqqara, there are two other pyramids. [10]The first began its construction with flat sides inclined at 52 degrees. [11]However, after it was more than two-thirds built, the builders changed the angle to 43½ degrees. [12]Note that this pyramid is called the "Bent Pyramid". [13]The entire second pyramid was built at 43½ degrees, but all later pyramids have the original face inclination of 52 degrees. [14]Interestingly, almost all of the large pyramids were built within a 150-year period, 2650–2500 B.C. [15]Since then, Egyptian society has called upon its engineers to do other things. [247 words]

Scan and Code

	QW_x	$LP(M_{n\,1})$	V_l	$M_{n\,2}$	N_p
1.	Who,	in human history,	were	the first	engineers?

$$A = (QW_x(X_s) * LP(M_{n\,1})) + V_l + (M_{n\,2} * N_p) + (?)$$
$$= B_{3\ \text{question}}$$

Note: Unscrambled, the core B_3 sentence's text is "*engineers were who*."

	V_l	X_s	$N_{p\,1}$	$LP(M_{n\,1})$	$N_{p\,2}$
2.	Were	they	the Chinese architects	of the Great Wall,	the Aztec temple builders,

$N_{p\,3}$	C_c	$N_{p\,4}$
the Roman aqueduct designers,	or	the primitive men and women

$RC(M_{n\,2})$
who invented the wheel?

$$A = V_l + X_s + (N_{p\,1} * LP(M_{n\,1}))\ |\ N_{p\,2}\ |\ N_{p\,3}\ |\ (N_{p\,4} * RC(M_{n\,2})) + (?)$$
$$= B_{3\ \text{question}}$$

Note: Here, there is no question word. The verb-first, subject-second syntax at the beginning and question mark at the end signal that the sentence is interrogative. Unscrambled, the core B_3 sentence's text is "*they were the Chinese architects, the Aztec temple builders, the Roman aqueduct designers, or the primitive men and women.*" Note also, the four predicate nouns are simplified expressions—adjective and noun combinations each coded as one compound noun.

X_s V_t $RC(N_o)$

3. Some say that it was Imhotep—a great physician, architect, and statesman in ancient Egypt.

$$B_2 = X_s + V_t + RC(N_o)$$

N_s V_t $M_{n\,1}$ $M_{n\,2}$ N_o $LP(M_{n\,3})$

4. Imhotep designed the first known pyramid at Saqqara

$LP(M_{n\,4})$ $LP(M_{n\,5})$

about 2650 B.C. as a tomb for King Zoser.

$$B_2 = N_s + V_t + (M_{n\,1} * M_{n\,2} * N_o * LP(M_{n\,3}) * LP(M_{n\,4}) * LP(M_{n\,5}))$$

$VP(M_{n\,1})$ N_s V_t

5. Built as a series of steps and known as "the Step Pyramid the project marked

$M_{n\,2}$ N_o $LP(M_{n\,3})$

the first documented use of hewn stone for a civil structure.

$$B_2 = (VP(M_{n\,1}) * (N_s)) + V_t + (M_{n\,2} * N_o * LP(M_{n\,3}))$$

$[LP(M_{v\,1})]^F$ $M_{n\,1}$ $M_{n\,2}$

6. Over the next century and a half, many other

N_s V_{pass} $LP(M_{v\,2})$

pyramids were built by the Egyptians.

$$A = [LP(M_{v\,1})]^F \ldots (M_{n\,1} * M_{n\,2} * N_s) + \left(V_{pass} * LP(M_{v\,2})\right)$$

$$= B_{2\ pass}$$

Note: Un-inverted, the core B_2 sentence's text is "*the Egyptians built pyramids*."

M_d N_s V_1 $M_{v\,1}$ M_p $LP(M_{v\,2})$ C_c

7. These structures were truly remarkable in both scope and size, and

$[LP(M_{v\,3})]^F$ X_s V_t X_o

for the engineering discipline, they marked one of the first applications

$LP(M_n)$

of arithmetic and geometry to guide and improve designs.

$$A = (M_d * N_s) + V_1 + \left(M_{v\,1} * M_p * LP(M_{v\,2})\right) \mid [LP(M_{v\,3})]^F \ldots X_s$$

$$+ V_t + (X_o * LP(M_n))$$

$$= B_3 \mid B_2$$

Note: The above compound sentence leads off with the subject term "These structures…" coded "$M_d * N_s$." Recall that *determiners* such as "this" and "these" are specialized adjectives represented as "M_d."

	$M_{n\,1}$	N_s	V_t	N_o	$LP(M_{n\,2})$	M_c	$LP(M_v)$
8.	One	innovation	was making	the sides	of the pyramids	smooth	instead of stepped.

$$B_5 = (M_{n\,1} * N_s) + V_t + (N_o * LP(M_{n\,2})) + (M_c * LP(M_v))$$

	$[LP(M_v)]^F$	X_e	V_l	$M_{n\,1}$	$M_{n\,2}$	N_o
9.	Nearby Saqqara,	there	are	two	other	pyramids.

$$A = [LP(M_v)]^F \ldots X_e + V_l + (M_{n\,1} * M_{n\,2} * N_o)$$

$$= B_{3\,\text{expletive}}$$

Note: Unscrambled, the core B_3 sentence's text is "*pyramids are nearby Saqqara*."

	N_s	V_t	X'	N_o	$LP(M_v)$
10.	The first	began	its	construction	with flat sides inclined at 52 degrees.

$$B_2 = N_s + V_t + (X'\text{-}N_o * LP(M_v))$$

	$[M_{v\,1}]^F$	C_s	$(X_o)_s$		$M_{v\,2}$	$M_{v\,3}$	V_{pass}
11.	However,	after	it	*was*	more than	two-thirds	*built*,

N_s	V_t	N_o	$[LP(M_{v\,4})]^E$
the builders	changed	the angle	to 43½ degrees.

$$A = ([M_{v\,1}]^F \ldots (X_o)_s + (M_{v\,2} * M_{v\,3} * V_{pass}))\backslash(N_s + V_t + N_o \ldots [LP(M_{v\,4})]^E)$$

$$= B_{2\,\text{pass}} \backslash B_2$$

Note: In this complex sentence, the first $B_{2\,\text{pass}}$ clause is subordinate, and the second B_2 standard clause is independent. Un-inverted, the first B_2 dependent clause is "*[the Egyptians] built it*"—if we assume the invisible (elliptical) doer in the original is "*the Egyptians*."

	$[X_s]$	V_t	$RC(N_o)$
12.	[You]	Note	that this pyramid is called the "Bent Pyramid."

$$A = [X_s] + V_t + RC(N_o)$$

$$= B_{2\,\text{command}}$$

13.

$M_{n\,1}$	$M_{n\,2}$	$(N_o)_s$	V_{pass}	$LP(M_v)$	C_c	$M_{n\,3}$
The entire	second	pyramid	was built	at 43½ degrees,	but	all

$M_{n\,4}$	N_s	V_t	$M_{n\,5}$	N_o	$LP(M_{n\,6})$
later	pyramids	have	the original	face inclination	of 52 degrees.

$$
\begin{aligned}
A &= \left(M_{n\,1} * M_{n\,2} * (N_o)_s\right) + \left(V_{pass} * LP(M_v)\right) \mid (M_{n\,3} * M_{n\,4} * N_s) + V_t \\
&\quad + (M_{n\,5} * N_o * LP(M_{n\,6})) \\
&= B_{2\,pass} \mid B_2
\end{aligned}
$$

Note: In this compound sentence, the first clause is a $B_{2\,pass}$. Here, similarly to example sentence 12, we will assume the invisible doer in the original is "the Egyptians," and that the un-inverted form's core elements are "*[the Egyptians] built pyramid*."

14.

$[M_{v\,1}]^F$	$M_{n\,1}$	$(X_o)_s$	$LP(M_{n\,2})$	V_{pass}
Interestingly,	almost	all	of the large pyramids	were built

$LP(M_{v\,2})$
within a 150-year period, 2650 – 2500 B.C.

$$
\begin{aligned}
A &= [M_{v\,1}]^F \ldots \left(M_{n\,1} * (X_o)_s * LP(M_{n\,2})\right) + V_{pass} * (LP(M_{v\,2})) \\
&= B_{2\,pass}
\end{aligned}
$$

15.

$[M_{v\,1}]^F$	N_s	V_t	X'	N_o
Since then,	Egyptian society	has called upon	its	engineers

$[VP(M_{v\,2})]^E$
to do other things.

$$
B_2 = [M_{v\,1}]^F \ldots N_s + V_t + (X'\text{-}N_o) \ldots [VP(M_{v\,2})]^E
$$

Sentence Stream Underneath the Text

(1) A $\{B_{3\,question}\}$—(2) A$\{B_{3\,question}\}$—(3) B_2—(4) B_2—(5) B_2—(6) A $\{B_{2\,pass}\}$—(7) A $\{B_3 \mid B_2\}$—(8) B_5—(9) A $\{B_{3\,expletive}\}$—(10) B_2—(11) A $\{B_{2\,pass} \setminus B_2\}$—(12) A $\{B_{2\,command}\}$—(13) A $\{B_{2\,pass} \mid B_2\}$—(14) A $\{B_{2\,pass}\}$—(15) B_2.

Frequency

B_n (all types)	B_n inversions/variations	Compound/complex	RCs and VPs
0× B_1	4× $B_{n\,pass}$	2× B_n I B_n	3× RC
13× B_2	2× $B_{n\ question}$	1× $B_n \setminus B_n$ or B_n / B_n	2× VP
4× B_3	1× $B_{n\ command/statement}$		
0× B_4	1× $B_{n\ expletive}$		
1× B_5			

Action Items

α	Select a short document (one page or less) *written by a peer or colleague* and play the role of editor. Print out a hard copy of this sample and read through it, underlining each verbal you come across (and hopefully you will come across some). Some of these verbals will lead phrases, and some will function solo. Note the part of speech that each verbal unit plays in the sentence by writing the corresponding sentence algebra symbol above it (keep it simple, "N," "Mn," or "Mv," rather than "VP(N)," etc.). See if you can also determine each verbal's type—infinitive (write "I"), past participle (write "P-ed"), present participle (write "P-ing"), or gerund (write "G").
α	The next time *you write* a short document (one page or less), play the role of editor again. Print out a hard copy of this sample and read through it underlining each verbal verb you come across (and hopefully you will come across some). Remember, some of these verbals will lead phrases, and some will function solo. Apply, once again, the analysis and annotation schema presented in the action item above. – If you did *both* of the above, consider your findings from examining the peer or colleague's sample versus your sample. What is different, and what's the same about the two writers' uses of verbals? Now make a list of three new ways you'd like to deploy verbal units in your own writing—that is, when doing so would be useful and productive. At some point, remember and apply your list.
α	Take a 5- to 10-min time-out during your work day, and write a declarative sentence that describes the most important thing you've accomplished thus far during the day: for example, "I completed my last round of research for my design project at 11:00 AM." – Now, using the question creating template presented in Chap. 8, rephrase your sample declarative sentence as a question five or more different ways: for example, "Did I complete my last round of research for my design project at 11:00 AM?" "What did I complete my last round of research for at 11:00 AM?" "When did I complete my last round of research for my design project?" – Consider (if you have time) also using the same sample declarative sentence you wrote for the above or a new one to practice commands/requests. Use the creating template presented in this chapter, and rephrase your sample sentence five or more different ways.

Recap

Key Learnings/Takeaways

- A verbal is a verb form appearing in a sentence (or clause) somewhere other than in the main verb slot. Verbals can appear either by themselves or at the head of verbal phrases. In sentences (or clauses), verbal units function (just like prepositional phrases and relative clauses) as word groups, each playing the role of a part of speech—noun, adjective, or adverb.
- There are four verbal types:
  - *Infinitive*: a form that combines the sign of the infinitive "to" with the verb's *root form*, or principal dictionary listing—for example, "to" + "invent" = "to invent." Infinitive verbals (and infinitive verbal phrases) play the roles of nouns, adjectives, and adverbs in sentence constructions.
  - *Past participle*: a form that, for regular verbs, duplicates the recipe for simple past tense, that is, combine the verb's root form with the suffix "-ed"—for example, "invent" + "-ed" = "invented." This form for irregular verbs takes a special spelling, for example, for the verb "fly," it's "flew," not "flied." Past participles (and past participle phrases) play the role of adjectives in sentence constructions.
  - *Present participle*: a form that, for both regular and irregular verbs, combines the verb's root form with the suffix "-ing"—for example, "invent" + "-ing" = "inventing." Present participles (and present participle phrases) play the role of adjectives in sentence constructions.
  - *Gerund*: a form that, for both regular and irregular verbs, also combines the verb's root form with the suffix "-ing"—for example, "do" + "-ing" = "doing." Gerunds (and gerundial phrases) play the roles of nouns in sentence constructions.

- In a sentence algebra equation, a verbal is distinguished by the subscript "v" for verbal—"V_v." A verbal generally heads up a verbal phrase, constructed using a basic predicate equation (P_1 through P_5 or some permutation thereof).
- As applied to the basic five sentence formulas, an *inversion* of one of those formulas means moving around a sentence equation's principal elements—particularly nouns, pronouns, and verbs—so that these elements are rearranged such that they deviate from standard, left-to-right word order (or syntax). For example, a passive sentence inverts the order of main verb and object from V_t + (N_o or X_o) to $(N_o \text{ or } X_o)_s + V_t$.
- Beginning with five basic sentence formulas (B_ns) as a starting place and framework on which to innovate, several inversions and variations of basic into advanced sentences (As) are possible:
  - *Passive voice*: a basic sentence rephrased from active (standard) into passive voice moves the main verb's object from the predicate (right of verb) to the subject area (left of verb). The original subject, the doer of the main action, is

either relocated to the sentence's predicate and nested within a prepositional phrase or removed altogether and moved "offstage." A passive sentence's main verb is conjugated as a form of "to be" combined with the original verb's past participle.

- *Questions*: to rephrase a standard basic sentence that makes a statement (declarative) to a sentence that asks a question (interrogative), the writer always signals this change by putting a question mark rather than a period at the sentence's end. In addition, questions are often formulated by putting a question word (QW) such as "did," "why," or "who"—at the head of the sentence and by rearranging nouns and verbs, in configurations similar to that of a passive voice sentence.
- *Commands/requests*: a sentence phrased as a command or request (imperative) generally intends to communicate an instruction, warning, or rule from the originator to the audience. Commands/requests are phrased almost identically to basic sentences, except that the subject of a command or request is always "you," stated directly or implied and elliptical.
- *Expletive introductions*: a B_n sentence with an expletive introduction begins with the expletive "it" or "there" followed by a form of the verb "to be." Similar to a passive voice sentence, a sentence that has an expletive introduction shifts important elements from one place to another, in this case from subject doer to predicate.

Reference

1. Steidel, R.F., Henderson, J.M.: The Graphic Languages of Engineering. Wiley, New York (1983)

Part II
Sentence Optimization

Chapter 9
Part II Primer: The Elements of Sentence Optimization

Learning Objectives

- *Know what "sentence optimization" is and what purpose it serves in applied engineering writing.*
- *Know what the initialism "OFI" stands for.*
- *Understand why this book's system stages its list of 20 opportunities for improving sentences into four progressive groupings.*
- *Understand how the system's progressive approach can benefit almost all learners and is especially advantageous for multilingual speakers of English who are still working on improvement opportunity #20, "use conventional phrasing."*
- *Understand how grammar theory and sentence algebra equations apply and don't apply to sentence optimization.*

Consider this scenario: You and your engineering team are moving along a large-scale, high-stakes project. An unexpected problem emerges and progress halts. You determine how to fix the problem, but the fix will be expensive. Now, suddenly, your project requires an additional $108K in unanticipated funding to reach completion.

You write a proposal memo to your division's on-site manager to procure the $108K. From past experience, you know that this manager reads fast, thinks quickly, and has zero tolerance for fluffy, meandering documents. Thus, you limit your memo to one sparse page of text comprising 270 words. How much, then, is each word worth? The answer is, on average, $400 per word ($108,000/270).

When engineers write important documents, especially short-form documents, they need to make certain that every word counts. There is no room for bloated phrasing or extraneous details. Each word should contribute to the document's purpose—which, in this case, is telling the manager specifically what the team needs, why the team needs it, and why granting the request for additional funding makes good business sense.

Beyond judiciously choosing words to build streams of sentences that are concise and clear, what about sentence-level correctness? What about appropriate writing

B. Henderson, *A Math-Based Writing System for Engineers*,
https://doi.org/10.1007/978-3-030-10756-7_9

style? What about effective formatting and document structure? For our hypothetical proposal memo's reader and the bulk of most other managerial audiences, all of these are important features of a successful workplace document.

The goal of Chaps. 9 through 13 is to move beyond sentence structure fundamentals and to refine your skills for writing optimal sentences and sentence streams known as paragraphs in workplace applications. Learning *sentence optimization* techniques enables engineering writers, as they write on-the-job, to engage in a continuous quality control process that strives to eliminate errors and ineffective constructions during sentence formulation—before they manifest on the page. Since mistakes still can and do happen, learning optimization techniques also equips the engineering writer with a keen set of "editorial eyes," enabling you to identify, troubleshoot, and repair a complete spectrum of 20 common sentence problems.

Commonplace errors have been invading and sabotaging sentences for many years and in many types of writing, and identifying these errors and discussing how to avoid them are by no means original to this book [1–3]. What's different about my treatment of common errors is simply that it selects 20 that I have observed to be top offenders in engineering writing; moreover, these chapters examine those errors using a math-based writing perspective and engineering examples.

Sentence Engineering Tip

Most writers do not regularly make 200 or 300 different types of sentence errors. Rather, they have a small number of bad habit errors that they commit frequently, even pervasively. Often, these are as few as 2 or 3 "favorites"—and often writers are unaware of them or maybe just lack the tools to identify and repair these types of mistakes. Thus, if you can identify what your heavy-hitter errors are and then learn sentence optimization techniques to address them, the quality of your documents might well improve significantly.

Twenty Opportunities for Improving Sentences

In this book, I refer to common sentence problems as *opportunities for improvement (OFIs)*. I divide the 20 OFIs into four progressive groupings:

- Sentence-level flaws that interrupt concision and clarity (Simplify and Clarify: OFIs #1 through #4)
- Critical errors resulting from grammar and punctuation mistakes (Eliminate Category I Errors: OFIs #5 through #10)
- Constructions that do not violate basic grammar and punctuation rules but nevertheless result in nonoptimal sentence style (Advance Optimal Style: OFIs #11 through #14)
- Less critical errors that interrupt smooth reading of sentences and mar their professional luster (Minimize Category II Errors: OFIs #15 through #20)

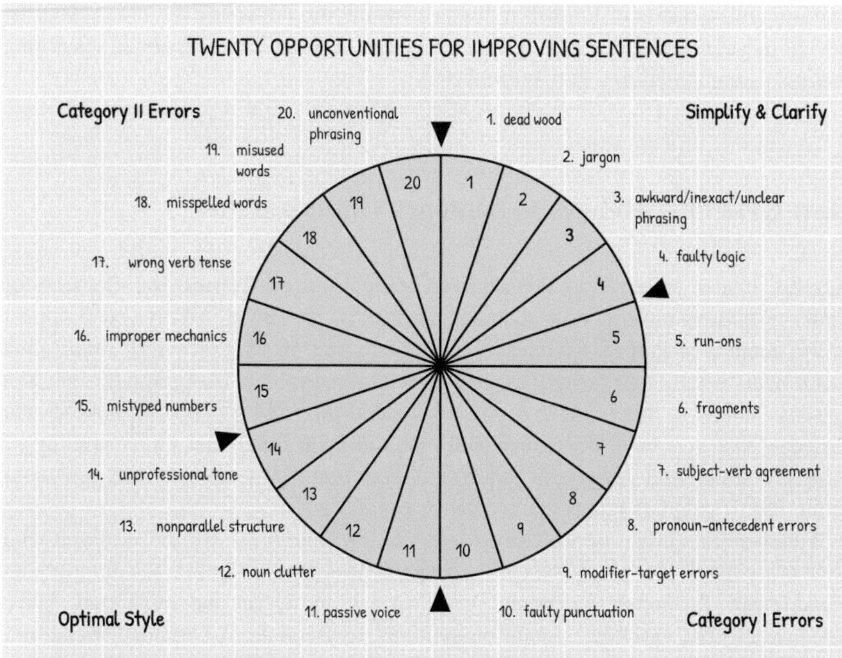

Fig. 9.1 Twenty techniques for improving non-ideal sentences on the optimization wheel

Figure 9.1 illustrates the full sequence of 20 OFIs positioned categorically around a sentence optimization wheel.

The optimization wheel's final grouping, the Category II Errors, deals more with nuanced native-speaker preferences and conventions than it does with grammar logic and the standard patterns (equations) that shape English sentence designs. This is especially true of *OFI #20—use conventional phrasing*, which calls upon the writer to go beyond the (mostly) objective rules that guide correct formation of sentences' structures into the more subjective terrain of being able to phrase text sentences "like a native speaker would say them."

Learning conventional phrasing, the final stage of mastering the English language, involves subtleties like knowing when (and when not) to insert articles (a, an, or the), choosing prepositions (should it be "on" a team or "in" a team?), and interpreting and applying idioms (why, for example, did someone say the new equipment costs "an arm and a leg" rather than something straightforward like "$250"?). What makes conventional phrasing even more complicated is that proficiency in it usually comes from immersion in a native-speaker community and practice over time (often several years), not from books and classes.

For this reason (as well as a general, most-to-least-interruptive-to-readers scheme), this book organizes its lineup of 20 OFIs with #20 last. While working on the long process of mastering OFI #20, this book's system encourages multilingual speakers of English to focus diligently on mastering the other 19 OFIs, which present fewer (or no) disadvantages to a multilingual person's learning process as compared to a native speaker's.

How Grammar Relates to Sentence Optimization

Another matter to cover in this chapter, aside from previewing the elements of sentence optimization, is how sentence algebra, as presented in Part I, will appear in the context of Part II. Chapters 9–13 focus more on the "how tos" rather than theoretical "whats" and "whys" associated with sentence and paragraph design. The lessons assume at this point that you understand how sentences work beneath the surface—that you have developed and internalized a functional understanding of basic grammar. And because of this assumption, you will see mostly text sentences in the Part II chapters rather than sentence equations.

To some, this might be welcome news. On the other hand, for those of you who have taken the time to study the earlier chapters on sentence algebra, this might seem paradoxical. To better understand why we will now, for the most part, leave sentence-algebra variables, operation symbols, and equations behind, I invite you to consider the following analogy. If you are an engineer, early in your preengineering education, you took an entry-level geometry class and learned the Pythagorean theorem—which, in math code, looks like this:

$$a^2 + b^2 = c^2$$

Most likely, your teacher began the lesson by taking you through a detailed, methodical proof to yield the equation above and then assigned you homework problems that applied the theorem, starting with easy problems and progressing, step by step, toward more advanced challenges. After a while, you achieved mastery, and then you were able to swiftly and easily apply the theorem to all types of problems.

Today, the Pythagorean equation remains in your engineering equation toolbox. Anytime a right triangle appears in your engineering work—in modeling a structure, locating something on an X–Y grid, or computing the grade of an incline—the theoretical equation behind the right triangle is implicit in your recognition and processing of right-triangle geometry, wherever it presents itself. It is a foundational element so deeply ingrained in you that you can call upon it and apply it (as the saying goes) *without even thinking about it.*

Likewise, the sentence optimization chapters assume that you have an internalized understanding of basic grammar and that this understanding (combined with your prior education and applied experience using English) allows you to compose sentences in word code (text), string them together, and produce routine documents, which are, if not always excellent, of serviceable quality.

How you acquired your core writing skills—through the approach in this book, through traditional English classes, or through extensive reading and writing practice—does not matter. If your sentence engineering skills present opportunities for improvement (just a few or many), and you are interested in consistently writing excellent documents, then you are likely to profit from the sentence optimization lessons. Even those of you who begin Part II without complete proficiency in grammar will likely profit from learning about sentence optimization, though the extent to which you will profit might be less.

Action Items

α	The next time you write a request for money—professional or personal—capture and save the document's text into a temporary Microsoft Word file (or equivalent) and use the word count feature to count total words in the document. Then calculate how much in dollars, on average, each word is worth.
α	Consider the sentence-level quality you observe in your own engineering writing and in your peers', and determine several common errors you regularly see. Note: You needn't know the official name for these errors. Your own descriptive labeling is fine—for example, "uses 'it's' when it should be 'its'" or "doesn't use commas correctly." Do you know how to troubleshoot and repair these errors? Revisit this observation later, after you have read all of the Part II chapters.
α	Estimate an approximate X-percent rating—for example, ~50%, ~60%, and ~85%—for how you would rate the sentence-level correctness of the final drafts of important, formal documents that you write as an engineer. Think of this number as a quality control guarantee. For example, "In general, I guarantee that engineering documents I write and circulate are built out of sentences that are X% grammatically correct." Are you satisfied with this percentage rating? If not, make a plan for improving it.

Recap

Key Learnings/Takeaways

- Chapters 9 through 13 examine *sentence optimization*, a set of techniques to help engineering writers eliminate or repair 20 common sentence imperfections.
- This book refers to sentence imperfections as *OFIs* or *opportunities for improvement* and divides the 20 OFIs into four progressive groupings:
  - Sentence-level flaws that interrupt concision and clarity (Simplify and Clarify: OFIs #1 through #4)
  - Critical errors related to grammar and punctuation mistakes (Category I Errors: OFIs #5 through #10)

- Constructions that do not necessarily violate basic grammar rules but result in nonoptimal sentence style (Advance Optimal Style: OFIs #11 through #14)
- Less critical errors that interrupt smooth reading of sentences and mar their professional luster (Minimize Category II Errors: OFI #15 through #20)

• Dividing the 20 OFIs into four progressive groupings is useful for all writers, because this scheme assigns first priority to learning how to identify and repair errors that are most interruptive to readers.
• The progressive sequencing can also be helpful for multilingual speakers of English. Whereas the first three OFI groupings cover techniques for sentence optimization that have to do with standard grammar and punctuation logic, and basic patterns that shape sentence style, the fourth grouping covers OFIs associated with native-speaker conventions and preferences, some of which (especially *OFI #20—use conventional phrasing*) are not easily learned from books and classes. While they may be still mastering OFI #20, the system encourages multilingual speakers to develop expertise in the other 19 OFIs.
• Chapters 9–13 assume that you have already learned and internalized an applied understanding of basic grammar and that, when you write, you typically generate sentences that, if not always ideal, are for the most part grammatical. How you acquired your applied understanding of fundamental grammar—sentence algebra, traditional grammar lessons, or extensive experience reading and writing in English—doesn't matter with respect to much of Part II's content. Thus, mastery of this book's chapters on sentence algebra is not a prerequisite for studying and benefitting from the chapters on sentence optimization; however, it certainly doesn't hurt.

References

1. Lunsford, A.: The Everyday Writer, 6th edn. Bedford/St. Martin's, Boston, MA (2016)
2. William Jr., S., White, E.B.: The Elements of Style, 4th edn. Longman, New York (1999)
3. Hacker, D., Sommers, N.: Rules for Writers, 7th edn. Bedford/St. Martin's, Boston, MA (2011)

Chapter 10
Simplify and Clarify

Learning Objectives

- *Understand why concision and clarity are more important than fancy language in engineering writing.*
- *Be able to identify and resolve four specific sentence-level errors or "Opportunities for Improvement" (OFIs) that unnecessarily complicate and obfuscate engineering writing.*

The engineering writer's first order of business regarding sentence-level quality control is to make certain that every sentence is concise and clear. This book calls the process *simplify and clarify*. This chapter reviews four specific techniques for eliminating words and phrasing that unnecessarily complicate and obfuscate a sentence's intended meaning. Figure 10.1 displays those techniques (OFIs #1 through #4) on sentence optimization's 20 OFI wheel.

If you've been told that short, simple sentences are too pedestrian or basic for serious, formal writing, it is high time that you unlearned this rule of thumb. This rule may sometimes apply to literary and scholarly writing. Certain genres do demand sentences that are stylish, nuanced, and steeped in allusions and esoteric subtext. The engineering world, however, does not praise fancy sentences; in fact, it demands the opposite.

B. Henderson, *A Math-Based Writing System for Engineers*,
https://doi.org/10.1007/978-3-030-10756-7_10

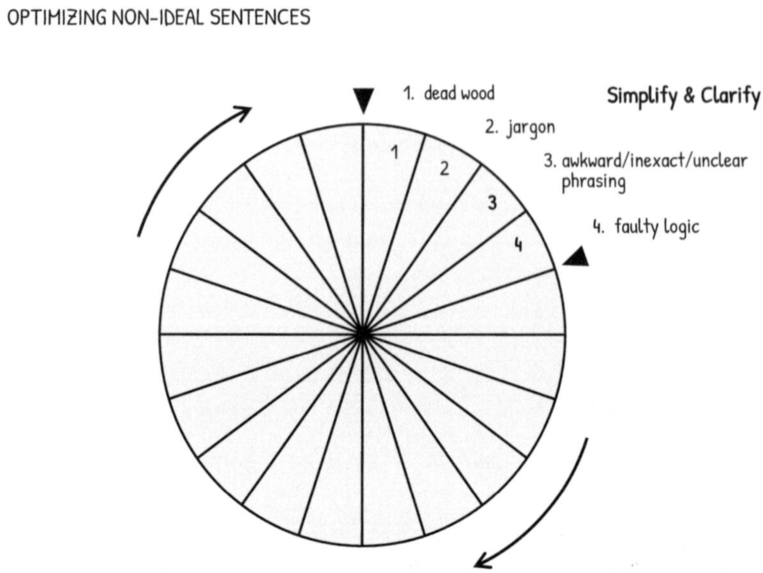

Fig. 10.1 The simplify and clarify errors on the OFI wheel

Example of *Simplifying and Clarifying* a Pompous, Bloated, and Comical Sentence

Weak:	My opinion regarding the matter is happily ardent—that is, with regard to my new engineering job at AmCoTech Corporation—in that I find both my assigned job duties and the employment venue itself keenly well-suited to my professional predilections and aspirations. (42 words)
Better:	I like what I do and where I work in my new engineering job at AmCoTech corporation. (17 words)
Shrinkage:	(1 − (17 words/42 words)) × 100% = 59.5%

Eliminate Deadwood

The first step of simplify and clarify is eliminating *deadwood*—words that don't earn their place. Below are three examples of original sentences with unnecessary words and their simplified, revised versions. Each example quantifies the level of improvement, from original to revised version, with a shrinkage calculation. If you can shrink a sentence's word count, even by one word, and retain the sentence's meaning, shrink it.

Original: The condition of the computer is such that it requires replacement. 10.1a

Revised: The computer needs to be replaced. 10.1b

(1 − (6 words/11 words)) × 100% = 45.5% shrinkage

Original: The nature of helium is such that it is a gas at room temperature. 10.2a

Revised: Helium is a gas at room temperature. 10.2b

(1 − (7 words/14 words)) × 100% = 50.0% shrinkage

Original: She performed the calculation of tensile stress to achieve a value. 10.3a

Revised: She calculated tensile stress. 10.3b

(1 − (4 words/11 words)) × 100% = 63.6% shrinkage

Eliminate Jargon

Since engineering writing conveys content that is complex and engaging, propping it up with *jargon*—buzz words, fancy expressions, and gratuitous foreign language terms—is a waste of time for both the writer and the reader. There are many lists of jargon words to avoid in print and on the Internet, and opinion varies from source to source as to what the top contenders are. At the end of this segment, I present a short, illustrative list of my favorites to avoid.

There are many types of jargon. Some jargon is generic and seemingly endemic, such as "utilize," the gobbledygook version of "use." Some is specific to the business world, such as "monetization" and "investiture." Other jargon is specific to high-tech industry—for example, "implement" and "facilitate."

Jargon words often sound "jargony" and are coined with suffixes such as "-ize." They can be colloquialisms like "basically" and "really." Although perhaps "cool" in conversation, such words do not play well in formal workplace writing. Oftentimes, jargon terms emphasize melodrama and hype—for example, "catastrophically slanderous" or "prodigiously colossal."

On the other hand, technical terms associated with a specific discipline or a company's proprietary products, processes, and technology are not jargon. Just because a word is exclusive, intricate, sophisticated, or multisyllabic does not make it jargon. A word that is necessary for concise, clear, and correct communication—as well as being the best choice for exact expression—is always the best word to use. Jargon words are unnecessary fancy or pretentious words that exist alongside plain English alternatives.

Although some jargon words have no place at all in engineering writing, others are not entirely replaceable. They become jargon when they're overused. In engineering, "implement" is one of those words. "Implement" lacks a simpler, precise replacement. "Start" and "initiate" are close, but not exact twins. Thus, engineers sometimes need to use "implement" to write precisely and accurately. But it's rarely necessary to use "implement" multiple times in a sentence, or even in a single paragraph or page. "Very" is another word that is potent but overused. "Very" efficiently and effectively emphasizes descriptions and assertions. However, when a writer uses it repeatedly, the reader becomes immune to it and filters it out.

In the next example series, 10.4 through 10.6, each sentence pair's original version contains jargon. The jargon terms are *italicized* in the original and removed or replaced with suitable alternatives in the revised version.

Original: Both MRI and X-ray imaging *facilitated implementation* of tissue monitoring around the insulin pump implant during the beta test. 10.4a

Revised: The beta test used both MRI and X-ray imaging to monitor tissue around the insulin pump implant. 10.4b

Original: *In view of the fact that* it is dangerous, *termination* of welding *in the immediate vicinity of* flammable liquids is *highly desirable.* 10.5a

Revised: Because it is dangerous, stop welding near flammable liquids. 10.5b

Original: *Basically*, it wasn't until we *utilized* social media in our marketing strategy that sales *really* escalated from ~50 K to ~500 K units/year. 10.6a

Revised: When we used social media in our marketing strategy, sales escalated from ~50 K to ~500 K units/year. 10.6b

Some Examples of Jargon Words and Phrases to Replace or Avoid

ascertain – find out
at the present time – now
during the course of – during
employ – use
endeavor – try
caveat – warning
facilitate – lead
in the immediate vicinity of – next to
in view of the fact that – because
terminate – stop, end
utilize – use
transpire – happen
until such time as – until
firstly, secondly . . . – first, second . . .
very – [use sparingly, once max. per pg.]
really – [delete]
basically – [delete]
as it were, if you will – [delete]
elucidate – explain
facetious – humorous
noteworthy – laudable

Note: the above list is representative, not comprehensive. Beware: there are more.

Revise Awkward, Inexact, or Vague Sentences

Awkward sentences sound strange when they're read aloud. When a sentence's phrasing is jumbled and rough, sometimes a reader can still tell what the writer means, but awkwardness makes the sentence sound unprofessional.

In contrast, sentences that are inexact may be smooth to read, but fail to communicate clearly what the writer intends to say. *Inexact sentences* are difficult for readers to decipher because they encode information using language that is imprecise or inaccurate. Sometimes the reader can infer what an inexact sentence is trying to say, based on context and piecemeal cues, but the analysis required slows down reading and assigns the reader extra work. Other times, inexact sentences are so imprecise that their intended meaning totally baffles the reader.

Here are several illustrations—first, an awkward sentence, original and revised (10.7), and then two inexact sentence examples, which display different degrees of the same problem, troublesome (10.8) and then catastrophic (10.9):

Original: I have in applying my engineering skills proficiency since they are supported by my statistician training in process control. 10.7a

Revised: My engineering and statistical training give me proficiency in process control. 10.7b

Problem: *awkward phrasing*
Repair tactic: *rephrase to enable smooth and rapid reading*

Original: Without contamination, be sure to apply latex gloves only after you put on booties, goggles, and a hood for sterile gowning in the clean room. 10.8a

Revised: When gowning for the clean room, put on booties, goggles, and a hood before you put on latex gloves to ensure that the gloves' outer surfaces remain uncontaminated. 10.8b

Problem: *inexact phrasing (decipherable)*
Repair tactic: *unscramble words and fix inaccuracies/ambiguities; for example, a worker should put on booties, goggles, and a hood before, not after, latex gloves*

Original: After resonance, the next reduction was based on the new fabrication, before reducing collateral input, and the sonic flange was compromised by the anodized aluminum stock. 10.9a

Revised: ??? 10.9b

Problem: *inexact phrasing (undecipherable)*
Repair tactic: *delete sentence; try again*

Aside from awkwardness and inexactness, another closely related sentence problem is vagueness. *Vague sentences* express a message that is incomplete, ambiguous, or both, because the message lacks sufficient detail. Whereas a sentence containing deadwood is troublesome to readers because it contains excessive, unnecessary details mixed in with the sentence's intended message, a vague sentence is just the opposite—not enough detail.

Vagueness (as well as awkwardness and inexactness) often happens when a writer relaxes quality control and allows a sentence that is imprecise and inaccurate to "slide by" and make it onto the page. To revise a vague sentence (or an awkward or inexact sentence), you must put more energy into writing that sentence—and then keep working on the sentence until it encodes your intended message smoothly, clearly, precisely, and completely. There's no shortcut, just effort and high standards.

To make a vague sentence's text more specific, I recommend using the 5W + H equation, our system's shorthand for the prompts: *who*, *what*, *when*, *where*, *why*, and *how?* To determine missing details, subject the text to the prompts, determine where the prompts apply, develop applicable responses to complete the sentence's message, and then revise accordingly. Example 10.10 illustrates the method, with vague text underlined and annotated in the sentence's marked up version (10.10b) and then exchanged for more specific text in the revision (10.10c).

Original: Near a mountain lake, we installed a unit on a high edifice to collect desired test data during a set period of time. 10.10a

Marked up: Near a mountain lake (what lake? where located?), we installed a unit (what kind of unit?) on a high edifice (what edifice? how high?) to collect desired test data (what type of data?) during a set period of time (how long and when is time period?). 10.10b

Revised: Near Lake Tahoe, California, we installed a solar panel on Mount Tallac's peak at 7200 ft above sea level to collect data on average kilowatts harvested per day during June, July, and August last summer. 10.10c

In the above example, the original sentence lacks some important *what? when? where?* and *how?* specifics. For instance, the markup and revision demonstrate that "mountain lake" can be more exactly expressed by naming the lake (what?) and its location (where?) and revising to "Lake Tahoe, California." Additionally, in the original, the nebulous expression "a set period of time" fails to disclose test date (when?) and duration (how long?). These details can be communicated precisely and accurately by expanding the text to "June, July, and August last summer." The example reveals some additional opportunities for improvement, as well.

You may have noticed that Example 10.10 does not apply all six prompts. It applies *what? when? where?* and *how?* but assumes that, in the original, the details associated with *who?* and *why?* are okay as is. In general, how to best apply the 5W + H equation varies from sentence to sentence. When you are doing sentence optimization and attempting to avoid or repair a vague stream of text, consider all six 5W + H prompts to begin with, determine which apply (and where), and then develop responses and corresponding revisions as you see fit.

Sentence Engineering Tip

When you're writing an engineering document, unless you maintain vigilant quality control, it's relatively easy to let an awkward, inexact, or vague sentence slip onto the page and blend in with the rest. This happens because, as the writer, you know *what you're trying to say*, and as you scan text as you write, this original intent can bias what you see. Unfortunately, your readers do not have access to your sentences' original intended meanings. They're not mind readers. All they have access to is what lies on the page before them, and what those word streams, now detached from your intent, communicate *exactly as written*. As an engineering writer, strive to not make compromises. Do not allow nonoptimal sentences to slide into a document's text because you're feeling lax and thinking, "Oh well, that sentence is a little rough, but it's good enough." Keep working on your sentences *until all of them* say exactly what you intend them to mean.

Avoid Faulty Logic

To conclude this chapter, let's examine one more simplify and clarify practice. Sometimes sentences "don't compute" for readers because those sentences have been contaminated with faulty logic. Below are three example pairs, 10.11 through 10.13, each displaying a sentence containing faulty logic and a second version revised to be more logical and representative of the engineering writer's intended meaning. A note beneath each pair briefly discusses the specific problem and how it has been repaired.

Original: If nitrogen is odorless and nitrogen is a key component in nitrous oxide, which contains oxygen which is also odorless, obviously, nitrous oxide would be odorless, too, because it's a mixture. 10.11a

Revised: Both nitrogen and oxygen are odorless gases and can join together to form nitrous oxide, N_2O. 10.11b

Note: Sentence 10.11a is illogical because it expresses a conclusion built upon a faulty premise. That the two elements are odorless and can join together to form a compound doesn't mean the resulting compound is odorless. Another term for this type of faulty logic is "non sequitur."

Original: The agricultural engineer claimed there were only two options for further development of an automated harvester for commercial crop soy beans: his two new designs, concept "A" and concept "B." 10.12a

Revised: The agricultural engineer presented the two most viable options for further developing an automated harvester for commercial crop soy beans: his two new designs, concept "A" and concept "B." 10.12b

Note: Sentence 10.12a is illogical because the engineer presents his two preferred choices as the only two options. In truth, there are probably more than two options. Here, instead, it would be reasonable for the engineer to present two options, qualified as being "the two most viable," not the only two. Another term for this type of faulty logic is "either-or" reasoning.

Original: The technical marketing survey collected over 1000 positive responses regarding the Orion 55 Cruise Craft, and this result proves that our new solar-powered house boat will be our best product and easily net millions in profits. 10.13a

Clarified: The technical marketing survey collected more than 1000 positive responses regarding the Orion 55 Cruise Craft, indicating that our new solar-powered house boat will be a popular and profitable product. 10.13b

Note: Sentence 10.13a is illogical because it presents absolute conclusions—"proves," "our best"—based upon a non-absolute and insufficient premise. Engineering writers should always be careful with the word "prove." Remember the Scientific Method. Even with a large set of affirming data, good science and good engineering can only determine a hypothesis to be "false" or "not false," not unreservedly "true." Another term for this type of faulty logic is "sweeping generalization." A similar form of faulty logic also present in 10.13a is "hyperbole"—that is, grandiose exaggeration.

Action Items

α	Eliminate as much dead wood and jargon as possible from the following bloated sentence to communicate the same message with fewer words. When you're done, count the number of words in the simplified version and calculate shrinkage (1 − (X words/Y words)) × 100% = Z% shrinkage).
	Hint #1: It's possible to shrink this sentence more than 50%.
	Hint #2: What this pompous, verbose speaker is talking about is being hungry and needing a fork.
	Due to the fact of my immediate necessity and ravenousness, would you please be so kind as to pass me one of those handy, multipronged stainless steel utensils that we humans use for transporting food solids from plate to mouth. (Y = 40 words)

(continued)

α	What are the most popular jargon terms in your engineering community? For the next few days, note when you or someone else uses jargon instead of plain English. Each time, think about whether the jargon added to or subtracted from the communication's efficiency and effectiveness. Remember: Technical terms associated with your technical discipline or a company's specific products/processes, technologies, and intellectual properties are not necessarily jargon.
α	The next time you have to write an important email, read it aloud before you press "send." If your ears detect places that sound awkward or muddled, these places probably need to be simplified and clarified. Revise them to sound better. Though not necessarily a complete fix, chances are that you will improve rough spots simply by adjusting them to flow smoothly and "sound good" when read aloud.

Recap

Key Learnings/Takeaways

- In engineering writing, concision and clarity are more important than fancy words because the content is already complex and engaging. There is no need to dress up engineering sentences with fancy words, phrases, and stylistic intricacies. In fact, such elements can slow down or even halt reading.
- This chapter covers four specific sentence-level problems that unnecessarily complicate and obfuscate engineering writing (OFIs #1, #2, #3, and #4):
  - *Deadwood*: Unnecessary words that don't earn their place in a sentence. Deleting deadwood optimizes sentences. When eliminating unnecessary words, one quantitative metric for progress is "shrinkage" = (1 − (# of words in sentence after)/(# of words in sentence before)) × 100%.
  - *Jargon*: Fancy words placed into a sentence "for show." Instead, use straightforward terminology instead of jargon to optimize sentences. For example, choose "use" instead of "utilize."
  - *Awkward, inexact, or vague sentences*: Low-quality sentences that are cumbersome to read and confusing (either a little or a lot) for readers to understand. For each writing task, an engineering writer's work is done only when every sentence, from the reader's perspective, says exactly what the writer intends it to say. Revise awkward or inexact sentences to read smoothly and clearly. To fix vague sentences, this book recommends that engineers apply the 5W + H equation (*who? what? when? where? why?* and *how?*) to locate missing details. Once identified, the writer can add these details to the sentence to complete its meaning.
  - *Faulty logic:* Exactly what the name says. Engineering writers should watch for and avoid sentences that contain faulty logic, such as sentences that make claims based upon insufficient or irrelevant evidence; claims that are exaggerated, untrue, or impossibly absolute; and claims that are based upon individual opinions and wishes rather than data and facts.

Chapter 11
Eliminate Category I Errors

Learning Objectives

- *Know two high-level errors that create imposter sentences and the process for scanning text to identify, avoid, and repair them.*
- *Know and be able to reconcile three types of grammatical bookkeeping errors.*
- *Understand how punctuation marks guide and optimize readers' cognitive uptake when they decode/read sentences.*
- *Know and be able to deploy several specific punctuation techniques that produce sentences that are concise, clear, and correct—and reader-friendly.*

Be on the Lookout for and Eliminate Imposter Sentences

The first two Category I errors we will discuss are run-on sentences (OFI #5) and sentence fragments (OFI #6). Figure 11.1 displays these two errors along with the other category I errors on the OFI wheel.

A *run-on sentence* occurs when a writer inadvertently links two grammatically complete sentences together with a faulty connection—either by *fusing* two independent sentences together without a proper connector or by joining two independent sentences together with a *faulty connection device*, oftentimes "weak-link" punctuation in the form of a "comma splice."

A *sentence fragment*, on the other hand, is an incompletely constructed sentence—a cluster of words that begins with a capital letter and ends with a terminal punctuation mark but that is only masquerading as a sentence because its construction fails to comply with the basic grammatical requirements for independent operation.

B. Henderson, *A Math-Based Writing System for Engineers*,
https://doi.org/10.1007/978-3-030-10756-7_11

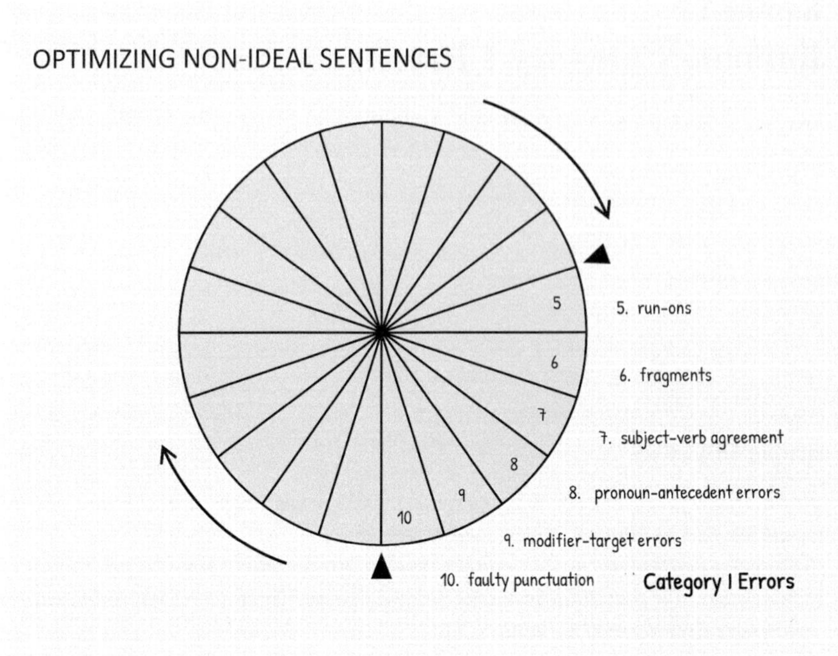

Fig. 11.1 The Category I errors on the OFI wheel

Even though run-ons and fragments look very similar to real sentences typographically, they are not. They are *imposter sentences*. Many readers in the professional world consider run-ons and fragments to be glaring, critical sentence-level flaws. Some readers even view run-ons and fragments as a type of mistake that indicates communication incompetence. In your writing as a professional engineer, you can get away with a few lower-level writing errors, but you should always eliminate imposter sentences from your formal workplace writing.

Example Imposter Sentences

Below are two examples that illustrate imposter sentences—one run-on (11.1) and one sentence fragment (11.2). Each example presents a flawed original version (the imposter) and a revised version underneath. *Italics* flags the location of the error in the original. A note below each example describes what the error is and why it is ungrammatical through the lens of traditional grammar. The note also explains the repair technique used and additional repair options.

Run-on Sentence

Original: Hydroelectric dams can efficiently harness potential energy from stored *water now* engineers are designing devices for harvesting kinetic energy from ocean waves. 11.1a

Revised: Hydroelectric dams can efficiently harness potential energy from stored water. Now engineers are designing devices for harvesting kinetic energy from ocean waves. 11.1b

Note: The original is two independent clauses (basic sentences) fused together and typed as one sentence. The result is a run-on imposter. An easy way to fix the error is to divide and punctuate the two fused sentences separately, so they stand independently as two structures, both legitimate basic sentences.

You can also repair this error by joining the two sentences together with a properly punctuated coordinating conjunction (e.g., *", and") or with a properly punctuated conjunctive adverb (such as "; in fact,"). Additionally, you could join the two independent clauses with a semicolon (";"). All three options form grammatically correct compound sentences.*

Sentence Fragment

Original: The invention of a solar-powered tomato harvester by agricultural engineers might revolutionize field crop farming. Although such a machine is not yet feasible with current technology. 11.2a

Revised: The invention of a solar-powered tomato harvester by agricultural engineers might revolutionize field crop farming, although such a machine is not yet feasible with current technology. 11.2b

Note: The original is two streams of text both typed as sentences. The first stream is an independent clause that can stand alone as a basic sentence. However, the second is a dependent clause that begins with the subordinating conjunction "although." Since, grammatically, a dependent clause cannot stand alone, the second stream is a fragment.

The standard repair here would be change the period "." between the two streams to a comma "," and then attach the imposter sentence (dependent clause) to the basic sentence that precedes it (independent clause). The result forms a legitimate complex sentence.

The Scanning Process

Identifying run-ons and fragments in the previous two examples (11.1 and 11.2) calls upon you to *scan the original text* for imposters and reconcile them. More specifically, the process requires three action steps:

1. Scan through the text and distinguish the boundaries of *typed sentences* that begin with a capital letter and end with a terminal punctuation mark (period or question mark).
2. Analyze and identify whether each typed sentence is a grammatical construction or an imposter.
3. Determine a repair technique for revising imposter sentences into legitimate sentences that conform with grammar rules.

The explanations given below the previous examples assume that you know how to identify basic sentence structures within word streams—such as independent clauses and dependent clauses. The examples also assume that you are familiar with a complete set of correct devices for joining clauses (basic sentence structures) together to form compound or complex sentences (what this book's system refers to as advanced sentence structures).

For those of you who already possess the ability to scan written text, avoiding imposter sentences is mostly a matter of vigilance. When writing something new, you must take care to avoid inadvertently creating run-ons or fragments. When revising and editing something already written, you must always be on the lookout for imposter sentences (and other high-level errors) and repair them without fail before you release a document into circulation.

But what if sometimes you have trouble recognizing what is and what is not an independent versus dependent clause within a word stream? What if you *do not* have the ability to scan text and identify the word patterns underneath the streams that distinguish them as grammatical sentences as opposed to imposters? Whether you struggle to see such things through the lens of traditional grammar; or through intuitive grammar skills acquired through speaking, reading, and writing practice; or through a combination of both, here is some good news for you.

As you read further into this book's chapters on sentence optimization, you will learn more about how sentences function at the component level—as well as how to make sentences operate properly and grammatically. This additional learning alone might close the gap. If not, you can, if you wish, also consult the chapters in Part I for more detailed, comprehensive lessons on grammar through the lens of math. Also, the Appendix demonstrates how to scan and repair a sample paragraph of text that contains faulty sentence equations (imposters) using the fundamental elements of sentence algebra.

Sentence Engineering Tip

Unlike coordinating conjunctions ("and," "but," "or," "nor," and so on), which can be punctuated with a single comma, conjunctive adverbs ("however," "in fact," "therefore," "consequently," and so on) require a *semicolon before* and *comma after* when used to join complete sentence structures. When punctuated otherwise, they create a faulty link and, as a result, an imposter sentence. For example, both of the following sentences are grammatical: "I like this, but you like that." and "I like this; however, you like that." On the other hand, the next sentence is a run-on: "I like this, however you like that."

Do Grammatical Bookkeeping and Reconcile Disagreements

Another ongoing quality control measure for engineers to apply to their writing is called *grammatical bookkeeping*. The process ensures that functional units working together in pairs within sentences (such as a subject noun and main verb), as well as in sentence sequences (such as a noun in one sentence paired with a pronoun replacement in a sentence that follows), join together correctly and clearly when a reader reads them. This section discusses three common pairings that are prone to faulty connections and disconnects, OFIs #7, #8, and #9.

Subject-Verb Agreement Errors

The first grammatical bookkeeping error is a basic subject-verb pair that fails to agree in number—that is, a plural subject mismatches with a singular verb or a singular subject mismatches with a plural verb. Engineering writers typically have less trouble with subject-verb agreement errors when writing short sentences, because few or no words stand in between the subject and verb. In this case, the engineer's "ear knowledge" of the language is more likely to guide verb choice to achieve a result that both "sounds" and is right.

On the other hand, longer sentences with multiple word units inserted between the subject and verb pair may sound less wrong when there is a mismatch. What might sound like an overt "clunk" in a short sentence becomes a small "clink" inside a longer, more layered sentence. From the reader's point of view, however, both bumps interfere with the decoding process. So beware of even tiny "clinks." The following three example sentences display subject-verb agreement errors that need to be reconciled. Note that as sentence length increases, the errors "sound" progressively less overt.

One astronaut *are* a biomedical engineer. ["CLUNK"] 11.3

The rigorous training program that prepares crew candidates for zero-gravity missions *yield* a 28% attrition rate. ["clunk"] 11.4

The New Generation Space Shuttle—with its unassisted takeoff and reentry capabilities, its maneuverability in space, and its highly skilled crew members—*exemplify* aerospace technology at its best. [little "clink"] 11.5

The main verbs for sentences 11.3, 11.4, and 11.5 are all incorrect in number. Their conjugations should be [astronaut] *is*, [program] *yields*, and [Shuttle] *exemplifies*. When read aloud, 11.3 sounds like a glaring clunker; 11.4 sounds odd, but in a manner that's not as obvious; and 11.5 *almost* sounds correct when read aloud, because of the presence of several other nouns—one singular, "maneuverability," and two plural, "capabilities" and "crew members"—located in between the sentence's true subject and main verb.

Because these other nouns are positioned immediately upstream from the verb, it's easy to mistake them as "imposter subjects" and let them dictate verb number—for example, pairing "crew members" (plural) with "exemplify" (plural). However, this sentence's true subject, "New Generation Space Shuttle," is singular and must pair with a singular verb. Beware: The difficulty of detecting subject-verb agreement errors is often directly proportional to the number of words between the true subject and verb.

Pronoun-Antecedent Reference Errors

The second grammatical bookkeeping error concerns the pairing of antecedent nouns (Ns) and their partner pronouns (Xs). Engineering writers, just like any type of writer, regularly find it useful to substitute pronouns for nouns. In a progress report, for example, the five people in Jeff's engineering work group can conveniently become one pronoun, "we." The three projects proposed for the next fiscal year by engineer Jillian might become "those." Using pronouns to stand in for nouns can streamline sentences—that is, as long as the reader can easily and clearly identify the noun being replaced.

Each time a reader encounters a pronoun in a sentence, the reader must "solve for X in reverse." In other words, starting with pronoun X, the reader then must glance (or think) backward to locate the most logical candidate for noun N paired with X—that is, the antecedent of X. From the reader's point of view, noun-pronoun pairings reveal themselves largely based on position. Thus, a best practice for engineering writers is to arrange antecedent noun-pronoun pairs so that the antecedent noun is physically close to its pronoun partner.

To reiterate using the math-based system of Part I of this book, when a writer assigns a pronoun X to replace a noun N, the process looks like this: N → X.

However, when a reader first encounters an X, the reader must look (or think) backward to search for an associated noun antecedent N. For readers, the process looks like this: N ← X.

Here are three examples of sentences with pronoun-reference problems. In each sentence, the pronoun is underlined. The most likely candidate it refers to, from a reader's point of view when doing a right-to-left hunt (←), is also underlined. As you review the examples, think about how you might revise them to make them more clear.

> After the technician milled 3 cm wide × 3 cm deep passes in each of the $6 \times 6 \times 6\ \text{cm}^3$ aluminum blocks and then drilled the resultant two <u>faces</u>, <u>they</u> went to the next batch station.

$$N \leftarrow X \qquad \text{“Faces”} \overset{?}{\leftarrow} \text{“they”} \tag{11.6}$$

> We drove to Truesdail labs and did a lengthy presentation. The mileage back and forth from our hotel in Cerritos was 46 miles in <u>total</u>. <u>It</u> was more time-consuming than previously anticipated.

$$N \leftarrow X \qquad \text{“Total”} \overset{?}{\leftarrow} \text{“it”} \tag{11.7}$$

> Last week at Beta biomedical, the neuro-oscilloscope was calibrated by the <u>technician</u> <u>that</u> usually maintained the robotic laser scalpel.

$$N \leftarrow X \qquad \text{“Technician”} \overset{?}{\leftarrow} \text{“that”} \tag{11.8}$$

In sentence 11.6, when the reader encounters the pronoun "they" and glances back to reconcile it, the first plural noun the reader sees is "faces." Yet here, the logical antecedent is "aluminum blocks." An easy fix for 11.6 would simply be to change "they" to "the aluminum blocks." Sometimes it's better to convert a pronoun back into its original noun form and let clarity trump repetition. Another fix would be rearranging the sentence's word order to place the antecedent and pronoun closer together.

Sentence sequence 11.7 shows a situation where there is no explicit antecedent. Logic implies that the antecedent is "driving." Here, for clarity, "driving" should replace the vague pronoun "it."

In sentence 11.8, the pairing of "technician" and "that" is clear. They're right next to each other. However, the writer has chosen the wrong pronoun to stand in for "technician." People prefer to be referred to with the people-type relative pronouns, "who" or "whom," rather than with the place or thing relative pronoun, "that." Here, the correct relative pronoun is "who," as in "who usually maintained the robotic laser scalpel."

Modifier-Target Location Errors

A third bookkeeping error deals with pairings between a word or word group functioning as a noun or verb and another word or word group functioning as an adjective (M_n) or adverb (M_v). Unlike antecedent-pronoun pairings, this type of bookkeeping requires readers to search for "specificity filters" on *either the left or the right* of the item modified, instead of only on the left. In sentence-algebra language, the shorthand diagrams for noun-modifier pairings and verb-modifier pairings are ($M_n \rightarrow N \leftarrow M_n$) and ($M_v \rightarrow V \leftarrow M_v$), respectively.

The possible applications of the above pairings are too numerous to explore comprehensively. The problem we choose to examine here is a modifier location error that occurs frequently when writers misplace adjectival phrases relative to their target partner nouns. The three examples below (with misplaced modifiers *italicized*) showcase how this error can create confusion for readers and can also cause embarrassing, unexpected results for the writer.

A management trainee stood in front of the cafeteria *leaning on a pole and wearing glasses.* [wacky] 11.9

My experiment blew up in front of the human resources director *going ka-boom and making a mess.* [absurd] 11.10

Increasing to maximum pressure and going into an explosive mode, the engineer conducted a burst test of the nitrogen gas regulator. [scary] 11.11

In the first two examples, the writer needs to move the adjectival phrases from the end to the beginning of the sentences, so the phrase in 11.9 modifies the noun "management trainee" rather than "cafeteria," and the phrase in 11.10 modifies "experiment" rather than "human resources director." With the third example 11.11, it is the opposite. The writer needs to move the adjectival phrase from the beginning of the sentence to the end, so the phrase modifies "nitrogen gas regulator" instead of "engineer."

English teachers generally refer to the above type of offense as either a *misplaced modifier* or a *dangling modifier*—where the latter is the more serious error because the modifier dangles alone in a sentence, and what it ostensibly modifies is either missing altogether or located in a remote location beyond easy, convenient association. In general, some excellent advice regarding modifiers and their targets is to follow the lead of the real-estate broker and think *location, location, location.*

Signal Process Points Within Sentences Using Commas and Other Devices

In streams of text, periods and question marks are important signal devices. They indicate key *process points*—places that cue the reader to complete the uptake and compiling of a unit of thought (that is, a sentence) before moving on to decode the next one. Just as with the layered hierarchy of input, output, and decision junctures in computer code, there are many punctuation levels in word code—punctuation *to end* a sentence; *to join* (or bond) basic sentences together to make compound or complex sentences; *to join* words, phrases, and clauses within sentences; and *to subordinate* and *separate* one sentence element from another within sentences.

Process point signals that occur within sentences include *commas*, *dashes*, *semicolons*, and *colons*. These *internal punctuation marks* organize sentence elements into logical pieces and direct how and when readers recognize them as they read a whole sentence—start to finish. This segment of this chapter presents specific strategies and tactics for using commas, dashes, semicolons, and colons in sentences to optimize structure and readability. The segment also highlights a series of sentence-level errors that are caused by missing or improper punctuation within sentences. These errors display collectively as OFI #10 on Fig. 11.1.

Set Off Introductory Elements

The first in-sentence signal device to be explored is the comma setting off an introductory element. The element can be a single word, a group of words, a prepositional phrase (a phrase led by a preposition, such as "At the office, ..."), or a verbal phrase (a phrase led by a verb, such as "Going to the office, ..."). As readers read sentence text, moving their eyes from left to right, when they encounter a comma placed immediately after a word or word group at the sentence's beginning, the comma denotes that element's introductory role and signals the reader to discern the element, and then carry its unit meaning forward, looking to pair it with the next logical partner or piece of the sentence. Oftentimes, introductory words and phrases work as adjectives that modify the sentence's subject noun or as adverbs that modify the sentence's main verb.

Even more important, aside from cueing upcoming associations, an introductory comma signals the reader to look for the sentence's subject to appear very soon after the introductory comma, not before. Thus, if the signal is missing and readers receive no explicit cue separating introduction words from subject words in a sentence's beginning, those words can blur together and communicate unintended meanings and confusion. In sentence design, leaving out an introductory comma on behalf of typographical economy does not pay off. Whether the introductory element is a single word, a few words, a phrase, or a subordinate clause, I advise engineering writers to insert a process-point comma.

Here are example sentences featuring several different introductory elements—a single word (11.12), a prepositional phrase (11.13), and a subordinate clause (11.14). In each "a" version sentence, the comma is missing, and in each "b" version, the comma is included with the insertion place (word before and word after) *italicized*. You are likely to find the "a" versions less user-friendly for a reader than the "b" versions.

Original: Next test the flip-over station drift relative to position two and the placement of the work piece. 11.12a

Revised: *Next, test* the flip-over station drift relative to position two and the placement of the work piece. 11.12b

Original: In the final iteration factor the wear of the stainless steel break plate into the Failure Modes and Effects Analysis (FMEA). 11.13a

Revised: In the final *iteration, factor* the wear of the stainless steel break plate into the Failure Modes and Effects Analysis (FMEA). 11.13b

Original: Rejecting marketing's advice for retargeting they presented plans aiming at the same customer base in their sales proposal. 11.14a

Revised: Rejecting marketing's advice for *retargeting*, *they* presented plans aiming at the same customer base in their sales proposal. 11.14b

Did you notice blur or confusion zones when you first read the "a" versions of 11.12, 11.13, and 11.14—that is, "next test," "iteration factor," and "retargeting they"? Most likely, the answer is yes, and just one comma fixes the problem in all three cases.

A similar, though not identical, case of attached sentence elements deserving mention here would be that of elements at the end of a sentence, rather than the beginning. These elements can also be distinguished with punctuation marks. Words and short phrases at the end of a sentence, however, do not usually benefit from a comma setting them off because sentences decode left to right. Thus, words and short phrases at the end of a sentence are easier for readers to discern and compile unambiguously. On the other hand, longer sentence elements can be usefully set off with dashes and colons (and often are). This section's segment on restatements, amplifications, expansions, and lists covers these scenarios.

Set Off Nested Elements

A second type of in-sentence punctuation signals the insertion of a discrete word or group of words into, rather than before (or after), a complete parent sentence. These insertions typically do one of the following:

- Promote smooth and clear *transitions* between sentences or parts of a sentence.
- Serve as a *restatement or renaming* of someone or something.

- Function as a *verb phrase* (that is, a phrase that begins with a verb) or *relative clause* (that is, a clause beginning with a relative pronoun such as "who," "whom," "that," or "which") that adds *adjectival* or *adverbial* modification.

Because all of these insertions are located somewhere after the first word and before the last word in the parent sentence, I call them "nested."

Since this is a case of sandwiching rather than tacking on, setting off a *nested element* requires *two* process points signaled by *two* punctuation marks—usually two commas and sometimes two dashes. In either case, the punctuation marks must match. If one of the punctuation marks is missing, the signal to the reader is faulty because it is incomplete. If neither punctuation mark is there at all, the signal is missing altogether.

With nested items, the parent sentence already stands alone and makes complete sense in and by itself before the insertion. Thus, it is helpful to conceptualize nested elements as being "parenthetical." Writers incorporate *parenthetical remarks* into sentences to add an aside, a note, or "food for thought." Technical documents regularly include parenthetical remarks to set off links to figures and tables, for instance, as well as references to other bits of text. The standard punctuation for a parenthetical remark is two parentheses—one on the left and one on the right. Likewise, we use either two commas or two dashes—one on the left and one on the right—to insert nested elements into already autonomous parent sentences.

The difference between a parenthetical remark and a nested element is a matter of degree. An item nested between two commas is "parenthetical" in nature, but its relative importance (from the point of view of the sentence's writer) is greater than text set off between parentheses. In turn, a writer can further increase the level of importance of a nested item by using more potent punctuation marks—two dashes—rather than two commas (default mode).

Here are some example sentences that include a variety of nested elements—words, phrases, and relative clauses. Each illustration is presented as a sentence pair. The "a" version has punctuation marks missing, and, immediately below it, a "b" version shows the same sentence punctuated correctly. The examples use *italics* to identify places where punctuation has been added to fix errors.

Original: The company in question Bilco Oil does not partner with other companies that produce toxic waste. It does however transport refined petroleum by rail in tanker cars that are noncompliant with OSHA standards. 11.15a

Revised: The company in *question, Bilco Oil, does* not partner with other companies that produce toxic waste. It *does, however, transport* refined petroleum by rail in tanker cars that are noncompliant with OSHA standards. 11.15b

Original: In designing the deep-well irrigation system which took him five work days and translated into 40 billable hours Hyram Dausen, our agricultural engineering consultant, specified a pump that would not cavitate, even with a draw depth of 270 meters. 11.16a

Revised: In designing the deep-well irrigation *system—which took him five work days and translated into 40 billable hours—Hyram Dausen*, our agricultural engineering consultant, specified a pump that would not cavitate, even with a draw depth of 270 meters. 11.16b

Original: The semiconductor manufacturing expert who was our CEO's brother-in-law and provided consulting for free gave an interesting overview for the process technicians; he stated that compared to modern transistors early transistors were enormous—about the size of a pea. 11.17a

Revised: The semiconductor manufacturing *expert—who was our CEO's brother-in-law and provided consulting for free—gave* an interesting overview for the process technicians; he stated *that, compared to modern transistors, early transistors* were enormous—about the size of a pea. 11.17b

The examples above display sentences that, for the most part, might be comprehensible to readers in both forms, "a" and "b." However, as far as ease of decoding is concerned, the "a" versions most likely present a slower, more tedious decoding process than the properly punctuated "b" versions.

Proper punctuation in sentence 11.15b clarifies that "Bilco Oil" is a restatement or renaming of the "company in question." Proper punctuation also helps the transition word "however" assert a more effective function. The dashes around the relative clauses in 11.16b, "which took him five work days and translated into 40 billable hours," and 11.17b, "who was our CEO's brother-in-law and provided consulting for free," make both more easily distinguished modifiers, the former adjectival (modifies the nearby verbal, the gerund "designing," in the phrase before it) and the latter also adjectival (modifies the compound noun preceding it, "semiconductor manufacturing expert"). Also in 11.17b, the two commas setting off "compared to modern transistors" help the reader distinguish this verb phrase as a modifier that qualifies the extent of early transistors' enormity (they were enormous).

Sentence Engineering Tip

Here's an easy way to remember the importance of correctly punctuating nested elements. Nearly any writer would agree that it makes no sense to insert a parenthetical aside into a sentence using just one parenthesis—either a "(" on the left or a ")" on the right. It is equally illogical to set off a nested item using a single comma or dash.

Use Commas to Separate All Items in Lists of Three or More

Since engineering products and processes are complex and multifaceted, when engineers write about engineering, they often modularize content into manageable pieces using *lists*. For lists of *two items*, there is a standard rule: join the items with the conjunction "and" or "or" (very rarely "nor") and do *not* use a comma. Here's an example:

Incorrect: My favorite undergraduate courses were Statics, and Dynamics. 11.18a

Correct: My favorite undergraduate courses were Statics and Dynamics. 11.18b

However, lists of *three or more items* require punctuation for precision and clarity. To correctly punctuate a set of three or more items, place a comma after each item up to and including the next-to-last item. Then, you also connect the second-to-last and last item with a conjunction, either "and" or "or" (very rarely "nor").

Although some people, in the case of lists of three or more, consider the comma between the second-to-last item and conjunction (also known as an *Oxford comma*) to be optional, if you view each listed item to be discrete and autonomously weighted (usually the case), and if you want your reader to interpret the list as you do, then place a comma after the next-to-last item to distinguish this item as being separate from the last. If you do not insert this final comma, you risk misinterpretation and blur. Here are two examples with trouble spots, with the fix *italicized* in the revised versions:

Original: Before the guests arrived, the ship's maintenance technician ate a breakfast of toast, ham and eggs. 11.19a

Revised: Before the guests arrived, the ship's maintenance technician ate a breakfast of toast, *ham, and eggs.* 11.19b

Original: The navy engineer boarded the diesel research vessel and inspected the top deck's layout, railing, specialized pedestal block and tackle.* 11.20a

Revised: The navy engineer boarded the diesel research vessel and inspected the top deck's layout, railing, *specialized pedestal block, and tackle.* 11.20b

* Note: A "specialized pedestal block" is a block-shaped equipment base bolted onto the ship's deck, whereas a "tackle" refers to the ship's rope, block, and pulley system.

For listed items that are not clearly distinguished within a sentence, the above two examples show how adding a comma can eliminate the ambiguity. In 11.19, the ambiguity is somewhat trivial—whether the technician ate ham and eggs mixed together versus whether the technician ate a slice of ham with eggs on the side for breakfast probably does not matter. In version 11.19b, we add a comma to clarify for clarity's sake. However, in the case of example 11.20, the ambiguity is whether a piece of nautical hardware is one thing or another. Here, adding a comma for clarity has a serious purpose.

Besides individual words, nested items can also be word clusters, phrases, or clauses. To punctuate these, if one of the items has a comma situated within it, the solution is to use both semicolons (first-order) and commas (second-order) to create a hierarchy within the list's structure. Example 11.21 below illustrates such a hierarchy. The revised version uses *italics* to highlight where a comma (or semicolon) can eliminate ambiguity or confusion.

Original: The intern manufactured the prototype swiftly yet methodically: procuring a new tool bit, safety goggles, and piece of stock 5150 steel, chucking the stock in the mill, which was located in Bld 7, to machine the stock's specified axial dimension, and then remounting the stock on the lathe, which was located in Bld 3, to turn the specified radial dimension. 11.21a

Revised: The intern manufactured the prototype swiftly yet methodically: procuring a new tool bit, safety goggles, and piece of *stock 5150 steel; chucking* the stock in the mill, which was located in Bld 7, to machine the stock's specified *axial dimension; and then* remounting the stock on the lathe, which was located in Bld 3, to turn the specified radial dimension. 11.21b

Use a Dash or Colon to Announce Restatements, Amplifications, Expansions, and Lists

Earlier in this segment, we discussed how (and how not) to attach introductory elements to the beginning of a sentence. In those scenarios, a comma set off the introductory element—a word or a word group.

A writer can also attach phrases or clauses onto the end of a parent sentence, as long as the attachment is logical and properly punctuated. For such constructions, we use a dash ("—") or colon (":"). Typically, there are four types of phrase/clause attachments: a *restatement*, *amplification*, *expansion*, or *list*.

While either a dash or colon can often work adequately as a signal device for these four constructions, the signal issued by a colon is more emphatic. By convention, engineering writers usually use a colon to set off lists and expansions, and they use dashes to set off restatements and amplifications (that is, an added item that elaborates with further details).

> **Sentence Engineering Tip**
>
> When typing a dash, be sure to hit the hyphen key twice, "--" (most word-processing software will automatically convert this entry into the long dash symbol, "—"). Otherwise, your reader might interpret what you, the writer, intend to be a dash as a hyphen, because it's typed as a hyphen "-".

Here are examples that show how to add the four types of phrase/clause attachments to the end of a parent sentence using proper punctuation (and best-choice conventions):

Restatement

They held the IEEE conference in Indianapolis, IN, in the Trumpet Tower Conference Center—a popular venue for large-audience presentations. 11.22

Note: In 11.22, the added item restates what the conference center is. Also note what would happen if the dash were mistakenly entered as a hyphen. What's a "Center-a"?

Amplification

The engineer was burst-testing the regulator—increasing the nitrogen gas feed to maximum pressure, inducing permanent deformation, and then continuing till rupture. 11.23

Note: In 11.23, the added item increases the specificity—and intensity—of the activity described in the preceding parent sentence.

Expansion

The genetic engineering team came up with a revolutionary idea for commercializing the production of cloned human cartilage: this was the breakthrough that upper management was waiting for to entice an additional $15M of venture capital to fund a new 5-year plan. 11.24

Note: In 11.24, the added item is a complete sentence that acknowledges what's described in the parent sentence and also strategizes about the next maneuver.

Horizontal List

Eclectic ABC Company has announced that next year's new products for sale will include the following and more: cars, nitroglycerin, and radios. 11.25

Note: In 11.25, the added item is a list of three new products that ABC company will offer for sale. The list is "horizontal," because it develops in standard, left-to-right, horizontal type.

Punctuating Vertical Lists

Engineering documents can, and often do, include *vertical lists*. Whereas horizontal lists (as illustrated above in 11.25) are typed in standard, left-to-right sentence text, *vertical lists* begin with a horizontally typed introduction, which oftentimes is a complete sentence ending with a colon. Then this introduction sets off a vertical column of items—tagged with *bullets* (symbols such as "•", "♦", and "*") or *numbers* ("1", "2", "3," and so on).

Along with being vertically arranged, the list is also punctuated differently from its horizontal equivalent. The examples below display two key conventions for typing vertical lists. When the list is composed of *words and phrases* (the case of 11.26), the writer places *no punctuation after the items* nor a conjunction ("and" or "or") between the second-to-last and last item. When the list is composed of *sentences* (the case of 11.27), the writer inserts *a period (".") after each item* and, here again, also does not follow the horizontal list convention of including a conjunction between the second-to-last and last items.

Vertical list with word or phrase-type items

Eclectic ABC company has announced that next year's new products for sale will include the following and more: 11.26

- Cars
- Nitroglycerin
- Radios

Vertical list with sentence-type items

Eclectic ABC Company's business model relies on the following policies: 11.27

- Create moderate net profits with low risk.
- Support a large, installed base of existing products.
- Provide excellent customer service.

Distinguish Compound Adjectives from Compound Nouns Using Hyphens

A final item to discuss in this chapter's punctuation signals segment is using hyphens to distinguish compound adjectives from compound nouns. Engineering content is intricate and sophisticated. Oftentimes, it takes not just a word but a group of words to articulate a specific type of adjective or specific type of noun. From "vacuum tubes" and "copper-wire rheostats" to the "state-of-the-art silicon wafer fabrication" and "nanoscale integrated circuits," the engineer's vocabulary embraces a wide variety of multi-word nouns and noun modifiers.

The punctuation convention for distinguishing compound adjectives versus compound nouns is as follows: hyphenate multi-word clusters when they function as adjectives, as in, for example, "high-speed-incubation"; and do not hyphenate multi-word clusters that function as compound nouns, as in, for example, "tissue regeneration protocol." In a sentence, an effective pairing of a hyphenated compound adjective and non-hyphenated compound noun could be this:

Effective Hyphenation

The biomedical engineer developed a high-speed-incubation tissue regeneration protocol.

Although the preceding sentence is technically dense, a technical reader would most likely be able to understand (decode) the sentence's message without struggling the first time through, whereas the next version (see below), which lacks hyphenation to signal the compound adjective versus compound noun, might require a technical reader to slow down and exert extra cognitive effort (perhaps even reread), to make sense of the stacked-up, six-word cluster, "high speed incubation tissue regeneration protocol," at the sentence's end.

No Hyphenation (Less Effective)

The biomedical engineer developed a high speed incubation tissue regeneration protocol.

With hyphenation, sometimes there is leeway regarding logical interpretation of how best to hyphenate a multi-word adjective. For instance, in the preceding example, maybe what the engineer developed could be more accurately described as a "high-speed-incubation-tissue regeneration protocol"—a term structured around the compound noun, "regeneration protocol" rather than "tissue regeneration protocol"? Or maybe not.

Ultimately, as an engineering writer, you must decide how you want your readers to interpret multiword adjectives and nouns, and then code them as such—*with signaled intention.* In fact, as an engineering writer, your writing will be optimal when you are decisive and intentional regarding all your punctuation choices.

Action Items

α	Using whatever grammar system you wish (traditional, intuitive, or sentence algebra), take a page or two of your own writing (or someone else's) and scan it for imposter sentences. Annotate your scan by underlining all word streams that represent basic sentence structures (independent and dependent clauses). Next, circle the connection devices—that is, the connector words (conjunctions) and/or connector punctuation marks (such as semi-colons ";")—between the basic sentence structures. Did you come across any imposters? If so, were they run-ons or fragments, and how would you fix them?
α	Pick a recent email that you (or a colleague) wrote. Print out a hardcopy and grab a pen or pencil. Now read through the text pretending that you're the intended reader, encountering the email's text for the first time and having to unpack the meaning of each sentence, one by one. As you read, try to underline the subject noun (or pronoun) and the main verb it pairs within every basic sentence you come across. Did you find any *subject-verb agreement* errors? How would you fix them? If you have time (now or later), continue onward as follows:

(continued)

	– *Pick another email and try the exercise again. This second time, instead of subject noun (or pronoun) and main verb pairs, underline every pronoun you come across and the antecedent that it pairs with (upstream). Did you find any instances where a pronoun-antecedent pairing might be clearer for readers if the two items were located more closely together? Are there any instances where the pronoun should probably be tossed out and the noun restated?* – *Pick yet another email and do the exercise a third time. This time, underline every modifier word (or word group) that functions as an adjective (M_n) or adverb (M_v) and the target noun (N) or verb (V) that it intends to make more specific (modifies). Are there any instances where modification would be clearer if the two items (modifier and target) were located more closely together? Are there any instances where the target is missing and there is a dangling modifier?*
α	For the next few days, scan your writing for missing commas. Look particularly for places in sentences where a comma(s) should be present to clearly set off and distinguish an introductory element or nested element. Keep a running tally, either in your head or on scratch paper, of how many missing commas you find. Hopefully, you will not encounter any. If you do, however, give yourself credit for demonstrating that you have learned enough in this chapter's segment on internal punctuation to identify where commas are supposed to go. And henceforth, be more vigilant in applying this knowledge while you are writing workplace documents—and especially while you are editing documents prior to releasing them for circulation.

Recap

Key Learnings/Takeaways

- Two Category I errors that create imposter sentences are run-on sentences (OFI #5) and sentence fragments (OFI #6).
  - A *run-on sentence* occurs when a writer inadvertently links two sentences together with a faulty connection—by either *fusing* two independent sentences together without a proper connector or joining two independent sentences together with a *faulty connection device*, oftentimes "weak-link" punctuation in the form of a "comma splice."
  - A *sentence fragment* is different from a run-on sentence. A fragment is an incompletely constructed sentence—a cluster of words that begins with a capital letter and ends with a terminal punctuation mark, but that is only masquerading as a sentence because its construction fails to comply with the basic grammatical requirements for independent operation.
- In this book, sentence scanning means (a) reading through streams of text and determining where streams are typed like sentences (they begin with a cap letter and end with a terminal punctuation mark), (b) examining each stream typed like a sentence to determine whether its structure is grammatically correct or whether it's an imposter sentence, and (c) determining an appropriate repair for all imposters.
- In general, "grammatical bookkeeping" is a quality control process that a writer undertakes to ensure that sentence elements work properly together in pairs and

clearly reveal their pairing and grammatical logic to a reader. More specifically, this chapter examines three types of grammatical bookkeeping problems, each classified as a Category I error (OFIs #7, #8, and #9):

- *Subject-verb agreement error:* occurs when a writer writes a sentence in which the number of the subject noun or pronoun (N or X) disagrees with the number of the main verb (V). The more words a writer places between subject and verb, the more likely this error is to occur.
- *Pronoun-antecedent reference error:* occurs when a writer uses a pronoun (X) in a sentence and positions the pronoun's antecedent noun (N) in a place that's difficult for the reader to find. This type of error results in the reader associating the wrong noun with the pronoun, obscuring the sentence's intended meaning.
- *Modifier-target position error:* occurs when a writer uses a modifier word or group of words (adjective = M_n or adverb = M_v) in a sentence and positions that modifier in a place that makes it difficult for the reader to pair it with its intended partner or "target." This type of error, just like a pronoun-antecedent reference error, causes confusion for the reader.

• Punctuation marks guide and optimize readers' cognitive uptake when they decode/read sentences by acting as *signals* that indicate *process points*—places where the reader recognizes, uptakes, and compiles a unit or subunit of dynamic human thought. Whereas a period or question mark signals the end of a sentence, *internal punctuation marks* signal elements (subunits) *within* a sentence that need to be identified and processed by the reader.
• Internal punctuation marks include commas, dashes, colons, and hyphens. In general, these marks *organize*, *separate*, *subordinate*, and *identify* elements within a sentence—ranging from single words and word clusters to phrases and clauses. This chapter covers five internal punctuation rules, all the source of a Category I error (OFI #10) when not followed properly:

- *Use commas to set off introductory elements.*
- *Use commas and dashes to distinguish/bracket nested elements.*
- *Use commas to organize/distinguish elements in lists of three or more.*
- *Use dashes and colons to set off elements (including horizontal and vertical lists) that are tacked onto sentence ends.*
- *Use hyphens to distinguish compound adjectives from compound nouns.*

Chapter 12
Advance Optimal Style

Learning Objectives

- *Understand how appropriate style and tone can contribute to excellence in engineering writing.*
- *Know how to identify passive voice and noun clutter and how to optimize sentence style by reducing occurrences of both of these non-optimal features.*
- *Know how to identify and avoid unsymmetrical sentence constructions.*
- *Know how to recognize and achieve appropriate, professional tone in sentences.*

Chapters 10 and 11 have presented a sequence of ten OFIs that focus on concision, clarity, and correctness as fundamental features of excellence in sentence design. Beyond these general attributes of high quality, there is another. A sentence that is lean, lucid, and technically correct can still be suboptimal at communicating its intended message if it lacks *appropriate style*.

Writing style is especially important in documents that must persuade and impress. Let's say that management mandates a new workplace policy that affects your group and assigns you to communicate this new policy to your team. You anticipate that some team members will find the policy objectionable. Not only will your announcement email need to precisely and accurately describe new policy logistics and details, but its message will also need to rally support for team-level buy-in. This can be accomplished, in part, with sentences that effectively blend authoritativeness, diplomacy, and professional polish and finesse—in other words, sentences crafted with appropriate writing style. This chapter examines four ways to achieve this (see Fig. 12.1).

B. Henderson, *A Math-Based Writing System for Engineers*,
https://doi.org/10.1007/978-3-030-10756-7_12

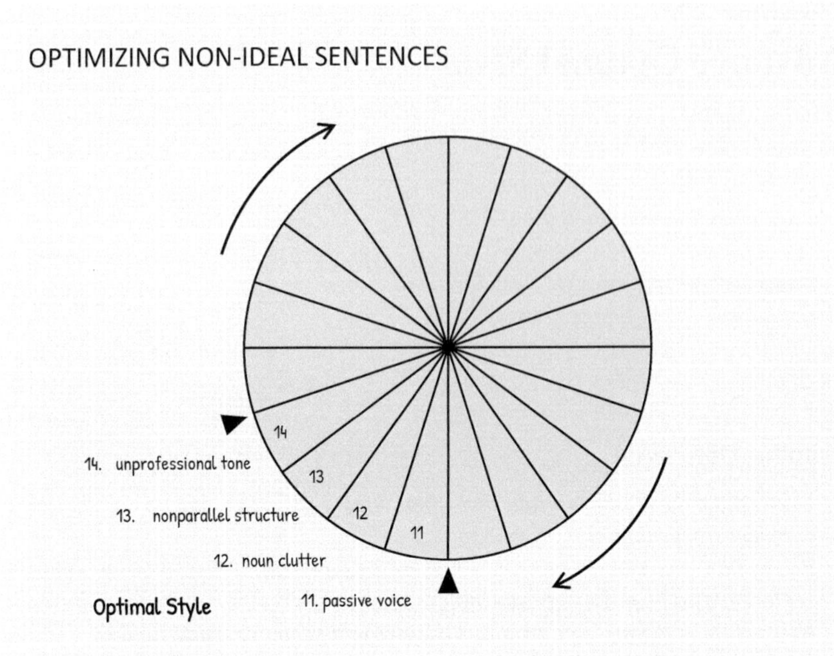

Fig. 12.1 Four opportunities for advancing optimal style displayed on the OFI wheel

Strive to Use Active Rather than Passive Voice

The first of four specific matters of style covered in this chapter is whether to phrase a sentence in *active* or *passive voice* (OFI #11). Voice does not impact a sentence's grammatical verb tense, correctness, or basic meaning. Voice is a matter of syntax—how the components in a sentence are arranged and emphasized.

To construct a sentence using the *active voice*, the writer builds the sentence around a basic sentence equation that includes a *subject noun* or *pronoun* (N_s or X_s), *transitive verb* (V_t) that transfers action to an object, and *object noun* or *pronoun* (N_o or X_o), arranged in that order. In sentence algebra, the resultant core equation is (N_s or X_s) + V_t + (N_o or X_o). Here, the *subject* is the *action "doer,"* and the *object* is the *action "receiver."*

To construct the same sentence using *passive voice*, the writer *inverts* the sentence equation: $[N_o$ or $X_o]_s$ + V_{pass} * (L + $[N_s$ or $X_s]_o$). What was previously the object in the active version now plays the role of subject—$[N_o$ or $X_o]_s$—in the passive version. The verb form changes from regular V_t to passive V_{pass}, and what was formerly the subject and "doer" in the active version relocates to the passive sentence's predicate. In this case, the doer is situated as the object in a prepositional phrase—(L + $[N_s$ or $X_s]_o$)—where L (the preposition leading the phrase) is usually the word "by." In some cases (as you will soon see), the action "doer" is not situated within a prepositional phrase, nor is it present within the passive sentence at all. Instead, the action "doer" moves "offstage."

Here is an example that demonstrates all of the preceding as the sentence switches from active into passive voice:

active sentence

Maria evaluates the solutions.

$$N_s + V_t + N_o$$

versus

passive sentence

The solutions are evaluated by Maria.

$$[N_o]_s + V_{pass} * (L + [N_s]_o)$$

In the above example, you might have noticed that in the passive sentence's text version, the passive verb form "V_{pass}" takes shape as two words, "are evaluated"—the auxiliary verb "be" joined together with the original verb's past participle, which, for regular verbs, is the root form + the suffix "-ed." If you are interested in reading more about the grammar underneath passive constructions, past participles, and expanded equations containing adjectives (M_n), adverbs (M_v), conjunctions (C), and so on, you can consult Chap. 8, which covers sentence inversions in more detail. However, there is another way to identify active versus passive constructions, and it does not involve analyzing the sentence algebra equations, or grammar, underneath sentences. In sentence optimization, we refer to this simplified approach as "the left-side versus right-side rule."

Left-Side Versus Right-Side Rule

This word-based (rather than math-based) rule observes that in sentences phrased using ***active voice***, main verb action flows from left to right. Thus, the ***action doer is on the left*** of the verb and the ***action receiver is on the right*** (→). Conversely, ***passive voice*** means that main verb action flows from right to left. Thus, the ***action receiver is on the left*** of the verb and the ***action doer is on the right*** (←), or sometimes the doer is missing from the sentence altogether and ***placed "offstage."***

The following example sequence demonstrates how the rule operates:

- The sequence begins with case "A," an active voice sentence (12.1a). The main verb action flows from left to right (→). "Bob" (underlined) is the action doer on the left, "solved" is the action (main verb) in the middle, and "the problem" (in *italics*) is the action receiver on the right.
- Next, voice shifts to passive in case "B" (12.1b), and the sentence (and flow) inverts (←). Action-doer "Bob" moves from left-side position to right-side

position, the action word remains in the middle but changes form to "was solved," and action-receiver "the problem" does the opposite, moving from right side to left.

- Finally, passive voice case "C" present an alternate passive voice option (12.1c), with the same flow (←), but with "Bob" disappearing "offstage."

active voice, case "A" →

Bob solved *the problem* 12.1a

Bob (action doer), left side | *the problem* (action receiver), right side

← ***passive voice, case "B"***

The problem was solved by Bob. 12.1b

The problem (action receiver), left side | Bob (action doer), right side

← ***passive voice, case "C" (doer "offstage")***

The problem was solved. 12.1c

The problem (receiver), left side | [doer is not mentioned]

Pros and Cons of Active Versus Passive Voice

What are some advantages of writing sentences that are phrased actively versus passively? Most importantly, using the active voice typically shortens sentences. Moreover, active sentences sound different (and most would say better). They sound *vital* and *direct*, while passive sentences tend to sound, well, *passive*.

Using passive voice also has advantages. In the scientific disciplines (more so in academe than industry), some researchers and research publications prefer passive voice for experimental reports and reviews because they consider passive voice to be more objective and formal. In addition, if you are trying to avoid gender bias and gendered personal pronouns, instead of using the cumbersome "she/he" or "he or she," one option is to switch to passive voice and leave the action doer unnamed and "offstage." Or if you need to communicate negative news and de-emphasize or mask the identity of a doer who has done something negative, you might write a sentence like this: "An inappropriate remark was made." When phrased this way, the sentence discloses the negative action (making an "inappropriate remark")—*but not who did it*. In situations like this, you, the writer, will need to decide whether or not it to include the doer by weighing strategic benefit against ethical consequence.

In general, when writing engineering documents in industry, I advise engineering writers (whenever possible and permissible) to use active rather than passive voice. Its vital and direct (rather than flat) tone makes active sentences more engaging to read and often more concise. Sometimes, of course, it helps to deviate from active voice. For instance, a long string of active sentences that are all phrased in exactly the same way can be just as monotonous to read as a long string of passives. In such situations, you can disrupt the repetition by occasionally inserting a passive sentence, with its different rhythm, sound, and result-first (rather than doer-first) syntax.

Sentence Engineering Tip

When writing engineering reports, especially the methods and procedures sections, instead of using passive voice to avoid direct references to yourself and team members, allow yourself to use personal pronouns such as "I," "we," etc. Thus, instead of

> The mixture was heated to 105 °F. It was then centrifuged by the experimenter into liquids X and Y. After cooling, the refined liquids were stored by personnel …

write

> I heated the mixture to 105 °F. Then I centrifuged the mixture into liquids X and Y. After X and Y cooled, my team and I stored the refined liquids …

Because of the stale tone and awkward phrasing that result from avoiding personal pronouns, many twenty-first-century engineers now write reports that are concise, clear, correct, and sufficiently "objective" using the active voice.

Three Additional Examples

Below are three more examples that demonstrate passive versus active voice. In sentence pairings 12.2, 12.3, and 12.4, all main verbs (both passive and active versions) display in *italics*. Additionally, in each example, the action doer is underlined to highlight how it follows the "left-side versus right-side rule" as phrasing shifts from passive (original) to active (revised). A shrinkage calculation underneath each example verifies that the active versions are slightly shorter than their passive counterparts. Finally, note that in the last example (12.4), the main verb is *irregular*. "Teach" + "-ed" = "Teached," which is not a standard English word. To form the passive verb, we combine "will be" and "taught."

Original: Petroleum *is burned* by Ukraine to heat most of its factories. 12.2a

Revised: Ukraine *burns* petroleum to heat most of its factories. 12.2b

(1 – (9 words/11 words)) × 100% = 18.2% shrinkage

Original: A costly mistake *was committed* by Jeff when he was doing end-of-fiscal-quarter accounting for the 3-mile bridge project. 12.3a

Revised: Jeff *committed* a costly mistake when he was doing end-of-fiscal-quarter accounting for the 3-mile bridge project. 12.3b

(1 – (16 words/18 words)) × 100% = 11.1% shrinkage

Original: Next week, the engineering interns at NASA *will be taught* by Veronica how to use sulfuric acid to remove impurities from lunar specimens. 12.4a

Revised: Next week, Veronica *will teach* the engineering interns at NASA how to use sulfuric acid to remove impurities from lunar specimens. 12.4b

(1 – (21 words/23 words)) × 100% = 8.7% shrinkage

Strive to Use Verbs as Verbs and Avoid Noun Clutter

Another stylistic technique for creating well-crafted sentences is *using verbs as verbs* (OFI #12). You can change many verbs into nouns by adding a suffix, such as *-ion*, *−tion*, *−ment*, *−ent*, *−ance*, *−ence*, *−ancy*, *−ency*, etc. One name for changing verbs into nouns is "nominalization." In sentence optimization, we refer to these constructions as *noun clutter*, because nominalized verbs can clutter up otherwise good engineering writing.

Just like passive voice, noun clutter does not make a sentence grammatically incorrect. Its presence or absence in sentences is a matter of style. I advise writers to avoid noun clutter because it's inefficient. Noun clutter occupies space in a sentence equation as a noun naming a person, place, or thing while at the same time expressing action indirectly.

For example, the verb "assess" *expresses* action. When it becomes "assessment" by adding the suffix "-ment," it is a word that *names* a type of action. Here are two versions of a simple sentence, one using a *verb as verb* and the other a *verb as noun* (noun clutter):

Bob assessed the jet engine.

versus

Bob did an assessment of the jet engine.

Both of these sentences describe Bob assessing something (the jet engine). The former does so by using "assessed" as the principal verb. The latter places "did" into the verb slot to complete the sentence (and make it grammatically correct) and then places what Bob was doing ("an assessment") somewhere else. Noun clutter constructions using verbs as nouns are almost always less concise than constructions that use verbs as verbs.

Below are three sentence pairs (12.5, 12.6, and 12.7) that each show an original sentence version with *noun clutter* (in *italics*) and a revised version. The revised versions either change instances of noun clutter back into *verbs* (also *italicized*) or delete the noun clutter words altogether (if they're extraneous). Like passive sentences revised into active sentences, the revised sentences here "sound" more vital and direct than the originals when read aloud. Shrinkage values for each example verify that reducing noun clutter makes these sentences more concise.

Original: John performed the standard deviation *calculation* for the data set. 12.5a

Revised: John *calculated* standard deviation for the data set. 12.5b

(1 – (8 words/10 words)) × 100% = 20% shrinkage

Original: The *summary* of the data in "Table 13b—Epogen Trials on One-Year-Old Rats" is a *presentation* of the first test trial for a new pharmaceutical product. 12.6a

Revised: "Table 13b—Epogen Trials on One-Year-Old Rats" *summarizes* first test trial results for a new pharmaceutical product. 12.6b

(1 – (17 words/26 words)) × 100% = 34.6% shrinkage

Original: *Resumption* of the oil refinery *operations* will occur next Tuesday, because union *negotiations* resulted in a new contract ending the strike. 12.7a

Revised: The oil refinery will *resume* operations on Tuesday, because the union *negotiated* a new contract ending the strike. 12.7b

(1 – (18 words/21 words)) × 100% = 14.3% shrinkage

In the above revisions, we observe one instance of noun clutter changed back into a verb (12.5b); one instance of noun clutter converted into verb, while another gets deleted (12.6b); and finally, two instances of noun clutter converted into verbs with another left as is (12.7b). In the last revision, if we were to swap out the noun clutter term "operation" and swap in a verb form such as "to operate" or "operating," the sentence's phrasing would change slightly, though it would not necessarily improve. When optimizing sentences, the best rule of thumb is not to eliminate all noun clutter, but rather to *reduce noun clutter to the extent that it improves style.*

Apply Symmetry to Sentence Design

Symmetry in phrasing—commonly called *parallel structure*—is another recommended best practice in engineering writing (OFI #13). As applied to sentence design, this means phrasing sequences of related sentence elements (related words, phrases, and clauses) in the same way. Doing so establishes logical, repeating patterns within sentences that are easy for readers to recognize and quickly understand.

The following three sentence pairs (12.8, 12.9, and 12.10) demonstrate symmetry applied to sentence design. The original versions *italicize* "out-of-parallel" or unsymmetrical elements. The revised versions show more efficient and effective designs that rephrase the elements in a sequence so they are all "in parallel." *Italics* also highlight the revised sequences.

Original: After removing the scalp, perform the following three surgical operations in the patient's skull: (1) the bone hole *is drilled*, (2) *suction* it, and (3) facilitate *insertion* of the electric probe. 12.8a

Revised: After removing the scalp, perform the following three surgical operations in the patient's skull: (1) *drill* the hole, (2) *suction* it, and (3) *insert* the electric probe. 12.8b

Original: A good test *would use* small quantities of cheap Teflon, *requires* little time, *accuracy,* and *be simple* to run. 12.9a

Revised: A good test *would use* small quantities of cheap Teflon, *require* little time, *be* accurate, and *be* simple to run. 12.9b

Original: Einstein *developed* the supremely elegant equation $E = mc^2$, with extrapolation of that theorem *determining* time as a variable, and yet his public image *was notably disheveled.* 12.10a

Revised: Einstein *developed* the supremely elegant equation $E = mc^2$, *extrapolated* that theorem to determine time as a variable, and yet *presented a notably disheveled* public image. 12.10b

The unsymmetrical (out-of-parallel) constructions in the above examples occur in advanced (rather than basic) sentences that contain a subject or subjects (doer or doers) paired with multiple verbs (actions). Although each unsymmetrical construction is out of parallel in slightly different ways, all revisions incorporate parallel structure using a common strategy: that is, adjusting the subject-and-multiple verb pairings so verb sequences conjugate in the same way.

- In 12.8, we accomplish this by rephrasing the list of procedural steps as commands "[you] drill . . .," "[you] suction . . .," and "[you] insert . . . ."
- In 12.9, we similarly link three main verb actions with a common subject noun "good test" and shared auxiliary verb "would."
- Finally, in 12.10, we partner a common subject—"Einstein"—with a series of three in-parallel verbs. To preserve the original sentence's message, in this case, we swap in "presented" for the original verb "was disheveled" in the third term.

Strike a Professional Tone

In effective engineering writing, the writer's voice need not be completely invisible. It is okay, occasionally, to thread a measure of charisma and wit into your sentences. Still, engineering writers should assign first priority to communicating engineering information—not being clever or engaging in emotional appeals. One way to achieve this principal goal is to keep the tone of sentences controlled, objective, and positive through prudent phrasing and word choices (OFI #14). This segment presents three stylistic tips for incorporating professional tone:

1. **Avoid gratuitous abstractions**. These include hype words such as “huge,” “best in the world,” and “phenomenal.” Instead, dial down to words such as “large,” “top-rated,” and “excellent.” **Also avoid excessively impassioned terminology**—for example, “love [it],” “[it’s] wonderful,” and “[I] personally guarantee.” Instead, tone it down to words such as “like [it],” “[it’s] excellent,” and “[I] strongly recommend.”
2. **Use positive rather than negative phrasing whenever possible**. If a sentence says the equivalent of “I have not ever done this, but I think I can do it because I have relevant experience doing something similar,” strike the negative lead-in, and rephrase it as “I think I can do this, because I have relevant experience doing something similar.” Even better, also strike the “I think” and say, “I can do this, because . . . .” Most of the time when you qualify a statement with “I think” or “I believe,” the reader views both to be already implied. If you’re using the first-person point of view, obviously you “think,” “believe,” or “feel” whatever the sentence says because you’re the one who wrote it.
3. **When given the choice, use “can” over “could” and “will” over “would” to be more assertive.** For example, when discussing a career opportunity in an email exchange with a recruiter, a graduating engineer might write:

   I *would* put together a viable cover statement and resume and send them to you, if you *could* consider waiving the “5 years of direct experience in industry” requirement that is stated in the job ad.

   or this more assertive version

   I *will* put together a viable cover statement and resume and send them to you, if you *can* consider waiving the “5 years of direct experience in industry” requirement that is stated in the job ad.”

Here are three additional examples applying this segment’s three tips. The tone-impacting words that adjust from (a) version to (b) version are highlighted in *italics*. As you review 12.11, 12.12, and 12.13, most likely, you will see the benefits of tone adjustment, and the ways that this stylistic device, applied to an otherwise unchanged sentence, can make that sentence’s message a little (and sometimes a lot) more acceptable and persuasive for a professional readership.

Original:	The reason that I *totally love* our new X3000 microprocessor chip and *always* recommend it to clients is that it’s the *absolutely perfect choice* for *all* applications.	12.11a
Revised:	The reason that I *like* our new X3000 microprocessor chip and *usually* recommend it to clients is that it’s *versatile and adaptable* for *most* applications.	12.11b

Original: My previous project was *not* like this one, given that I developed software for micro-core USA *instead of* its overseas partner, micro-core UK, but I guess that their software needs are nearly identical. 12.12a

Revised: In my previous project, I developed software for micro-core USA to meet software needs that are nearly identical to the new request from its overseas partner company, micro-core UK. 12.12b

Original: Taking on the manufacturing assignment at our commercial plant in Singapore *would* provide technical experience that I *would* not get at our research facility in Iowa, and I *could* also enjoy some sightseeing. 12.13a

Revised: Taking on the manufacturing assignment at our commercial plant in Singapore *will* provide technical experience that I *will* not get at our research facility in Iowa, and I *can* also enjoy some sightseeing. 12.13b

Action Items

α	When it comes to personal development, sometimes part of learning how to do something well means experiencing how it feels to do it poorly. Therefore, as a part of developing your sentence optimization skills, find a low stakes situation for practice—for example, writing a social email to a friend, writing a note on a cocktail napkin, or even conversing at a party (not related to work). In this context, compose some sentences that contain any number of the following: passive voice, noun clutter, unsymmetrical constructions, and inappropriate tone. Have fun writing (or saying) these examples of what *not to do*. Later, when you resume high-stakes engineering communication at work, remain conscious of and avoid the stylistic flaws and tone blunders that you have just explored.
α	Spend a moment to think about this: When you read "passive voice," what example of a passive verb pops into your head? One stipulation is that your example must come from your own writing or a colleague's, *not* from this book. What is this passive verb? How often do you (or they) use it? Do this same thought exercise for the other three matters of style covered in this chapter. Consider "noun clutter"—what example comes to mind? Consider "symmetry"—what example comes to mind? And finally, consider "tone"—what example comes to mind?

Recap

Key Learnings/Takeaways

- *Style* is an attribute of professional writing that goes beyond basic grammar and sentence precision. In general, optimal style enhances ease and speed (and even enjoyment) of reading for readers. The payback to writers for optimizing sentence style is that it helps your writing exhibit a level of excellence above and beyond sentence-level conciseness, clarity, and correctness. Additionally, style is often viewed as a reflection of a writer's professional confidence, credibility, polish, and finesse.

- This chapter covers four matters of style (OFIs #11, #12, #13, and #14):
  - *Passive voice* results in structuring a sentence with the *action receiver* on the *left side* of the main verb and the *action doer* on the *right side* of the main verb, nested in a phrase, or sometimes located "offstage." The opposite is *active voice*, which places the *action doer on the left* of the main verb and the *action receiver on the right*. This book prefers the active voice because it generally shortens sentences and gives them more vitality as compared to passive voice.
  - *Noun clutter* refers to the non-recommended practice of making verbs into nouns by adding suffixes like *"-tion," "-ment," or "-ance."* For example, "calculate" (verb) becomes "calculation" (noun). This book recommends *using verbs as verbs*, not as nouns, whenever possible because this practice simplifies sentence structure.
  - *Parallel structure* in a sentence occurs when multiple items in a sequence of elements playing common roles are phrased similarly—in repeating patterns. We can refer to parallel structure as being "symmetrical" and out-of-parallel structure as being "unsymmetrical." This stylistic device makes a sequence of elements in a sentence easy for readers to distinguish and process.
  - *Tone* is a style attribute that characterizes a writer's voice in a sentence. Generally, for engineering professionals, a tone that is controlled, objective, and positive works best. To achieve appropriate tone:
    - Avoid gratuitous abstractions.
    - Use positive rather than negative phrasing whenever possible.
    - When applicable, use "can" instead of "could" and "will" instead of "would" when phrasing a sentence to add assertiveness.

Chapter 13
Minimize Category II Errors

Learning Objectives

- *Understand the difference between Category I and Category II sentence errors.*
- *Know how to recognize all six Category II errors and how to repair them—or better yet, avoid them—in your engineering writing.*
- *Know some practical techniques for maximizing the sentence-level quality of your workplace writing that are particularly useful for multilingual speakers who are not yet completely fluent in English (i.e., still working on the long process of mastering OFI #20, "use conventional phrasing").*

The fourth and final set of sentence optimization techniques addresses the *Category II errors*. These are shown in Fig. 13.1 as *OFIs #15 through #20*. While Category II errors do impair sentence quality, they typically do not violate basic grammar and punctuation rules. As a result, and because of their tendency to be less interruptive to readers than Category I errors, this book ranks them as less severe than Category I errors. Nevertheless, when Category II errors are present in sentences, these imperfections can still irk readers, occasionally confuse them, and nearly always degrade a document's quality, credibility, and professional excellence.

Here is an illustration of a sentence plagued with Category I errors (13.1a), versus a second version of the same sentence containing only Category II errors (13.1b), and then, finally, a third version with no errors at all (13.1c).

Example sentence with Category I errors

Rated 75% more efficient than their competitors' engineers at Florida Sun Inc. has developed a new solar panel technology next year they anticipate product release to consumers. 13.1a

B. Henderson, *A Math-Based Writing System for Engineers*,
https://doi.org/10.1007/978-3-030-10756-7_13

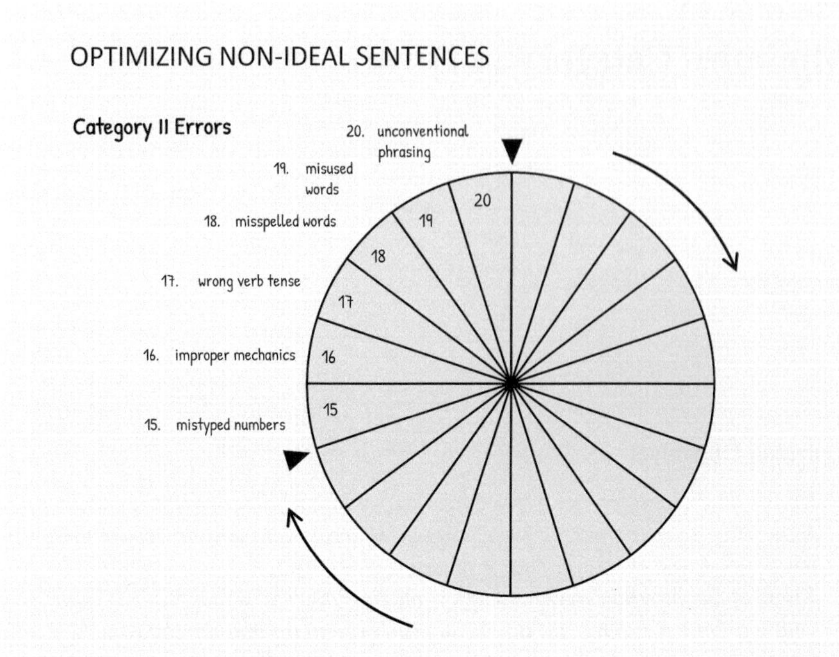

Fig. 13.1 The Category II errors on the OFI wheel

Example sentence with Category II errors

Engineers at florida sun inc. developed new solar panel technology rated 75% more efficient then their competitor, and next year, the company anticipate a product release onto consumer. 13.1b

Corrected sentence with no errors

Engineers at Florida Sun Inc. have developed a new solar panel technology rated 75% more efficient than their competitors', and next year, the company anticipates product release to consumers. 13.1c

In reading through the above sentence versions, you probably found the first version with Category I errors (13.1a) to be confusing. Most likely, you had to slow your normal reading speed or even reread all or part of the sentence to fully understand (or attempt to understand) its message. When you read the next version (13.1b), the Category II errors probably were a nuisance factor and source of some misinterpretation. On the other hand, these flaws probably did not obscure your understanding of the sentence's basic message, nor appreciably slow down your reading speed. But I'm guessing that you found the third error-free version (13.1c) to be the easiest to understand and read. Chapter 11 equipped you with a set of tools for avoiding and repairing critical Category I errors in your engineering writing. Chapter 13 will complete your sentence optimization tool kit so that you can avoid and repair less-critical, though still undesirable and unprofessional, Category II errors.

Numbers in Engineering Writing

In engineering, numbers are of paramount importance. They appear frequently in engineering documents, and there are many established conventions and standards for inserting numbers into document text. These conventions and standards can vary from company to company and organization to organization. For example, the *Institute of Electrical and Electronics Engineers (IEEE) Style Guide* for editing IEEE's numerous publications contains 11 guidelines on how to write mathematical equations in articles and books [1].

This segment reviews several basic conventions for writing numbers that usually apply in engineering writing and documents. In your own applied writing beyond this book, however, I recommend that you consider what is presented here to be generic advice and that you proactively research and incorporate all additional conventions and standards that may apply to your company's documentation.

Numbers at the Beginning of Sentences

Do not begin a sentence with an Arabic (or Roman) numeral. Using numbers in this way can be jolting to a reader's eyes. Instead, as illustrated below, either write the number as a word or revise the sentence so the number does not come first.

Eleven

Example: ~~11~~ board members, all engineers with M.B.A.degrees, advise our CEO regarding our company's research and development plan.

Or: Our 11 board members, all engineers with M.B.A. degrees, advise our CEO regarding our company's research and development plan.

Numbers Within Sentences

Within a sentence, spell out single-digit numbers—whole numbers zero through nine—and use Arabic numerals for whole numbers with two or more digits—10 or greater—except when the number partners with a standard scientific (or monetary) unit. This is often the case in engineering documents, since numbers and units go hand in hand. For instance, write "3 ml," not "three ml." Here is a sentence with numbers corrected so they conform with the preceding conventions. "Liters" is a scientific unit, and "steps" is not.

nine 2

Example: The distillation process had ~~9~~ steps and produced approximately ~~two~~ liters of ethyl alcohol every hour.

Combinations of One Digit and More Than One Digit Numbers

If a sentence has a whole number representing a nonscientific (or monetary) quantity that is less than 10 and also a number that is 10 or greater, use Arabic numerals for both numbers.

Example: We purchased 12 plane tickets and reserved 4 ~~four~~ rental cars for our team's trip to the corporate meeting in Singapore.

Leading Zero

For decimal numbers less than 1.0, *always* use a leading zero before the decimal point to ensure that the reader sees it: e.g., write "0.1," *not* ".1."

Example: If you are targeting beam intensity of > 0.01 ~~.01~~ lux, a sample size of 125.5 mg should be sufficient for testing the lithium X-ray purification technique.

Equations

You can insert an equation into an engineering document either immediately *after* the text sentence that introduces it or *inside* the sentence that refers to it.

The standard practice for inserting the equation following the sentence that introduces it is to end the introductory sentence with a colon (":") and then center the equation in a horizontal band of blank space immediately beneath. The text (the next sentence) resumes in a new text block *after* the equation, as illustrated below:

> The *Henderson equation*, developed by S. Milton Henderson, was first used to estimate the drying time of harvested grain crops:
>
> $$1 - \mathrm{rh} = \mathrm{e}^{-cTM_{\mathrm{E}}^{n}} \qquad (13.2)$$
>
> Later, researchers further refined the equation and developed it into a computer program. . . .[2]

When you insert an equation *within* a text sentence, the text ends (mid-sentence) immediately before the equation, the equation comes next, and the interrupted sentence resumes in the text block underneath. There is no special punctuation, before or after the equation, as illustrated next:

The *Henderson equation*, developed by S. Milton Henderson

$$1 - \mathrm{rh} = \mathrm{e}^{-cTM_{\mathrm{E}}^{n}} \qquad (13.2)$$

was first used to estimate the drying time of harvested grain crops. Later, researchers further refined the equation and developed it into a computer program. . . .[2]

Note also that if you have not defined an equation's variables earlier in the document's text (and they need to be defined for clarity), the standard convention is to define terms immediately after the equation. Thus, as applied to Eq. (13.2) (either case), the text immediately beneath the equation would read as follows:

where rh = equilibrium relative humidity
e = e
M_E = equilibrium moisture content
T = temperature, R
c, n = constants

Finally, standard conventions also call for you to *number all equations*. Doing so makes it easy for readers to quickly and correctly refer to equations mentioned in a document's text, such as the *Henderson equation*, labeled as Eq. (13.2), in the above demonstrations.

Basic Mechanics: Capitalization, Bold Font, Italic Font, and Abbreviations

The conventions and standards that apply to the typographical features of words in sentences are known as sentence *mechanics*. Just as they do for numbers, these conventions and standards can vary from company to company and organization to organization. However, the following guidelines generally apply:

Capitalization

Aside from capitalizing the first letter of proper nouns and the first letter of a sentence, an additional point to learn about capitalization in sentences is this: unless you are typing a *safety notice* in a document, such as "WARNING" or "STOP," an *acronym* like "NASA" (National Aeronautics and Space Administration), or an *initialism* such as "LCD" (liquid crystal display), do not type text in all capital (all caps) letters for the sake of emphasis. Instead, use *italic* or **bold** font to emphasize words and phrases in sentences—as discussed and illustrated in the next segment. In addition, even though it might seem easier to compose informal emails while leaving your keyboard's caps lock key perpetually *on* (or *off*), sentences typed in all caps (or all lower case) are a distraction and tedious to read—as well as unprofessional. Here are several text samples, with the ineffective typography corrected.

Please note that attendance
Example: ~~PLEASE NOTE THAT ATTENDANCE~~ at team meetings is
mandatory.
~~MANDATORY.~~

Here's a quick note to say thanks for the
Example: ~~HERE'S A QUICK NOTE TO SAY THANKS FOR THE~~
meeting. I appreciated your input.
~~MEETING. I APPRECIATED YOUR INPUT.~~

H I
Example: ~~h~~ere's a quick note to say thanks for the meeting. ~~i~~ appreciated your input.

Bold and Italic Font

Conventionally formatted documents use **bold** and *italic* font to highlight document titles, labels, section headings, and heading hierarchies. Chaps. 15–20 will discuss document-level typography in more detail. In sentence-level writing, an effective use of *italic* font and **bold** font (or ***both*** together) is either to emphasize words and phrases (and occasionally even whole sentences) or to refer to print and digital media titles—such as print books and e-books, print journals and e-journals, and videos—by typing titles in *italics*.

Using italics to distinguish the title of a book, journal, video, and so on is straightforward: simply type the title in italics. This labeling practice provides a universal cue for readers. (Note: for subunits within a book, journal, video, etc., distinguish the labeled unit with double quotation marks—e.g., "chapter," "article," "episode," etc.) On the other hand, there is no one universally accepted practice for using italics and bold font to emphasize words in sentences. For this, you must use your own discretion. My only general advice would be as follows: exercise restraint and don't overuse this technique. Here are some examples of what I consider to be effective uses of bold and italic font:

Using Italics, Bold Font, and Both for Emphasis

Example: Please note that attendance at team meetings is *mandatory*.

Or: **Please note** that attendance at team meetings is ***mandatory***.

Using Italics to Denote Titles

Example: After reading the *IEEE Transactions on Nanotechnology's* recent issue on nanoscale robots, I felt compelled to purchase Jeb Treestone's new book, *The Amazing Shrinking MEM*.

Abbreviations

Just as numbers abound in engineering writing text, so do *scientific units* that define the type of quantity that a given number represents. Most scientific units of measure have well-established abbreviations such as meter (*m*), kilogram (*kg*), second (*s*), and Newton (*N*). Whenever you are writing a document and are certain that your readers know the abbreviation for the unit of measure you are referring to, it is both useful and economical to immediately (and thereafter consistently) use the unit's abbreviation instead of the unit's full name.

However, if some of your readers might not be familiar with the abbreviation, you should spell out the unit the first time you use it, and then define the abbreviation in parentheses after the unit's full spelling. Thereafter, you can use the unit's abbreviation when referring to it.

Also, when using an abbreviation, be certain that you know its proper form, and if you don't, refer to a style guide or suitable reference and find out. For instance, scientific units named after historic persons, such as "Pa" for the pascal (mathematician, physicist, and philosopher Blaise Pascal, 1623–1662), are generally capitalized. Units derived otherwise, such as "rad" for the radian, are generally not.

The following example displays correct insertions of abbreviations for two scientific units in a sentence. The original, with the units spelled out in full, is correct, as well, though slightly less economical in its phrasing.

Example: About 65 ~~kilometers~~ km west of the relay tower, there is a 0.25 ~~hectare~~ ha plot of higher-elevation land on which we can build the transmission center.

Selecting Verb Tense

When you are writing an engineering document, being smooth and consistent when choosing verb tense is a good best practice that your readers will likely appreciate. However, no grammatical rule requires that, once you have established verb tense in a sentence, you must stay in that tense. The same holds true for a succession of sentences. The determiner is always this: the *when* and *how* of a given verb's action must match the *when* and *how* of the verb tense chosen to record that action. Ultimately, matching a verb word with correct verb tense is a matter of case-by-case precision.

Here are two examples that show a writer making *exact* instead of inexact verb tense choices. Some complementary examples of inexact choices are shown in brackets. The first example shows a multi-clause sentence that, to be logical, must communicate more than one time frame among its several verbs. The second example displays a chain of linked sentences, all related to an experimental procedure.

Example: Although we *have tested* [not "tested"] the apparatus several times already, if we *are entering* it in the design competition, we *must test* it again to be certain that it *is* robust.

Example: First, we *measured* current (I). Next, we *took* a reading of voltage (V) across the circuit. Since Ohm's law *tells* us [not "told us" or "will tell us"] that $R = V/I$, we then *calculated* resistance.

Maintaining one consistent tense does not work for either of the above examples. In the first case, it does not work because the sentence describes action that has already occurred in the present ("have tested"), action that is occurring in the present ("are entering"), and an assertion made in the simple present tense ("must test"). In the second example, several sentences describe procedural actions that have occurred in the past. The final sentence, however, includes the present-tense verb "tells," which describes a universal principle. We assume that universal principles are true at all times (past, present, and future). Thus, here and elsewhere, we always describe the action of a universal principle in the present tense.

Sentence Engineering Refresher

To review, a verb's tense—that is, its situation-specific spelling (or conjugation)—communicates *when and how a verb's action occurs*. There are three main types of tenses—*simple* (indicates simply whether a verb's action occurs in the past, present, or future), *perfect* (refers to an action that has already occurred within the general time frame of either the past, present, or future), and *progressive* (indicates ongoing action occurring sometime within the general past, present, or future). If you'd like more detailed coverage on the grammatical functions and tenses of "V" = verb, consult Chap. 3 on sentence algebra's core variables.

Spelling Errors

Spelling errors introduce inaccuracies into sentences and reduce a document's professional appeal. The example below illustrates two common spelling errors and an oddly spelled word "neuw" (most likely a typo) being corrected to improve a sentence's quality:

accommodate new separate

Example: We cannot ~~accomodate~~ our ~~neuw~~ customer's request for a ~~seperate~~ pricing scheme for the first 1,000 units—namely, $5 per part rather than our standard of $8.

Since many useful lists of common English word misspellings are available in other books and online web pages, and also because modern spellcheckers are very good at identifying misspelled words in word-processed text (and immediately offering correct spelling alternatives), this segment provides some general advice, in lieu of cataloging its own list of culprit words. When you are preparing the final draft of a formal engineering document for release (not necessarily the earlier drafts that you write during document development), and you are doing an editorial sweep through the final draft's text, here are several suggestions:

- Verify that your word processor's spellchecker is turned on. Then, address and repair (as necessary) every identified word.
- Bear in mind that spellchecker software generally links to a standard, mainstream dictionary and that such dictionaries do not include code names and coined terms associated with new engineering inventions, proprietary secrets, and so on.
- Remember that ultimately, you—not your computer—are accountable for making certain that every word is spelled (and used) correctly in a workplace document released with your name as author.
- If you are uncertain about any word's spelling, even just a little, further investigate and resolve the uncertainty using a dictionary or appropriate reference/source. (Note: There are several excellent dictionaries available online, so there's no excuse for not using one.)

Word Usage Errors

Spellcheckers are excellent tools for catching a variety of word spelling errors, but they are not reliable for checking word usage errors. Word usage errors, in addition to introducing imprecision into your documents, can also yield both unprofessional and highly embarrassing outcomes.

For example, let's say you are a biomedical engineer writing a status report email that describes a batch of test cultures of bacterial strains, and your spellchecker indicates an instance of the typo "organizm." To fix the problem, you swap in a correctly spelled alternative from a list offered in a pop-up menu. Unfortunately, in a rush to get the email sent, you inadvertently select "orgasm" rather than "organism" from the menu's choices and press send. In the preceding scenario, here is the sentence that passed spellcheck, along with the appropriate correction:

organism

Example: This week, we identified and categorized a mutated ~~orgasm~~ in the new batch of test cultures that used radioactive micro fertilizer.

Since spellcheckers offer less helpful assistance for misinterpretations than they do for misspellings, writers have no computerized safety net for word usage problems. This segment presents its own list of top word usage offenders. The list focuses on 20 word pairs—look-alike words that cannot be used interchangeably—that I have observed being confused in engineering writing. As an engineering writer, you should be able to use all words on this list correctly:

1. **Accept vs. except**: To *accept* (verb) means to willingly receive or acknowledge something, whereas *except* (as a verb) means to exclude something from a field of choices, or alternately, *except* (as a preposition) signals that something is omitted or excluded.

   "He *accepted* the job bid to resurface his company's asphalt parking lot, but *excepted* the contractor's additional request for payment-in-full upfront."

   or

   "He *accepted* the job bid to resurface his company's asphalt parking lot, *except* for the contractor's additional request for payment-in-full upfront."

2. **Affect vs. effect**: To *affect* (verb) means to change or transform something, some way or somehow; however, the result of a change-type action is the action's *effect* (noun). When used as a verb, to *effect* (verb) means to cause or to bring about something.

   "We needed to *affect* the fertilizer's water-absorption coefficient, so to *effect* reformulation, we decided to add sulfur and proprietary agent G. Ultimately, the long-term *effect* of this change was substantial profits from reformulated fertilizer sales."

3. **Bare vs. bear**: *Bare* (verb) means to uncover, and *bare* (adjective) means to be uncovered. However, to *bear* (verb) means to support or carry, and a *bear* (noun) is an animal.

   "The bridge over the creek can *bear* the weight of a hungry *bear*, so make certain the cupboards are *bare* when you leave the cabin."

4. **Complement vs. compliment**: To *complement* (verb) means to join (complete) one thing with another to achieve a harmonious union; however, to *compliment* (verb) means to praise someone for an action or quality. Both of these words have noun forms, too.

   "He proposed that we *complement* the heat pump unit with an auxiliary solar panel to improve the system's efficiency, and we *complimented* him for being environmentally conscious."

5. **Compose vs. comprise**: To *compose* (verb) means to make something, whereas to *comprise* (verb) means to include or embrace something.

   "A committee of factory supervisors *composed* a set of safety regulations that *comprised* emergency response plans for floods, earthquakes, and tornados."

6. **Deduce vs. induce**: To *deduce* (verb) means to reason from general to specific; however, to *induce* (verb) means to reason from specific to general.

   "Since our former CEO possessed an executive MBA from a prestigious university, we mistakenly *induced* that he had the ability to successfully manage our company. When we hired our current CEO, we reviewed all of her past executive appointments and verified whether, during those appointments, the annual reports to shareholders indicated trends of fiscal growth and profits (which they did). Thus, we *deduced* that she would make an excellent CEO for our company (which she did)."

7. **Discreet vs. discrete**: *Discreet* (adjective) means being private or quietly polite about something; however, *discrete* (adjective) means that something is separate from or distinguished as different from the rest.

   "Worried about the competition stealing her idea before she could apply for a patent, the inventor of the touchless laser toothbrush held a *discreet* meeting with potential investors and showed them the *discrete* pieces of a disassembled prototype."

8. **Eager vs. anxious**: To be *eager* (adjective) is to anticipate something enthusiastically, even impatiently; however, to be *anxious* (adjective) is to be nervous or uncomfortable about something.

   "He was *eager* to receive the results of his Professional Engineering (P.E.) Exam, because he wanted to put this accomplishment on his resume; however, when he received an email with the subject line 'PE Exam Score,' he felt *anxious*, as he was worried that he may have failed."

9. **Eminent vs. imminent**: *Eminent* (adjective) is a term that attributes prominence to people and things, whereas *imminent* (adjective) means something (usually negative) is about to occur.

   "Given his *eminent* ability to solve complex problems, we knew steady-state operation would occur soon, even though, at present, a temporary plant shutdown was *imminent* because of the reactor leak."

10. **Farther vs. further**: Both *farther* and *further* (adverbs) can be used to refer to distance; *further* (adjective or adverb) can also be used to refer to a passage of time, either literally or abstractly—but *farther* cannot.

    "In ancient times, a catapult could hurl stones *farther* [or *further*] than humans could throw them, yet *further* engineering eventually produced cannons, whose projectiles outdistanced rock hurling machines."

11. **Fewer vs. less**: To express the reduction in quantity of something countable that divides into a number of pieces, use *fewer* (adjective); however, to express the reduction in quantity for something that is not countable, or is measured in portions, use *less* (adjective).

"For the antique one-cylinder engine to operate more efficiently, the carburetor intake needs *less* air and more gasoline, and the ratcheted governor lever needs to be advanced a *few* more clicks."

12. **i.e. vs. e.g.**: An abbreviation of the Latin term "id est," *i.e.* means "that is"; *e.g.* is short for the Latin term "exempli gratia," which means "for example."

    "We subject alloy samples—*e.g.*, 2124 aluminum—to a standard test procedure for atmosphere endurance—*i.e.*, we isolate samples in a sealed vessel containing a hot, acidic, high-pressure mixture for about six weeks."

13. **Imply vs. infer**: When referring to something indirectly, a speaker or writer (the sender) *implies* it; however, a listener or reader (the receiver) *infers* it.

    "When Timothy presented the results of the transgenic trout study at the conference, he *implied* that his data plotted as a normal distribution; therefore, some members of the audience *inferred* incorrectly that his conclusions were credible rather than random."

14. **Its vs. it's**: The word *its* is a possessive pronoun meaning "of it," whereas *it's* stands for the contraction "it is" or "it has." These two terms *cannot* be used interchangeably.

    "*It's* estimated the satellite will establish geosynchronous orbit 20 minutes after *its* launch."

15. **Lie vs. lay**: *To lie* (verb) means to position oneself into a horizontal (or sometimes reclined) position relative to a surface. However, *to lay* (transitive verb) means to place someone or something onto a surface.

    "After the radioactive chickens *lay* 12 eggs, the research technician on nightshift will update the database and then *lie* down and take a nap."

16. **Meet vs. mete**: *To meet* (verb) means for someone or something to encounter (or come together with) another, whereas *to mete* (verb, often combined with *out*, as in *mete out*) means to portion out a substance or thing, usually in a controlled, measured fashion.

    "After I *meet* with the company's efficiency officer, I will then be able to *mete* out vacation-day approvals for the month of July."

17. **Principle vs. principal**: A *principle* (noun) is a fundamental truth, concept, or rule; however, a *principal* (noun or adjective) refers to the predominant (or the predominance of an) element or person among a specific set of elements or people (note: in finance, *principal* is also used to describe an invested amount of capital on which interest is either earned or paid).

    "During the 21st century's first decade, the *principal* reason for Kiwi Computer Inc.'s phenomenal success was that the company's *principal*, namely, the CEO and founder Fiona Willis, insisted on one key operation *principle*: 'value our employees, in order to add value to our products (and net revenues)'."

18. **Prove vs. conclude**: Although a mathematician can, in theory, *prove* (verb) a mathematical theorem and an attorney can *prove* someone's innocence in a court of law, in engineering and science rarely do we prove anything to be absolutely 100% true or false. Instead, we use experimentation and data to *conclude*, *indicate*, or *validate* (all verbs) that something works in a repeatable and reproducible manner.

    "Given that our test engineers have now logged over 800K hours of flying time for our Model III Sky Car without crashes or unplanned landings, we can *conclude* that the Model III demonstrates a safer operation profile than previous prototypes. However, this initial success does not *prove* that the Model III is safe for commercialization and sale to consumers."

19. **Respectfully vs. respectively**: *Respectfully* (adverb) indicates that something is being done with respect or reverence, whereas *respectively* (adverb) qualifies that the order of things in a sequence or list correlates with the order of things in a subsequent sequence or list that closely follows after the first.

    "Two bicyclists—the Swede, who is in the lead and *respectfully* wearing her national colors of blue and yellow, and the Nigerian, closing in from behind wearing green and white—raced by the checkpoint, where our rpm radar gun measured pedaling speeds of 100 rpm and 104 rpm, *respectively*."

20. **Shear** vs. **sheer**: *Shear* (adjective or verb) refers to opposing coplanar forces breaking or severing (or conspiring to break or sever) a material in cross section; however, *sheer* (adjective) characterizes something as being absolute or unmitigated, and it can also characterize a material as being almost transparent or a precipice or cliff as having a very steep decline.

    "To ensure that the nano-cooling fins will not *shear* off when loaded from the side, her proposal to heat treat the synthetic metal substance as if it were an alloy was *sheer* genius."

In general, as with spelling, the same bottom-line advice holds true: if you are unsure of the proper usage of a word, even slightly, *always consult a printed or online dictionary* to check the word's definition and proper usage.

Conventional Phrasing

How to correctly assemble an English sentence is determined by grammar logic and identifiable components and structures (or sentence equations), yet it is also determined, in part, by what this book calls *conventional phrasing*—that is, how a native speaker would "naturally say it." Sentences that lack conventional phrasing, even if otherwise well-designed and understandable by a reader, can sound "off" and unconventional to a native speaker. And in workplace writing (though not necessarily writing outside of the workplace), whether the business language is English,

Chinese, Spanish, Swahili, German, or French, native speaker preferences do guide sentence-level quality control standards in documents.

Sometimes writing (and speaking) that contains flaws in conventional phrasing is characterized as "having an accent." However, this is not an appropriate label for communication in the twenty-first-century global workplace, which includes (and relies upon) engineers from different countries and cultures, all working together and using common business languages for exchanges. Our focus here is one of these key business languages—English.

When a person learns a new language (second, third, or more) beyond their principal language (or languages), the final stage of mastering (acquiring) that language is becoming fully fluent—that is, able to speak (and write) in a manner that is indistinguishable from that of a native speaker. If you are multilingual and not yet fully fluent in English, what makes improvement opportunity #20—use conventional phrasing—particularly challenging is that the determiners for what is "conventional" and what is not are in many ways subjective.

They are defined by common usage and consensus among native English speaker communities, and not by objective, logical, readily teachable principles. The elements of conventional phrasing in English are also tied to the language's unique syntax (or "natural syntax," as discussed in Chaps. 2–8), vocabulary, and repertoire of idiomatic expressions—features, some of which have no equivalents in other languages' frameworks [3].

Some of the more perplexing "insider" rules and conventions that native speakers of English assign to English and that apply to formal workplace writing include the following:

- *Properly inserting articles*
- *Using idiomatic expressions*
- *Selecting prepositions heading up prepositional phrases*
- *Determining how and when an "s" goes on the end of a plural noun*

We can use the preceding four practices to examine how unconventional phrasing errors occur in English sentences, as well as how they can be repaired.

Inserting Articles

Although an explicit set of linguistic principles define *when* and *which one* of the articles—"a," "an," or "the"—should be placed before a noun, native speakers of English usually choose how and when to insert articles *intuitively* rather than by conscious reasoning. This intuitive process associates with what this book calls *ear knowledge* (as opposed to head/intellectual knowledge) of the English language. Here are two corrected sentences containing OFI #20 errors related to the improper insertion of articles:

the next sample a process operator an excimer laser ablation
Example: Give ~~next sample~~ to ~~process operator~~ to perform ~~excimer laser ablation~~.

[delete] the crystalline structures
Example: We use ~~the~~ electron microscopes to examine ~~a crystalline structures~~
[delete]
of ~~a~~ heat-treated samples.

As you can see in the above illustration, reconciling article errors involves both proper insertion as well as deletion.

Using Idiomatic Expressions

When a native speaker of English says, for example, "her new car cost *an arm and a leg*," this sentence means "her car cost *a lot of money*." Yet to a multilingual speaker, this idiomatic expression, if interpreted literally using dictionary definitions, might mean "the car was priced at a cost of two amputations." Here are two examples of idiomatic expressions that have been incorrectly interpreted and then re-expressed in a sentence, along with an appropriate repair:

Example: Introducing a new product into the marketplace can
attract the public eye
~~attract more publics' eyes~~.

you can't judge a book by its cover
Example: Since she knew that ~~you shouldn't decide on a book cover~~, the design engineer remained uncertain about the new self-adjusting office chair's ergonomic correctness, even though the chair looked sleek and functional.

Choosing Prepositions

Prepositions are words like "in," "on," "after," "between," and so on. They are used as construction devices in sentences and are not used independently. However, prepositions serve a very useful purpose when they partner with other words to form a prepositional phrase. To use prepositions properly, speakers/writers must know which type of preposition goes *naturally* with which type of phrase—and this requirement can be a dilemma for multilingual speakers because native speaker preference (above and beyond dictionary definition) frequently establishes what is "correct."

In the following examples, the original constructions contain prepositions that are, for the most part, logically selected using dictionary definitions, yet phrased in unconventional ways:

On of

Example: ~~Of~~ average, this method has an error rate ~~in~~ 1.5%.

for petroleum

Example: Sludge Oil Inc. fulfills consumers' needs ~~of petroleum~~ using deep-well technology and tundra fracking.

Adding "s" to the End of a Noun to Form a Plural

Knowing when and when not to put an "s" (or "es") on the end of a noun to form a plural can also present challenges. It has to do with whether the noun is singular or plural. For example, consider one "scientist," "laboratory," or "experiment" (singular) versus two "scientists," "laboratories," or "experiments" (plural). It also has to do with whether the noun is a count or noncount noun. A noncount noun refers to an indistinct set of items, such as "equipment," which could be "one piece of equipment," "10 pieces of equipment," or "200 pieces of equipment." Below are example sentences illustrating unconventional phrasing relative to "s" on the ends of nouns:

scientists equipment

Example: The two ~~scientist~~ purchased a variety of new test ~~equipments~~ for their lab.

jargon deadwood

Example: Having now been edited to remove ~~jargons~~ and ~~deadwoods~~, the operations manual is user-friendly for both technical and nontechnical readers alike.

The research and theories connected with language acquisition and the translation of one language into another are beyond the scope of this book. Expert linguists and language scholars can best explain such topics. Instead, this book takes the following approach in its coverage of conventional versus unconventional phrasing:

- First, identify and name it as a sentence-level problem, which is prevalent in twenty-first-century workplace writing because our international workforce includes large numbers of multilingual professionals.
- Second, characterize this type of error as a Category II and not a Category I infraction, and note that its presence in an engineering document is *non-optimal*—nothing more and nothing less.
- Third, recommend that multilingual engineering writers who are not yet fully fluent in English be on the lookout for unconventional phrasings in their

workplace writing and strive to minimize—or ideally eliminate—them as a sentence-optimization practice.

Being able to distinguish what an unconventional phrasing error is and what it is not is also useful for native speakers. Ideally, this knowledge will lead them to fairly and respectfully calibrate their individual quality control standards so that when they encounter phrasing slips and glitches in the writing and speaking of others, they will respond with proper perspective, understanding, and tolerance. Again, this book distinguishes 20 OFIs. OFI #20 is but one, and it presents a relatively benign interruption to readers as compared to some of the others.

Advice on Building Sentence Optimization Skills to Avoid Unconventional Phrasing Errors

I have taught engineering writing to internationally diverse audiences of engineering and science students for nearly two decades. Most of the class rosters have included a diverse mixture of native speakers and others for whom English is a second or third or fourth language. Through this experience, I have observed numerous attributes of quality as well as numerous avenues to failure in engineering documents. I have also observed many students from all types of backgrounds significantly improve their engineering writing skill sets and then go on to become successful engineering writers in industry locations all around the world.

If you are a multilingual engineering writer in English and yet are still working toward 100% fluency, here are some observations and advice that come from me as well as from my students (past and present) on how to minimize (and eventually eliminate) unconventional phrasing in engineering documents:

- When you are writing in English, strive to *think in English* during the process, rather than in your principal language, and then translate from that language into English as you write.
- Remember to *call upon your ear knowledge of English* in addition to your head knowledge, especially when editing final document drafts. Do so by reading your working drafts aloud—slowly, surely, and at ample volume (or have a friend or colleague—a writing buddy—read the draft to you while you listen). Each time you "hear" a possible phrasing imperfection, work on rephrasing it, doing so more than once if necessary to make the spot "sound natural"—the way that you think a native speaker would say it.
- *Practice conversing in English with native speakers and listening to native speakers speak English* as much as possible. This applied activity will advance your ear knowledge (and head knowledge) of English.
- *Watch movies and television shows in English* (at first, with subtitles in your native language), and *read articles and books written in English* as much as possible.

- In your workplace, *find and establish a win-win editing partnership with a writing buddy* who has already mastered conventional phrasing. Meanwhile, be sure that you have worked (or are working) diligently at mastering OFIs #1 through #19. That way, you and your writing buddy can *reciprocally exchange* document drafts as a quality control best practice. Your writing buddy can assist you with OFI #20 errors, and you can assist your writing buddy with the other 19 aspects of sentence optimization.
- *Keep working on sentences until each represents a personal best effort.* If you are composing a sentence and find yourself struggling to find the right words and right way to express what you intend, don't give up, move on, and leave the sentence "as is." If you know a sentence is non-optimal, keep working on it—try this, try that, maybe leave the sentence alone awhile, and then return and try again—until the sentence achieves the communication purpose *you intend for it.* If you release a document that contains sentences that you know are inexact or unclear, those faulty sentences will, almost invariably, cause trouble for your readers.
- *Be persistent yet patient.* Developing complete fluency and the ability to eliminate unconventional phrasing errors in your writing takes time, often several years or more of applied speaking and writing practice in English among native speakers (i.e., *immersion in a native speaker community*).
- Meanwhile, take heart in knowing that even if a sentence is not phrased perfectly and contains one or two instances of OFI #20 or other noncritical Category II errors—if it is otherwise exact and clear in its intended message, the sentence can still serve a useful purpose in workplace communication. It can help transfer important information from one person (you, the writer) to another (your reader).

Some Additional Resources

Here are some additional recommended references, for you to consult when you encounter engineering writing questions not addressed in this book. The English language is complex, and, like most languages, it contains oddities and anomalies. I am certain that you will encounter some of these in your engineering writing on the job. I am also certain that you will find answers to them if you approach them not as mysteries but as engineering problems that any good engineer like yourself can solve.

ESL Students, English as a Second Language, Purdue Online Writing Lab (OWL), Purdue University, Indiana, U.S.A.
https://owl.purdue.edu/owl/english_as_a_second_language/esl_students/index.html

ESL Resources, The Writing Center at UNC-Chapel Hill, North Carolina, U.S.A.
http://writingcenter.unc.edu/esl/resources/

English as a Second Language Resources, Doyle Online Writing Lab, Reed College, Oregon, U.S.A. http://www.reed.edu/writing/esl_resources.html

Action Items

α	Retrieve your "My Engineering Writing" file from Chap. 1's action items, or choose another sample of your workplace writing, a page or two long. Now that you have studied the fourth and final set of sentence optimization techniques (along with the other three), review the sample text and see if you discover any sentence-level errors. If so, what are they? Henceforth, be on the lookout for these errors in future documents that you write.
α	Make a list of pairs of words that you occasionally mix up, such as "effect" vs. "affect" or "lie" vs. "lay." Are these pairs in this chapter's list of 20 examples? If a pair is not on the list, consult a dictionary and discern the difference in meaning between them.
α	Identify an idiomatic expression that you like to use and that does not mean what the words literally suggest. How would you explain this expression to a multilingual speaker of English?

Recap

Key Learnings/Takeaways

- Category II errors (OFIs #15 through #20) have to do with flaws that degrade sentence quality but do not necessarily violate the rules of basic grammar and punctuation. This book considers Category II errors to be less critical and less interruptive for readers than the higher-level Category I errors (OFIs #5 through #10). Both error types affect sentence quality and make sentences non-optimal engineering writing. Ultimately, engineering writers should strive to eliminate both Category I and Category II errors from their workplace writing.
- The Category II errors are defined by native speaker conventions and consensus, as well as by grammar logic and rules. Therefore, when learning this book's progression of sentence optimization techniques, it may be advantageous for multilingual speakers of English to study Category II errors last, because learning OFIs #1 through #14 is often easier (and more objective) than mastering (more subjectively defined) OFIs #15 through #20 . This view is especially true for OFI #20: use conventional phrasing.
- Here is a list of Category II errors and a brief definition of each (OFIs #15 through #20):
  - *Use numbers correctly:* Although specific conventions for using numbers in engineering text can vary from company to company, there is a general set of conventions for using numbers that engineering writers should know how to apply—such as when and when not to write numbers as words (e.g., "eleven") rather than as Arabic numerals (e.g., "11") and how to properly insert equations into text.
  - *Use proper mechanics:* A set of general mechanics for typing text defines when and where to use italics, bold, or both for emphasis—as well as for writing titles.
  - *Use correct verb tense:* Action type determines the choice of verb to use, and when and how action occurs determine the verb's tense—simple present,

future perfect, past progressive, and so on. Verb tenses within a sentence and within sentence streams can vary and are determined on a case-by-case basis.

– *Spell words correctly:* On behalf of accuracy, precision, and professional excellence, engineering writers should always strive to spell words correctly.
– *Use words correctly:* Also on behalf of accuracy, precision, and professional excellence, engineering writers should always strive to use words correctly. Doing so is often challenging in applied writing because a word can be spelled correctly yet still be used incorrectly in a sentence. Spellchecker and grammar checker programs often will not recognize word usage errors.
– *Use conventional phrasing:* How sentences—or the elements within a sentence—are assembled is determined, in part, by conventional phrasing—that is, how a native speaker would "naturally say it." If you are a multilingual speaker of English, mastering the ability to phrase English sentences like a native speaker (OFI #20) takes time, often years of speaking and writing within a native speaker community. On the other hand, mastering OFIs #1 through #14, and in large part OFIs #15 through #19 as well, requires diligent effort, practice, and high sentence-level quality control standards. This is equally true for native speakers.

Review Box: 20 Sentence-Level Opportunities for Improvement for Engineering Writers

Simplify and Clarify

OFI #1: Eliminate deadwood.
OFI #2: Eliminate jargon.
OFI #3: Revise awkward/inexact/vague sentences.
OFI #4: Avoid faulty logic.

Eliminate Category I Errors

OFI #5: Avoid run-on sentences.

Use coordinating conjunctions, subordinating conjunctions, conjunctive adverbs, semicolons, colons, or dashes to connect basic sentences together to form compound, complex, or compound-complex sentences.

OFI #6: Avoid sentence fragments.
OFI #7: Reconcile subject-verb agreement errors.
OFI #8: Reconcile pronoun-antecedent reference errors.
OFI #9: Reconcile modifier-target location errors.
OFI #10: Signal process points within sentences using proper punctuation.

- *Set off introductory elements with a comma.*
- *Set off nested elements with commas, dashes, or parentheses.*
- *Use commas to separate all items in lists of three or more.*
- *Use a dash or colon to announce tacked-on restatements, amplifications, expansions, and lists.*

(continued)

Advance Optimal Style
- OFI #11: Reduce passive sentences.
- OFI #12: Reduce noun clutter.
- OFI #13: Incorporate parallel structure.
- OFI #14: Incorporate a professional tone.

Minimize Category II Errors
- OFI #15: Use numbers correctly.
- OFI #16: Use proper mechanics.
- OFI #17: Use correct verb tense.
- OFI #18: Spell words correctly.
- OFI #19: Use words correctly.
- OFI #20: Use conventional phrasing.

References

1. *IEEE Editorial Style Manual*, IEEE Periodicals Transactions/Journals Department, version 8, 2014
2. Henderson, S.M., Perry, R.L.: *Agricultural Process Engineering*, 3rd edn. The AVI Publishing, Westport, CT (1976)
3. Lunsford, A.: *The Everyday Writer*, 3rd edn. Bedford/St. Martin's, Boston, MA (2004)

Part III
Document Algorithms

Chapter 14
Part III Primer: The Elements of Document Algorithms

Learning Objectives

- *Know what a document algorithm is and how it works.*
- *Know what document features are independent from document algorithms and what is meant by document geometry.*
- *Know the difference between short-form and long-form engineering documents.*
- *Know five essential engineering documents and the reasons why professional engineers should be able to write them.*

What Is a Document Algorithm?

In Chaps. 14 through 21, we find our math touchstone in the world of software engineering, where flowcharted algorithms made of symbols, functions, arrows, and labels provide software designers with a top-level view of how a program moves through its operations—start to finish, input to output. A *document algorithm* is analogous. It defines how the operative flow of a human language message develops; how and when the message's language stream articulates descriptions, claims, and evidence; and how these elements aggregate and synthesize into a coherent, cohesive, and convincing message output.

As a preview of things to come in these chapters, Fig. 14.1 below displays a rudimentary algorithm applied to guide the writing of an evaluation of two design alternatives—in this case, two choices for glass thickness for a deep-sea submarine's observation window. The resultant short stream of text documents the engineering writer's decision-making process for selecting optimal glass thickness (best of two choices) based on stress and strain parameters (two criteria).

The beauty of using algorithms to define document structures is that algorithms eliminate some of the guesswork (and consequent anxiety) from document writing and instead make the writing process tangible and objective. Once you define an intended document's communication purpose, target audience, and situation-driven

B. Henderson, *A Math-Based Writing System for Engineers*,
https://doi.org/10.1007/978-3-030-10756-7_14

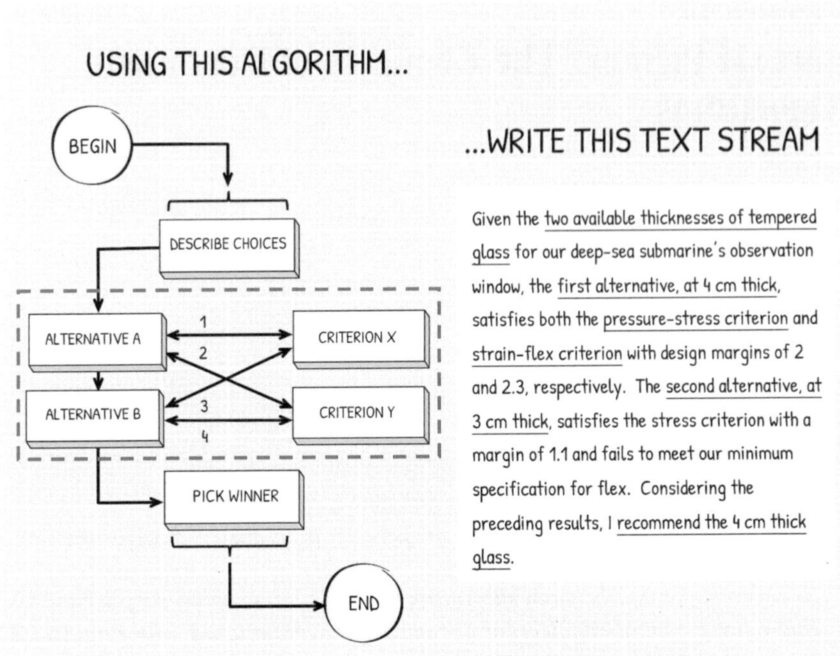

Fig. 14.1 Sample algorithm guiding the development of a short stream of text

format—which in twenty-first-century documents includes page design and typography, as well as communication medium (print or digital)—you can select an appropriate algorithm.

The remainder of the job will, of course, require effort, yet now it will be guided by a straightforward plan. To begin, instead of confronting a blank first page nagging the question "What should I write?," the algorithm tells you exactly what to include in an effective introduction. To compose the rest, you continue onward, guided by a flowcharted map, which prompts you to generate text that integrates smoothly and effectively into a known and tested structural design.

Writing excellent engineering documents is never easy, but with document algorithms, you might find that it gets easier. They prescribe standard recipes that work. Yet they are meant to be helpful tools, not creativity-smothering rules. If you wish, you can use algorithms as launching pads, frameworks upon which to innovate. But there's nothing wrong with keeping it simple and applying them "as is."

Part III examines the document algorithms and corresponding application processes for producing *five essential engineering documents*:

1. ***Project proposal***—a front-end document that engineers write to solicit authorization and funding for a proposed project from management or a prospective client.

2. ***Status report***—an interim document that engineers write to report the status of a project in progress to management and project team members.
3. ***Project report***—a back-end document that engineers write for management or the client to present project results and record activities that produced those results.
4. ***Tech to nontech brief***—a "how it works" document that engineers write to explain a technical concept to a nontechnical or lay audience.
5. ***Instructional job aid***—a "how to do it" document that engineers write to teach a technical job task to someone who needs step-by-step procedures for performing that task efficiently, effectively, and safely.

Although there are certainly more than five noteworthy engineering document types, this book focuses on the above list because they are "go-to" standards that almost all engineers will be called upon to write on the job, sooner or later.

What Does a Document Algorithm Do and Not Do?

When applying the frameworks that I recommend, both *independent* and *dependent elements* should be considered. For example, these frameworks do not—and cannot—control organization-specific formatting protocols. Placed side by side, two documents built around the same algorithm can look very different from each other. For example, Company Y might require all of its engineering project reports to be printed single-sided, in double-spaced text, and have the company name and logo on every page, whereas Company Z might require the same type of reports to be created only as digital uploads—entered section by section into text entry boxes on the company's web interface and then accessed and read online.

For shaping and arranging text and graphics on the page or screen, there are, however, generic formats and practices that are useful for engineering writers to apply. In math-based writing, these are called *document geometry*. For instance, a standard email's layout consists of routing information (top), subject line (next), and body text (remainder). A best practice for typing emails is to paragraph frequently to create short, rectangular blocks of text, which are quick and easy to read. The following chapters' lessons on document algorithms also contain tutorials on basic page design.

Document *length* is another element independent from algorithm type. The document's complexity and amount of content, not its algorithm, determine appropriate length. Once a target audience, purpose, and context dictate a writing situation's best choice of document algorithm, and the engineering writer adopts that algorithm to write a report, the report could span 1 page, 10 pages, 100 pages, or more, depending on the complexity and amount of content that must be communicated.

That said, however, as short as possible is always best. Twenty-first-century engineers predominantly write and transmit documents via email and online forms/collaborations, and these media demand brevity. Accordingly, all of the five essential document types and associated examples presented here are *short-form documents* (1 to 4 pages maximum). The document algorithms underneath the essentials are universal, however. Each short-form version can expand into a corresponding *long-form document* (5 pages or more).

The process for short-form into long-form expansion—which entails some alteration of format and modularization scheme and the addition of what is known as *front matter* and *back matter*—is the topic of Chap. 20.

Beyond the Five Essential Algorithms and Document Types

There are, of course, more than five important document types characteristic of the engineering profession—for example, the project plan, R&D report, design review, and technical brief to a technical audience (also called a white paper). My author web site discusses some of these. It also provides a tutorial and associated algorithm for another short-form document that is critically essential for beginning an engineer's career as well as advancing it—the job application packet (cover message and resume).

Additionally, a number of good engineering (and science) writing books in print focus more on long-form documents. These can provide engineering writers with a broader, alternative perspective and access to other useful document types and reference materials.

In general, whether a document is short or long, prudent choice of content (what to say) as well as arrangement or sequencing of that content (when and how to say it) has a great deal of impact on how effectively and efficiently that document achieves its intended communication purpose. This is why, beyond the five essentials, understanding how to look at document structure through the lens of an algorithm—and to recognize what is, and what is not, a successful developmental path—is a universally relevant skill for engineering writers.

Action Items

α	Take a moment to consider the last engineering report that you wrote. What was the report's purpose? Who was its target audience? What was the format (its layout, typography, graphics, as well as communication medium—print or digital)? For instance, it may have been a project update for your immediate supervisor written as an email.
α	After you've selected a personal example, consider how you went about structuring this report *as you wrote it*. Did you have a predetermined plan for what content you needed to include and how best to arrange it? Did you use a report template as a guide? Did you use a similar document that you or someone else had written as a model? Did you first sketch an outline or do a brainstorm blast? Or did you just "make it up as you went along"? If you had an appropriate document algorithm to follow, might that have helped you write a better report or

(continued)

	write the report more quickly and efficiently? The answer to the latter question could be *yes*, *maybe*, or *no*. If not now, but surely after reading the Chaps. 14–21, I hope that this book's math-based writing system will inspire you to say *yes*.
α	No matter whether your answer to the above action item was *yes*, *maybe*, or *no*, draw a simple flowchart composed of circles, boxes, diamonds, arrows, and text labeling that traces the developmental structure of the same report example you recalled for the above activities. The goal here is to produce a quick sketch. Be creative and don't worry about the flowchart's aesthetic quality or fancy-phrased labels. After you study Chaps. 14–21, refer back to your free-form sketch, and see how well it aligns with this book's approach to flowcharting documents' structures. There will likely be similarities between your sketch and the drawings presented here, because as an engineer, you already possess expertise in math-based thinking and conceptualizing systems with block diagrams and flowcharts.

Recap

Key Learnings/Takeaways

- Like an algorithmic flowchart used by a computer programmer to guide the logical development of a program's computer coding in lines and subroutines, a document algorithm can be used by an engineering writer to structure the logical development of an engineering document written in sentences and paragraphs.
- When assigned a writing task, an engineering writer must select the appropriate algorithm for the job. The writer accomplishes this by first defining the purpose, target audience, and situation-driven format and medium for the document and then using these characterizations to identify a best-fit algorithm. Thereafter, although writing the document will require effort, the algorithm will ease the writing process because it is now guided by a straightforward plan.
- Some document elements are independent from algorithmic structures. Format, typography, and medium, as well as document length, can vary depending on a number of factors. For instance, whether or not a document is a *short-form document* (1–4 pp.) or a *long-form document* (≥5 pp.) depends, quite simply, upon how many words are needed to get the job done—no more, no fewer. Chapters 14–21 focus their instruction on short-form documents because short is generally preferred to long in today's workplace.
- In math-based writing, *document geometry* refers to document design elements: the placement and arrangement of text and graphics on the page.
- Although there are more than five, Part III focuses on presenting the algorithms and associated application processes for five essential "go-to" documents that professional engineers are likely to write on-the-job in industry:

  – *Project proposal*—a front-end document for soliciting funding and authorization.
  – *Status report*—an interim document that reports status of a project in progress.
  – *Project report*—a back-end document presenting project results and the associated activities that produced those results.

- *Tech to nontech brief*—a "how it works" document that explains a technical concept to a nontechnical audience.
- *Job aid*—a "how to do it" document that teaches a technical job task to someone who needs step-by-step procedures for correctly performing that task.

Selected Reading

1. Beer, D.F., McMurray, D.A.: A Guide to Writing as an Engineer, 4th edn. John Wiley (2013)
2. Alley, M.: The Craft of Scientific Writing, 4th edn. Springer Nature (2018)
3. Day, R.A., Gastel, B.: How to Write and Publish a Scientific Paper, 7th edn. Greenwood (2001)
4. Tebeaux, E., Dragga, S.: The Essential of Technical Communication, 4th edn. Oxford University Press (2017)

Chapter 15
Project Proposals

Learning Objectives

- *Know the document algorithm for proposals structured around the mode of persuasion.*
- *Strategically design and tactically advance a win-win argument.*
- *Frontload a proposal with a statement of the bottom line and the critical components for argument setup.*
- *Use proper typographical conventions for formatting a short-form proposal as a memo.*

Engineers write proposals *to identify engineering opportunities* and then obtain *a charter to do* and the *resources to support* an associated engineering project. Proposals generate technical work.

Proposals can be short or long, digital or hardcopy, and formatted in a variety of ways. Proposals can be either *internal* or *external* workplace documents, where *internal* means that the document is confined to circulate along communication channels *inside* a company, while *external* means that the document is aimed at *outside* audiences, such as external clients, vendors, technical partners, financial partners, and any number of other outside organizations that might represent a business opportunity or possess a needed resource or market outlet.

A proposal can be *independently initiated* or may respond to a *request for proposals (RFP)*. An example of the first case might be an engineer who invents a new integrated circuit design for a nanoscale transistor. She thinks that the design aligns well with her work group's strategic mission and business plan, so she seeks approval to develop the idea into a chartered project. To get project approval, she writes and *independently initiates* an unsolicited proposal to management. A positive reply from management grants the engineer authorization to move forward with her new project. A rejection means that her design idea will, at least for now, remain undeveloped.

B. Henderson, *A Math-Based Writing System for Engineers*,
https://doi.org/10.1007/978-3-030-10756-7_15

A proposal that responds to an *RFP* might result from a situation where an outside source, often referred to as either the *client* or the *customer*, puts out a call to potential proposal writers (the *bidders* or *contractor*). The call is typically packaged and transmitted as an RFP document. For example, the city of San Francisco wants to outsource a renovation job. So the city puts out an RFP call to prospective bidders. One of these is a California-based civil engineer who specializes in retrofitting out-of-specification buildings to make them earthquake-proof. He writes a solicited proposal to respond to the RFP, along with several other bidders. The best proposal wins the job.

In general, whether 1 page or 100, internal or external, solicited or unsolicited, *proposals* are *front-end* documents written by engineers who are seeking support. In industry, this support is often authorization and/or funding from management *to begin* an engineering project. Engineers also write proposals to secure contracts from clients, to initiate business partnerships, and to advance professional development. Being a competent proposal writer is a critical skill for engineers, because winning proposals enable engineers to work on the projects they truly want to work on. Conversely, poorly written proposals lead to missed opportunities and dull consolation prizes, such as being assigned to work as a technical staff person, rather than project lead, on another engineer's design enterprise, even though you are quite certain that your initial concept for the same design problem was superior.

How Does an Engineering Writer Structure a Successful Proposal?

Like almost any successfully written engineering document, an excellent proposal is built out of concise, clear, and correct sentences, cohesive and coherent paragraphs, and accurate and precise data. An excellent proposal is aimed at (and tailored to) a specific audience, and it is also built with a textual geometry that enables fast and easy content uptake by readers.

What makes proposal writing particularly challenging, above and beyond the preceding, is that *proposals demand that engineering writers not only present new ideas and solutions but also sell them to an audience that initially is willing to look but not necessarily buy.* In proposals, argumentation strategy is supremely important, and it demands awareness of the audience as well as the prudent choice and arrangement of document content. In proposals, *what* the writer says as well as *how* and *when* the writer says it are all important.

In a proposal, from start to finish, the writer must champion a proposed idea. Yet the audience, typically, is initially neutral, uncommitted toward the idea. It is the writer's job, then, to provide that audience with sufficient reason(s) to say yes to the proposed idea.

An additional challenge is that arguments invite counterarguments. In a proposal, counterarguments take shape as little questions that pop into readers' heads as they read and encounter places that cause them to feel the impulse to argue back—*Wait, did you also consider…? Is that legitimate…? What about…? Sounds good, but do you have proof…?* Competent proposal writers anticipate their audiences' potential objections and proactively defuse them by writing responses directly into a proposal's text. Addressing and resolving audience questions when (or better yet, before) they are likely to arise prevents questions from being unresolved and thereafter developing into stubborn objections that can hinder (or worse, stop) the persuasion process. A rule of thumb in proposal writing is to deploy, wherever possible, a strategy of preemptive strike.

For instance, the engineer who is proposing a new nanotransistor foresees management concern about collateral costs such as specialized test equipment, so she mentions early on in her proposal that the company's R&D lab is already adequately equipped for building and testing her very small-scale prototype. Similarly, our civil engineer who is seeking a contract with the city of San Francisco predicts potential objections from elected city officials, because he knows that they have pledged to support a community initiative to preserve historic landmark buildings. Accordingly, he assures in his proposal that his retrofit process will not in any way change the building's original exterior design.

At this point, incorporating all of the above into your proposals might seem overwhelming. The good news is that this chapter's document algorithm can help you to generate and arrange a stream of effective proposal content that delivers enough reasons to persuade your audience to say yes. The algorithm's design also guides you to anticipate and negate routine counterarguments that are likely to arise in readers' minds as they read.

For instance, the algorithm assumes that you, as proposal writer, are making a request for project authorization and funding from management, or for a job contract and compensation from a client. Therefore, it is highly likely that your proposal audience will want to know *what*, specifically, the requested funds/compensation will buy, as well as *how*, exactly, the proposed project's expenses will equal the amount requested. Accordingly, the proposal-writing flowchart contains operational blocks (elements) that prompt you, as proposal writer, to respond to both of these items: the *what*, by directing you to include a list of concrete *deliverables*, and the *how*, by requiring that you also include a *line item budget*. The ultimate responsibility for satisfying a target audience's complete spectrum of wants/needs, questions, and concerns falls on you, the proposal writer, not an annotated flowchart. However, using this chapter's algorithm will definitely help you in preparing successful proposal arguments and documents.

Figure 15.1 illustrates our first document algorithm, the "win-win" project proposal. Built around *persuasion*, the algorithm achieves its overall purpose by advancing through six key operational blocks or key elements: (1) *problem/opportunity*, (2) *solution method and means*, (3) *deliverables* (Ds), (4) *project plan*, (5) *budget*, and (6) *outcomes* (Os).

The algorithm is intended to produce a "win-win" proposal because a successful proposal brokers a mutually beneficial transaction between two parties, the proposal reader and the proposal writer, each with different perspectives, agendas, and end

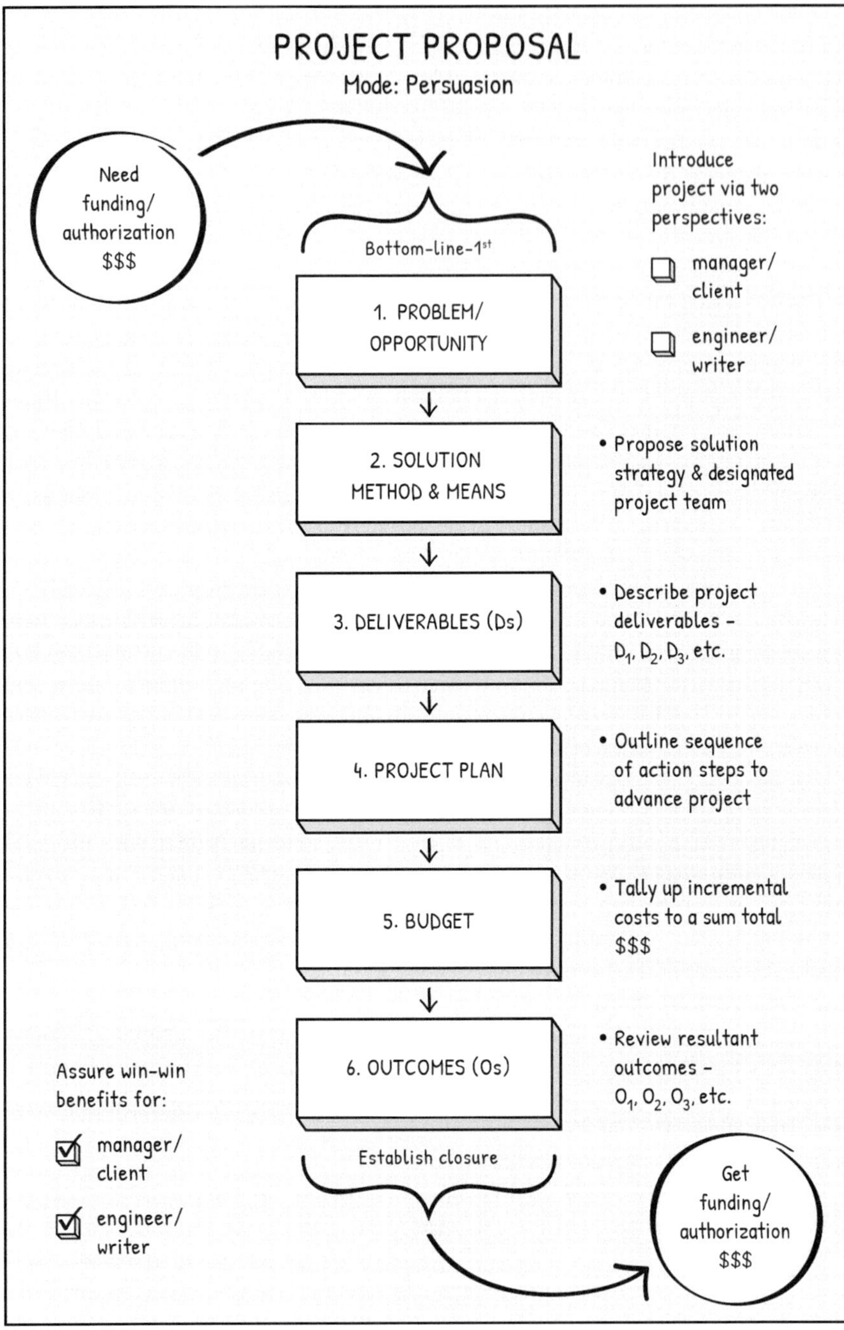

Fig. 15.1 Document algorithm for a win-win proposal

goals associated with the proposed project/activity. The upper-right corner of Fig. 15.1 acknowledges these two perspectives—first, the *manager/client's* (the person who reads the proposal and decides whether or not to fund/approve it) and, second, the *engineer/writer's* (the person writing the proposal who will undertake the project if the proposal is approved).

As noted in the algorithm's lower-left corner, the end goal is to achieve a "win" for both parties. The win for the audience is having an engineering problem solved or opportunity addressed in an effective way, at an acceptable cost on an acceptable schedule. The win for the writer is demonstrating the superiority of the proposed project idea and then procuring project support, usually money, to move the project forward.

How, Specifically, Does an Engineering Writer Apply the Proposal Algorithm?

The Framework

Figure 15.1's algorithmic flowchart leads off with a start operation labeled *need funding/authorization $$$* and concludes with the end operation labeled *get funding/ authorization $$$*. Between start and end, the algorithm's six key elements are further sandwiched between two bracketing operations: *bottom-line-first* and *establish closure*. These brackets remind the writer to frame the document's text as follows:

> *Frontload the bottom line:* Like nearly all engineering documents, a successful proposal must *state the bottom line up front (first)* by immediately and swiftly telling the reader—without prelude or preface—*what the document is about* and *what it offers and/or requests* from the reader. In a long-form document, this communication occurs in the document's summary or abstract. In a short-form document, the form emphasized in this book, the writer must state the bottom line in a sentence or two at the beginning of the document's text.

> *Establishing closure:* A successful proposal ends by letting the audience know that the document has concluded. Sometimes, in addition to wrapping up the proposal's pitch, the ending includes a "call to action"—wherein the proposal writer recommends a first step (or series of steps) to launch the proposed project, provided that the reader is now on board. In a long-form document, a proposal's ending often consists of a paragraph or two (or sometimes more) of text offering a retrospective review of the document's contents, interwoven with a set of concluding remarks. In a short-form document, this approach is too long-winded and plagued with what a busy reader is likely to consider redundancies. To end a short-form proposal, a sentence or two will do: signal *end of message* to your readers, and release them to go back to their busy workdays (and, ideally, to attend to your proposal's call to action).

The Six Key Elements

Now that we have examined the front and rear bracketing operations, here are the proposal algorithm's six key operational elements in recommended order:

1. *Problem/opportunity:* The proposal algorithm's first element functions as the launch pad for your proposal's argument. It aims the document at its target audience (the manager/client) as well as at its end goal—a "win-win" outcome for both the manager/client (target audience and reader) and you (the engineer/writer). The first element begins content development by introducing the proposal's topic/scope (a specific engineering problem/opportunity that needs to be addressed) and the proposal's purpose (to get support for an engineering project that responds to the problem/opportunity).

   The bottom-line statement that frontloads the proposal's beginning does some of the initial communication work. For effective argument setup, however, you must complete this task by adding a sufficient level of detail—no more, no less. What makes writing a proposal's introduction (as well as the rest of a proposal) particularly challenging is that the principal gauge for what is a "sufficient level" (i.e., what are the relevant details to cover and in what proportion) is not you and your (the engineer/writer's) technical interest, enthusiasm for, or bias toward the proposal topic—but rather *them*: your target audience, the manager/client (and their perspective).

   The proposal is trying to persuade them, not you; you are already convinced that your proposal is significant and worth supporting. (If you didn't, you wouldn't be writing it.) To persuade, your proposal must build empathy between you and your audience. To accomplish this goal, its message must speak to your audience's perspective first—and yours, second.

   Therefore, in the introduction, be sure to consider and acknowledge how, specifically, the problem/opportunity addressed in the proposal impacts the manager/client, as well as the people they are responsible for—employees, customers, product end users, investors, and so on. Make sure that the manager/client understands the true root cause of the problem. If applicable, make them aware of some facet of the problem/opportunity that might not have occurred to them. Also, if applicable, make them aware of other parties with similar problems. After completing the introduction, your target audience, the manager/client, should be both sufficiently informed and sufficiently engaged by the proposal topic—and ready and eager for the next segment.

Example: Two Perspectives for a Proposed Project

Here is an example of two project perspectives:

1. *The perspective of a project manager/client at Boeing Commercial Airplanes* who needs to outsource a job redesigning and upgrading the cabin air circulation and filtration system for one of Boeing's commercial jet airframes
2. *The perspective of an aerospace subcontractor engineer/writer* who responds to the RFP posted by Boeing

2. *Solution method and means:* Now that the first element has laid contextual groundwork, captured the reader's attention, and developed an initial level of engagement and empathy, the second element must go on to present the writer's proposed method for solving the problem. Element two focuses on presenting an overall engineering strategy for responding to the problem, rather than a step-by-step tactical plan. This is also the appropriate place for the writer to address means (or agents) of action, particularly the credibility and qualifications of the proposed project team. The proposal reader must be convinced that you, the proposal writer, and your group are the right people for the job. The people whom your proposal designates as project researchers, designers, builders, and so on are a part of the solution.

3. *Deliverables (Ds):* The third element now begins to drill down from strategy to *tactics*. This operation must further convince the reader to accept the proposal by providing the reader with more details about what the proposal writer and team pledge to produce. Here, the writer commits to delivering a set of concrete, tangible results—called the project deliverables or Ds (labeled D_1, D_2, D_3, etc. on Fig. 15.1). The paramount deliverable, of course, is the project's end result. However, oftentimes other deliverables precede and precipitate the end result. A good proposal writer spells out all of the important deliverables—initial results, interim results, as well as the final product or end result. Results, not action leading to results, are what the manager/client wants most from you.

Example: Deliverables for the Boeing Subcontract Project

An example of an *interim deliverable* would be D_1 = the subcontractor completes and submits a CAD drawing of a fully designed, upgraded air circulation filtration system for a jet airframe for manager/client approval. After concept approval and further action, an example of a *final deliverable* is D_2 = the subcontractor delivers a fully built system with all system components ready for use.

4. *Project plan:* Once the reader is sold on the proposed solution's concept, solution team, and resultant deliverables, the reader needs to know how and when the project deliverables will be delivered. Regarding the business aspect of engineering, time is money. Providing a project plan, the fourth element, gives the reader the clarity of a timetable—a sequence of action steps outlining what will happen and when.

   A Gantt chart is a useful graphic vehicle for illustrating a proposed sequence of action steps over time. It consists of a grid of rows and columns that shows the start, duration, and end of each project activity as a series of "do increments." This chapter's applied example shows a sample Gantt chart (see Fig. 15.2b). In general, a project plan creates accountability and an additional layer of commitment from writer to reader. When a proposal is accepted and the proposed project begins, the project plan remains useful for both the engineer and the client. The plan is a way to steer the project, avoid bottleneck stalls, and continually monitor whether project activity stays on course.

5. *Budget:* Beyond the visibility of deliverables and action steps, the fifth element provides the reader with financial transparency. This operation specifies how project funding will be spent in increments. Even though the proposal might state a bottom-line lump sum figure upfront, the budget segment usually provides a line-item breakdown of all individual "key" expenditures that total up to the lump sum.

   A line-item breakdown is best presented as a vertical list or table, not as a paragraph of text. For a short-form proposal, an appropriate level of detail is usually around three to ten items. A final "sum total" line should verify that the line items total up to the requested amount (see Fig. 15.2b for an example). Since a proposal is a front-end document, all budgetary figures are estimates. Consequently, a challenge for proposal writers is projecting cost figures. Strive to make your estimates achievable and safely margined for you (the engineer/writer), as well as reasonable and believable to them (the manager/client).

6. *Outcomes:* The sixth and final element of a proposal is a review of project outcomes or Os (labeled O_1, O_2, O_3, etc. on Fig. 15.1). Project outcomes are closely related to project deliverables, and sometimes distinguishing one from the other can get a bit tricky. To avoid confusion, think cause and effect. When a project produces (causes) an agreed-upon deliverable (a tangible concrete result), there is a corresponding outcome (benefit or effect).

   By this stage in your proposal (if you've followed the algorithm), you have already named your project's deliverables—D_1, D_2, D_3, etc.—and in doing so, it is likely you have already asserted or implied some benefits associated with the deliverables. The purpose of the algorithm's final operation is to complete the sales process. This is accomplished by highlighting and further elaborating upon the proposed project's final set of outcomes—particularly those that positively impact the reader (the manager/client).

   Project outcomes that benefit a manager/client often have to do with profitability. So, if your proposal projects future sales or cost savings, review those key numbers. Benefits for a manager/client might also include fulfilling a larger company business objective or mission, generating positive PR (public relations), meeting an environmental or safety regulation, or qualifying for a tax break. This final section of your proposal is there to ensure that, in the end, your proposal argument has given your readers enough (ideally, more than enough) reasons to say yes. When they finish reading the document, you want them convinced that your proposed solution/opportunity is just as solid a win for them as it is for you and that whatever you're asking for is worth that cost.

Example: Outcomes for the Boeing Subcontract Project

1. An example of a project outcome (benefit) from the manager/client's perspective is O_1 = once installed, the upgraded air circulation filtration system (deliverable) will make the new jet airframe more marketable to commercial airline companies (profits).
2. Another is O_2 = the new filtration system allows Boeing to meet new federal safety regulations for passenger cabin air quality (legality).
3. An example project outcome from the engineer/writer's perspective is O_3 = winning the contract and getting to do a technically challenging and meaningful project (professional success).

Application Example

Let's move on to an application example that demonstrates how to apply the algorithm to a short-form proposal document typical of what engineers write in industry.

- The application example presents a real-world scenario.
- It presents an algorithm-guided, pre-write sketch prepared by the scenario's engineer/writer.
- Finally, the example showcases an effective proposal memo, as well as an ineffective proposal memo for the scenario.

A brief tutorial on how to format the memo (document geometry) follows the application example. For readers already familiar with the form, the tutorial may serve as a useful refresher. It might also offer a few new useful tips on memo writing that are worth adding to your engineering writing toolbox.

The Project Scenario

Fenili Digital Mechanisms (FDM) is a large-scale computer products company that specializes in designing and manufacturing inkjet printers. Aside from chassis, internal components, and software drivers for the company's popular and profitable inkjet printer family, FDM engineering also designs the replaceable inkjet print cartridges (pens) that go into the printers, as well as the high-speed manufacturing lines that mass-produce the pens.

Because inkjet pens are short-life, consumable accessories that sell at a 7X markup, this product nets substantial profits for FDM. Not surprisingly, FDM management wants all inkjet pen manufacturing lines to be well supported and maintained by factory staff, so the lines can achieve maximum production yield and minimum downtime and scrap. When engineering develops a new and better printer model and complementary inkjet pen, and the new model is ready for release, the management wants the printer's pen manufacturing line debugged and operating in a steady-state mode as quickly as possible.

The project scenario assumes that FDM engineering has recently completed development of a next-generation printer (code name "Trinity") with a high-resolution inkjet cartridge that prints 2400 dots per inch (dpi). This printer supersedes another model (code name "Koala") that prints at 1200 dpi. In the scenario, the proposal writer is Linda Zhang, a manufacturing engineer, who took the lead role in designing the new manufacturing line for Trinity pens. Because of her technical expertise and experience, Linda will oversee line installation and start-up, as well as initial technical training for Trinity's factory floor staff. Ms. Zhang needs funding and authorization from management to initiate a vendor service contract (VSC) to hire an outside vendor (Take2 Limited) to film and produce a technical training video. The video will fulfill a critical training objective for her line start-up team.

Table 15.1 presents a pre-write sketch of Ms. Zhang's short-form proposal memo to management. She uses the Fig. 15.1 algorithm to select and arrange a content outline. Then, Figs. 15.2a, 15.2b, and 15.3 present effective and ineffective examples of Ms. Zhang's final proposal memo.

Table 15.1 Pre-write sketch of key proposal content guided by proposal algorithm

Elements of proposal algorithm	Sketch of key content for proposal elements
Bottom-Line-1st Start	• I (engineer/writer) request \$35.5 K from management (*what I want*) for contracting production of a technical training video for the new trinity pen line's start-up team
#1 Problem/ Opportunity	• *Management's perspective (them)*—management needs a fast line start-up for trinity pen manufacturing, because steady-state production is profitable and downtime/delays are costly • *Engineer/writer's perspective (me)*—based on my experience as project lead for the Koala line start-up, I know that workers assigned to specific production stations/areas can provide better support if they understand the whole line's workflow

(continued)

Table 15.1 (continued)

Elements of proposal algorithm	Sketch of key content for proposal elements
#2 Solution Method and Means	• I propose that we hire a vendor (Take2 Limited) to produce a workflow training video for the new Trinity pen line's staff. This training method and this vendor proved successful for the Koala line start-up, which I worked on as lead manufacturing engineer. Having all line staff understand line workflow is a critical training objective for the start-up manager • *Note: I anticipate the counterargument,* "Why not shoot the workflow video ourselves? Wouldn't that be cheaper and just as effective for training?" and I will need to address/refute this concern in this segment
#3 Deliverables (Ds)	• After filming, Take2 will deliver draft copies of the video for customer review and changes/edits (up to three review cycles) • After approval, the vendor will deliver a final version and transfer a master digital copy of the video to the information technology department for distribution • The vendor will provide postproduction support and service for minor edits/updates for 1 year
#4 Project Plan	• I will include a Gantt chart with an 8 week project timeline to display project activity divided into approximately 6 action steps
#5 Budget	• I will detail the $35.5 K request in a line-item budget
#6 Outcomes (Os)	• *Management's perspective*—having a fast line start-up to quickly reach steady-state production mode maximizes profits for the Trinity pen line (revenue). Fast start-up also ensures that Trinity inkjet printer owners/users will have an ample supply of replacement pens (customer satisfaction) • *Engineer/writer's perspective*—When funded/approved, I get to lead another engineering project to success at FDM (professional fulfillment)
Establish Closure	• I will indicate that I am sending necessary paperwork directly to management for approval/signature and say thank you

Effective Memo Example

TO: Doug Kepner,
Vice President Engineering Operations

DATE: *<month-day-year>*

RE: Funding Request for Trinity Line Workflow Video

FROM: Linda Zhang, *LZ*
Manufacturing Engineer

CC: Bill Torres,
Production Supervisor

The purpose of this memo is to request authorization for a $35.5K Vendor Service Contract (VSC). The VSC will expense production of a workflow training video for engineers, technicians, and operators assigned to our new high-speed manufacturing line for Trinity inkjet cartridges.

A top-level objective for our technical training program is to develop manufacturing personnel who can achieve quick line start-ups and then maintain steady-state operation resulting in maximum product yield, and minimum downtime and scrap. To achieve this objective, all factory-floor support staff—direct labor through salaried—need to understand the Trinity line's workflow—the sequence of 68 high-speed, robotic operations that assemble each Trinity cartridge.

FDM Marketing predicts large-scale worldwide demand for the new Trinity inkjet printer and its 2400-dpi print cartridges. Given that current downtime cost for our existing Koala product line is $30/second, we anticipate an equivalent or even higher downtime cost for Trinity. Line workers who understand upstream and downstream processes, in addition to knowing about their individual stations, are better-suited to engage in effective system-level thinking, troubleshooting, and repair.

Our company learned the benefit of using video training two years ago during start-up of our Koala pen line. We met our 45-day ramp-up goal, and then once up, held scrap rate to 5.4% (undercutting our 8% target) and pushed an impressive 250,000-unit monthly production rate for fiscal quarter one. Focus groups and post start-up assessment indicated that the Koala workflow video was an integral part of this success. I developed the technical script for the video and served as project lead. I will head up the Trinity video project as well.

My team will use Take2 Limited to film and produce the proposed video. This video production vendor provided responsive, quality service on the Koala project and offers competitive rates. Current industrial pricing for broadcast quality video production is $1.5K to $3K per minute. Take2's project bid for our 20-minute workflow video is $35K ($1.75K/min). A cheaper option that I considered, but rejected, was producing the video in-house using company-owned video equipment. However, my research indicates that, prior to Koala, FDM engineering did try producing in-house training videos for our prototype lab technicians. This project achieved unsatisfactory results, and engineering concluded it was more cost-effective to outsource the task to a vendor specializing in video production.

Page 1 of 2

- *Memo begins with a complete and properly formatted header.*
- *Note the blue ink initials.*
- *The body text begins by stating in a nutshell what the writer wants and why.*
- *Next, the text states the problem/opportunity from multiple points of view to build empathy and set up a final win-win outcome.*
- *Then, the memo presents a viable solution strategy.*
- *The text also establishes the engineer-writer's credibility and resolves an anticipated counterargument.*

Fig. 15.2a Effective proposal memo, page one

Once a VSC is in place, Take2 has agreed to work with my team to complete the project in eight weeks. Take2's chief deliverables are as follows:

- After filming and producing the 20-minute video, present proof copies for our review(s) and do changes/edits (up to 3 review cycles).
- After review, deliver an approved final video, and transfer a master digital copy to FDM's IT Department for secure, internal distribution.
- Provide post-production support and service for minor edits/updates for one year.

Figure 1 outlines our project plan:

	wk 1	wk 2	wk 3	wk 4	wk 5	wk 6	wk 7	wk 8
phase one = develop work-flow video script	o		Δ					
phase two = schedule filming and on-camera interviews								
phase three = shoot footage and record voiceovers			o					
phase four = produce, review, edit, and perfect							Δ	
phase five = receive final video and transfer to FDM IT								
phase six = post-project assessment								

Figure 1 – Gantt Chart for Trinity Video Project

As you can see in the Gantt chart, a critical juncture occurs at week 3 where scriptwriting ends (**Δ**) and filming begins (**O**). I will use vigilant project management to avoid a bottleneck here, as well as delays elsewhere.

Here is a line item budget for services to be provided by Take2:

Script consulting, filming, editing, and production, (\$1.75K/min for 20-minute video)	\$35,000
Light refreshments for FDM employee who participate in off- and on-camera interviews	250
Reserve for emergent issues	250
Total	\$35,500

The Trinity workflow video is a critical training tool to help FDM manufacturing meet its aggressive production targets; to improve troubleshooting response time and acuity for Trinity operators, technicians, and engineers; and to maximize revenue from the new line. Fast start-up will also foster customer satisfaction by ensuring that new owners/users of our Trinity inkjet printers have ample market supply of replacement pens. My department's logistics coordinator will submit a new VSC for Take2 Productions into your monthly signature cycle. Thank you.

Page 2 of 2

- *The text spells out the deliverables.*
- *The memo lays out a project plan using a Gantt chart.*
- *Here, the text splits vertically to let the graphic do its visual work with minimal labeling.*
- *Then, the memo presents a line-item budget of key expenses totaling up to the requested amount.*
- *The last segment "closes the deal" with a discussion of win-win outcomes.*

Fig. 15.2b Effective proposal memo, page two

Ineffective Memo Example

MEMORANDUM *<month-day-year>*

TO: Doug Kepner, Vice President Engineering Operations

FROM: Linda Zhang, Manufacturing Engineer

RE: TRAINING ISSUE

CC: Bill Torres, Production Supervisor

Doug;

The main thing about using video as a medium for teaching our personnel about new manufacturing equipment technologies is that it's visual in its format, rather than just about "hitting the books." Video lends itself well for seeing, efficiently and effectively, how something works overall, as well as in its increments. In terms of a manufacturing line, this means macroscopic workflow and more micro-level intricacies of individual stations. When it comes to training, I have never recommended our employees simply read specification manuals written by equipment builders, who don't understand the key training issues like my team does, from FDM's point of view. Our previous video training attempt in the form of the Koala line video is probably not going to win an Oscar, because it's not that type of movie, but it has seriously proved to be a topnotch education mechanism for line start-ups. I can say this because of the impressive production numbers we have achieved for the Koala line in our facility. Not surprisingly, we'd like to produce a workflow training video for our new Trinity cartridge manufacturing line. We need a $35.5K VSC and would very much like you to sign off on it. The vendor will be Take2 Limited, which is a video production company. I strongly recommend you approve this requested VSC as soon as it is possibly convenient for you, if you want us to keep "on-schedule" with our Trinity project timeline, which I'm sure you do. Please have your administrative assistant come to my work area and pick up the necessary paperwork for you to sign, as FDM's number one manufacturing engineering executive. Again, I stress that time and ultimate value of the project's results are of the essence.

Sincerely, *Linda Zhang*

- *The memo begins with an acceptable alternate format for the header (good), but memo's title is vague (bad).*
- *The body text begins with a salutation that contains a punctuation error—uses a semicolon instead of a colon.*
- *The text then launches into a "free-floating" discussion of the solution and doesn't clearly establish the problem upfront.*
- *The sentences use the wrong tone to assert credibility—negative and disrespectful, instead of positive and diplomatic.*
- *Finally, the text states the "bottom line" and some details to achieve it (good), but too superficially and too late (bad).*
- *To end, the memo assigns tedious response details and closes like a letter.*

Fig. 15.3 Ineffective proposal memo

Document Geometry: How to Format a Memo

Today, email is a prolific and highly effective communication medium for internal and external message documents. Prior to the digital explosion of the 1990s, however, the most frequently used person-to-person media for short-length textual communications in the professional world were (1) the *memorandum* or *memo* and (2) the *business letter*. Here, we will discuss *memoranda (memos)*. The memo is a format that continues to be useful for short engineering reports, policy statements, and, as we have just seen in the application example, short internal proposals, for which a hardcopy format adds a degree of formality and official-ness.

Memos are categorized as *internal* documents, because they travel between desks, departments, buildings, and divisions inside a company, using several channels including company mail and private hand delivery. Memos circulate among employees and are not meant to be routed outside of a company like a business letter using "snail mail" or traditional post services.

Regarding typography and layout, at its top, a *memo* begins with a text zone known as the *header* (see Fig. 15.4). In the header, the writer records all necessary information for internal routing. The header notes:

- To whom (**To:**) the memo is being sent (the target audience/reader).
- From whom (**From:**) the memo is being sent (the memo writer).
- A descriptive title indicating what the memo is regarding (**Re:** or **Subject:**)
- The date (**Date:**) when the writer writes the memo (the timestamp).

To indicate authenticity, the original master copy is often initialed in blue ink, immediately next to the writer's typed name. Optionally, the writer can add a listing (**cc:**) of secondary readers who will receive a copy of the original memo. To identify individual copies, a common convention is highlighting the recipient's name in yellow (or another color), using a highlighter pen, for each copy/name on the "cc:" list.

Underneath the header is a second, larger text zone that presents the memo's body text. Unlike an email or business letter, the memo's body text begins with a sentence and ends with a sentence. That is, the body text **does not include a salutation** (Dear So-and-So:) or complimentary closure (Sincerely,/ [vertical space for signature] / Writer's Name). Otherwise, typographically, a memo's body text is formatted much like a standard email or business letter—in single-spaced blocks, flush left, ragged right, with no tab at the beginning of paragraphs. The single-spaced text is double-spaced between paragraphs.

As is true of most short-length documents, memos are meant to be read quickly and easily. This goal is accomplished with a quick-scan layout that, top-to-bottom (vertically), is interspersed with small chunks of black text with white bands of space to provide optical relief. To accomplish successful chunking, the memo writer worries less about "topic sentences" and more about "logical breaks." Ideal paragraph length is one to eight lines. As for the body text's overall length, just like its cousins, the email and business letter, memos should be kept as short as possible. Whenever possible, one page is preferred.

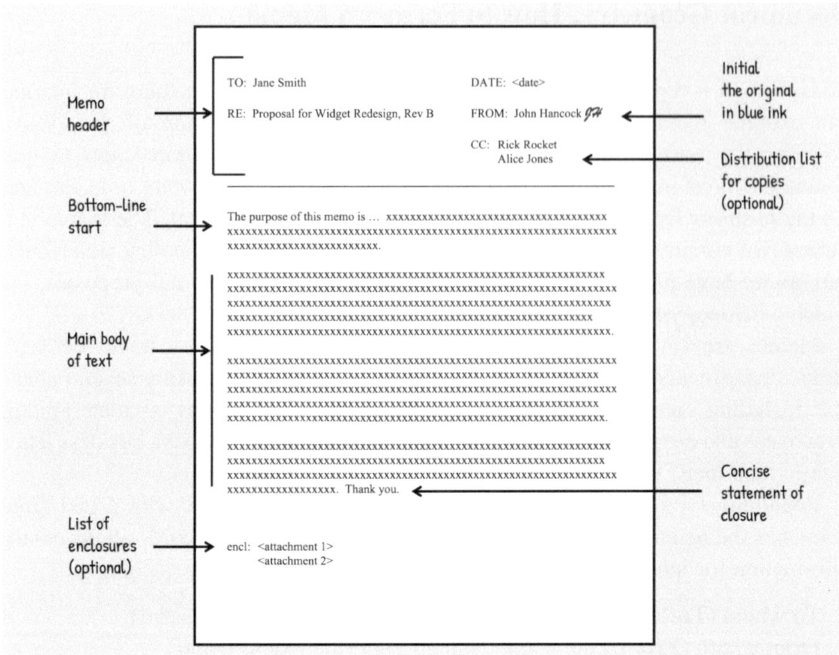

Fig. 15.4 Plan view of a memo

Figure 15.4 showcases the document geometry of a memo in top-down, or *plan view*, perspective. Because memos are relatively short documents, composing a memo requires a writer to produce significantly fewer sentences than required by a formal report or proposal. That's the good news. The bad news is that more responsibility and individual emphasis are placed upon each sentence in a memo. Thus, when sentence-level mistakes occur, they stand out. This book recommends that, as the final step, the engineering writer should read the memo slowly out loud before initialing and delivering it. Reading a memo aloud puts into play the writer's "ear knowledge" in addition to the writer's "head knowledge" of English. This double-edged quality control practice can often catch errors that might otherwise be missed when a writer quietly skims over his or her own work.

Document Engineering Tip

When you're doing a final read-aloud test of a memo (or any other engineering document) and you call upon your "ear knowledge" to assist you with quality control decisions, **always assume** that anything that sounds clunky **is clunky** and needs to be fixed. Even if you don't know theoretically (grammar-wise) what's wrong and why, if you work with the sentence using trial and error and get the sentence to sound better when voiced aloud, then chances are that the new version **is** better. Trust your "ear knowledge," and go with it.

Action Items

α	The next time that you are called upon to present an idea in a conversation or dialogue at a meeting, think about and make a note of how you incorporated (or did not incorporate) the "win-win" persuasive strategy, as presented by our first document algorithm. In your scenario, can you identify the multiple perspectives—yours as well as the client/manager's? How is the "what's in it for them" different from the "what's in it for you"? How does your argument address the specific wants/needs of your target audience and their perspective?
α	Within the confines of your own engineering activities, what's the biggest lump sum you've ever netted from a proposal? What's the biggest lump sum you've asked for but failed to get? For the latter, if you could somehow receive a "do-over," how might you use this chapter's document algorithm to your advantage?

Recap

Key Learnings/Takeaways

- The book's first document algorithm provides a recommended structure for crafting a project proposal for funding and/or authorization of an engineering project. The algorithm is built around *persuasion*. Like most engineering documents, a proposal begins by stating the document's "bottom line" (what the document is about and what it offers and/or requests from the reader) and ends with a brief statement of closure. Within this outer framework, the algorithm contains six major elements:
  - *Problem/opportunity*—Introduces the proposal topic from two perspectives: "you," the engineer/writer, and "them," the manager/client. The introduction must establish empathy between the writer and reader by addressing (above and beyond the writer's interests/insights) how the problem/opportunity impacts the manager/client's (reader's) business needs, employees, customers, investors, and so on.
  - *Solution method and means*—Proposes the overall strategy or path forward, as well as the means for project action, particularly the identity and credibility of the proposal project team (you and those who will work with you).
  - *Deliverables*—Describes and commits to (pledges) a set of project deliverables, that is, what the project team intends to produce as tangible results. These deliverables are what the requested project funding/authorization will "buy."
  - *Project plan*—Outlines a sequence of action steps and their timeline for moving the project from start to finish and producing all tangible results (interim and final) spelled out in the deliverables segment.
  - *Budget*—Discloses the project's financial details, that is, the discrete expenses/costs that you will incur/charge, and shows these expenses/costs as line items that tally up to the lump sum requested.

  - *Outcomes*—Reviews the *outcomes* (benefits) that the project aims to generate. Here, just as with element one, it is particularly important to highlight outcomes (benefits) to be gained by the manager/client. This segment ends your proposal's argument and *seals the deal*.

- Designing and advancing a win-win argument means that you are moving toward a "win" for your target audience, the "manager/client," and a "win" for you, the engineer/writer.
- In general, whether short or long, solicited or unsolicited, *proposals* are *front-end* documents written by engineers who are seeking support. In industry, this support is often authorization and/or funding from company management or an outside client *to begin* an engineering project. Also, proposals can either be *independently initiated* or may respond to a *request for proposal (RFP)* from a prospective client.
- A memo should begin with a header that records the document's "traveling information" (who is the audience? Who wrote it and when? What is its purpose?). The rest of the memo is the body text, which contains the document's message. Typographically, memos are designed to be read quickly, and "chunking" text helps with this. Also, unlike emails and business letters, the memo's message begins with a sentence and ends with a sentence, not a salutation and complimentary close.

Chapter 16
Status Reports

Learning Objectives

- *Know the document algorithm for status reports.*
- *Understand what managerial audiences want to encounter first in the body text of a short-form status report.*
- *Know the difference between linear (beginning to middle to end) document structure versus bottom-line-first structure, and know how to flip the former to reconfigure it into the latter.*
- *Know the proper typographical conventions for formatting a short-form status report as an email.*

Engineering projects often begin with the engineer writing a project proposal to gain funding and authorization (see Chap. 15) and end with the engineer preparing a final project report (see Chap. 17). While a project is "in progress," if you are the project engineer, you will most likely need to provide your manager with project updates.

Sometimes project updates are informal. For small, daily updates, a manager might prefer communications via in-person dialogue, voicemail, text message, or ultra-short casual email not intended for further sharing or preservation. For weekly or longer time spans, managers usually prefer more formal, comprehensive emails or memos for status reports. These documents must be suitably polished so that they can be circulated to other audiences beyond the immediate manager. This chapter covers how to write a status report as an email that incorporates "flipped," bottom-line-first organization—a document structure that can, at first, be counterintuitive to engineering writers but that is almost always optimal for managerial-type readers.

B. Henderson, *A Math-Based Writing System for Engineers*,
https://doi.org/10.1007/978-3-030-10756-7_16

Linear Versus "Flipped," Bottom-Line-First Structure

Chapter 15 briefly introduced project plans, which break project work into a series of "to-do" items, also known as "action steps." Project work can involve a wide variety of engineering tasks such as *observing*, *researching*, *analyzing*, *calculating*, *designing*, *iterating*, *maintaining*, and more. And an adequate engineering response to most of these tasks requires the engineer to undertake multiple action steps rather than one. Thus, because of its sequential nature, an easy, straightforward method for documenting engineering work is *linear description*. This means telling "the story" of how engineers march their way through a project step by step—responding to one assigned task, then another, and then the next over a period of time.

In a linear status report, the first line might be *On Monday, I was assigned to do ______*. Then, there might follow a cascade of sentence-level descriptions: *First, I ______. Next, on Tuesday, I ______. And thereafter, on Wednesday and Thursday, I did ______ ....* Finally, the ending of a linear project status report might be, *In conclusion, to complete the assigned task, as a last step, I did ______ and, therefore, recommend that ....*

For most engineering writers, describing a work activity step by step comes naturally. Storytelling is a natural way of speaking and writing for most people. It is how we describe and share events in our lives—both recent and long past. As an engineering writer, you must keep in mind that, from a manager's point of view, the most important part of a status report is how the story ends. Status report readers are chiefly interested in what progress you have made and what results you have produced as of now. Although it provides a user-friendly structure for writers, a linear status report means that readers get last what they want (and need) the most: the bottom line. When the delivery of the bottom line is delayed or buried among secondary details, status report readers become frustrated and impatient, and this reaction can prevent them from hearing part—or even all—of your report's key messages.

You can avoid this problem by using a two-phase document drafting process, a successful strategy for writing excellent status reports (and similar descriptive documents). Two-phase drafting *strikes a compromise* between what comes naturally for writers and what bottom-line, results-oriented readers naturally want. This process encourages engineering writers first to allow themselves to develop an initial rough draft in linear storytelling mode and then to *revise* and reconfigure the linear draft by "flipping" its structure.

Phase II calls upon you, the writer, to locate the linear draft's concluding sentence, delete any tag phrase such as "In conclusion ..." (or the equivalent), and then cut and paste the remainder to the second (flipped) draft's beginning, so its lead sentence states succinctly what the document reports overall—the interim results for some stage of an engineering project. Thereafter, you can further edit the second draft's middle and add a non-bottom-line ending—a simple statement of closure.

The Status Report Algorithm

The bottom-line-first progress report or status report algorithm is shown in Fig. 16.1. The bottom-line-first algorithm begins with a project in progress that has advanced through incremental tasks, and achieved one or more task results that need to be reported to the manager. The status report writing algorithm develops around the *mode of inversion* and ultimately guides the production of an audience-focused status report—here, as often is the case, an email. Between beginning and ending operations, the algorithm's middle moves through two key phases:

1. Generating a *linear draft* to record *step-by-step project action* leading to a *result* (top half of figure)
2. *Revising the draft* to produce a "flipped" *status report email* and to complete the algorithm with a finished document that is ready to send (bottom half of figure)

The Fig. 16.1 algorithm uses triangles to symbolize both parts of the two-draft process. The first triangle (apex up) depicts and labels the elements of a linear draft. Organized around physical action and time, this draft narrates the sequence of steps done by the engineer to complete an assigned project task(s) and produce a result.

Underneath, there is an inverted (base up) triangle. This triangle depicts a bottom-line-first revision that leads off with what previously came at the linear narration's end: the action steps' final result. Here also, the algorithm signals the writer to revise the second draft's middle, so it presents pertinent details that support the bottom line and cuts extraneous details that do not. Oftentimes, leaving pertinent details extracted from the first draft in their original order (linear) works just fine. However, since this is a nonlinear draft, rather than one mapped against time, you can also re-sequence pertinent details around other indices such as priority (i.e., each detail's relative importance toward supporting the bottom line). All in all, always keep in mind that the first draft generates raw content, and the second draft reorganizes content to best meet the reader's needs. To complete the inverted draft, since the original ending (bottom-line result) has been moved to the beginning, you need to insert a new ending (a brief sentence, or at most two) that establishes closure.

Document Engineering Tip

When you, the engineer/writer, present project results, any results (interim or final), think of those results as being a **claim**. Like an attorney in a courtroom, to ensure that you adequately convince your manager/reader that your claim is valid, you must support your claim with **evidence**—in a report, the pertinent details (facts, findings, data, and so on) that you offer to support your bottom-line (request, conclusion, and/or recommendation).

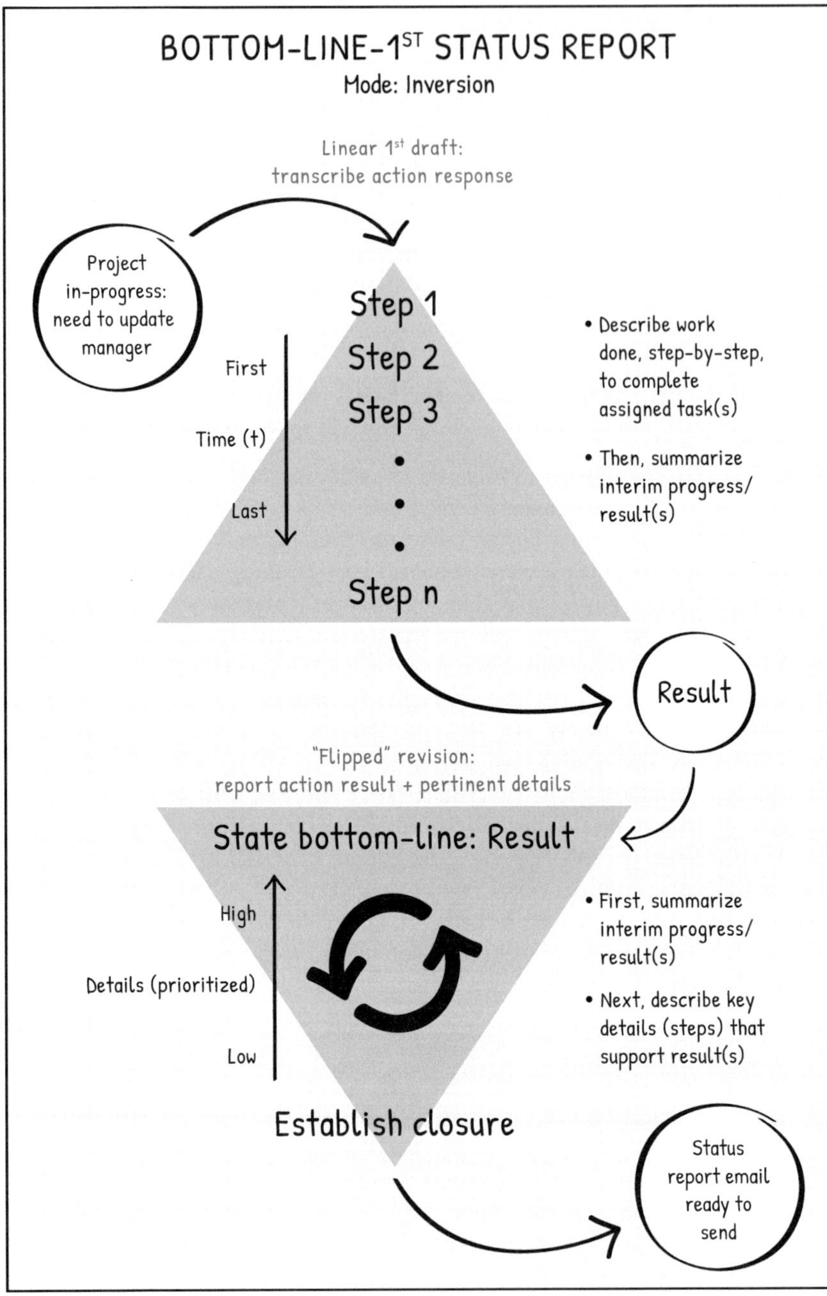

Fig. 16.1 Document algorithm for a bottom-line-first status report

Application Example

To become better acquainted with this chapter's algorithm, let's examine an application example built around a scenario in which an engineer has begun a project, is working on it, and is about halfway done. This is a long-term project that is *in progress* and has a project timeline that spans several months, rather than days or weeks. The engineer's manager requires weekly updates in the form of status report emails. The example demonstrates how the engineer uses the algorithm described in this chapter to help him write a successful status report.

Like the previous chapter, this chapter provides both an effective and an ineffective example document. Following the instructional segment and document models, you will find a brief tutorial on best practices for formatting an email's text. Though I assume that most readers—from entry-level engineers to seasoned CEOs—already have experience composing emails, I nevertheless encourage all to visit the end segment to examine how document geometry impacts the readability of an email and to review the several recommended optimization techniques.

The Project Scenario

Bui Aerospace's Control Systems Unit (Bui CSU) supplies custom-designed hydraulic control valves for commercial and military air and spacecraft manufacturers. Pete Knox is a CSU design engineer who reports to a divisional engineering team managed by Elaine Mayfield. Ms. Mayfield's team is currently working on a project for Green Planet Rocketry (GPR).

This client has contracted Bui CSU to supply a series of precision valves for a new communication satellite. GPR's contract specifies that one of these valves must be a fully mechanical unit that monitors the temperature of a system fluid (F1), an auxiliary thruster fuel, and then controls the flow rate of a second system fluid (F2), a coolant, based on temperature indexing. Ms. Mayfield has assigned Pete to be the project lead on developing this valve for GPR.

Pete has come up with a viable design concept. Pete's conceptual sketch (see Fig. 16.2) shows a cutaway view of the valve. The valve's index control mechanism is a mechanical bellows, filled with a third fluid, labeled "F3" in Pete's sketch. In his design, the temperature level of the auxiliary thruster fuel (F1) causes the fill fluid (F3) inside a corrugated bellows to expand or contract, opening or closing a poppet valve controlling coolant flow rate (F2).

At their last weekly meeting, Ms. Mayfield approved Pete's design concept, and she has cleared Pete to further develop a set of design specifications and then advance the project toward the prototype and test phase. This week, Pete's assigned tasks are to select a temperature-responsive fill fluid for the bellows as well as material for the prototype's valve body, actuating poppet, and corrugated bellows chamber. Our scenario assumes that Pete has successfully completed the preceding tasks and must now communicate his results to his boss in a status report email.

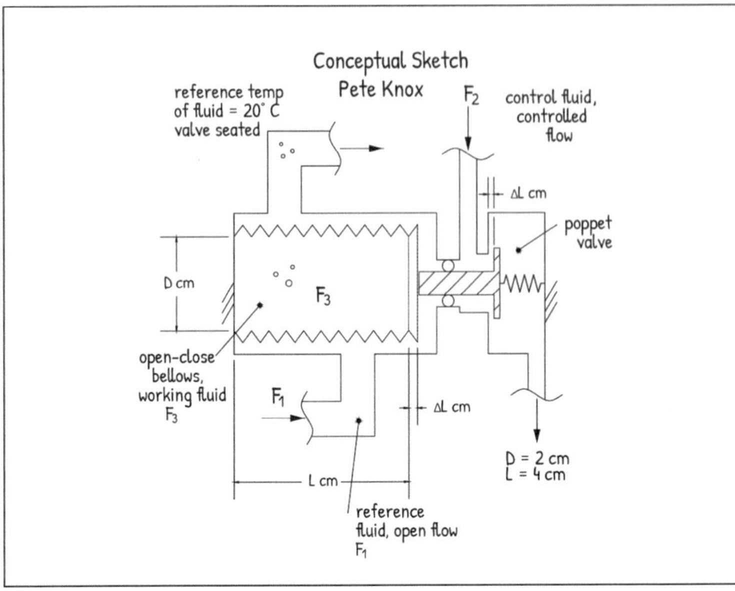

Fig. 16.2 Conceptual sketch of mechanical-bellows-actuated, fluid-temperature-referenced control valve

Writing the Status Report Email

After consulting the Fig. 16.1 algorithm to guide the production of his email's text, Pete writes a linear draft as shown below. This first draft is easy for Pete to produce, and it accomplishes the initial writing task of getting his weekly report's content out of his head and onto the page. Before pressing "send," however, Pete knows that he will need to revise the text of this first draft, which is 12 sentences long. Here, for instructional purposes, each of the sentences is tagged with a superscript number. This linear draft records a sequence of actions (both physical and intellectual) that progresses in real time, beginning with "After our last meeting . . ." and ending with a recommendation for proto fabrication materials and fill fluid for the bellows. (Note: Some of the linear draft's text is in italics. We will discuss this text later. Ignore it for now.)

Linear Draft of Weekly Status Report Email to Elaine Mayfield

Subject: *Progress Report*

Dear Elaine:

After our last meeting, my first action was to seek input on material selection from our team's stress analysis and materials expert, Van Nguyen.[1] Van recommended 316L stainless steel for the valve body and poppet, and C95410 bronze alloy for the bellows.[2] He verified these choices by plugging specifications for 316L and C95410 into a SolidWorks 3D virtual model of my design, and the analysis indicates that these materials satisfy strength and fatigue requirements with an adequate safety margin.[3]

We originally talked about filling the bellows with liquid benzene, which has a 0.00125 (1/°C) volumetric coefficient of expansion.[4] I further evaluated the suitability of this choice using MATLAB; unfortunately, it did not fare well.[5] My analysis indicated that, for our specified temperature change of +20 °C, liquid benzene expanded and pushed the poppet 1.0 ± 0.05 mm.[6] We are targeting a delta-L stroke length of 1.5 ± 0.05 mm.[7]

I did further research on potential fill fluids and determined that we can double the stroke length and achieve our target with GenAm synthetic ammonia, which has an expansion coefficient of 0.00247.[8] Synthetic ammonia, unlike natural ammonia, is inert when it wets bronze alloy/steel.[9] *Although not applicable to this project, but certainly of technical interest to me, I also discovered that GenAm is developing a synthetic form of acetone solvent.*[10] I have attached PDF copies of principal calculations, Van's simulation results, and my MATLAB printouts.[11]

In conclusion, I have completed the next step of the bellows-actuated valve project and recommend that we begin building a prototype with a 316L stainless steel body and poppet, C95410 bronze alloy bellows, and synthetic ammonia as a working fill fluid inside the bellows.[12]

Sincerely,
Pete

The above email text aligns with the top half of Fig. 16.1. It tells the story of how engineer Pete Knox responded to several tasks assigned by his manager. An illustration of the text's 12 sentences in the form of a sentence stream appears below. The text from sentence one (S^1) through sentence eleven (S^{11}) explicitly documents the steps Pete has taken to achieve a final set of results—what his manager, Elaine, would consider the bottom line. The last sentence, S^{12}, articulates the results, thus ending the story of Pete's latest chapter of project work.

$$S^1—S^2—S^3—S^4—S^5—S^6—S^7—S^8—S^9—S^{10}—S^{11}—S^{12}$$

Sentence stream: storyline email text (12 sentences, chronological order)

Though our engineering writer has drafted a text block that could function as the body of a status report email as is, the text is not optimal because it remains writer- rather than reader-focused. Elaine would prefer a "flipped" version that frontloads the final results (S^{12}) and places pertinent details second.

Pete continues by using the lower half of Fig. 16.1 algorithm to guide further revision (see revised draft below). First, he frontloads the bottom line by moving sentence twelve (S^{12}) to the beginning of the text. Next, he further studies the text and decides that ten of the original body-text sentences (S^{1} through S^{9}, as well as S^{11}) capture pertinent details. However, Pete concludes that the details captured in sentence ten (S^{10}), though pertinent to him, are not likely to be of interest to his manager Elaine. Thus, Pete eliminates the extraneous sentence, which is highlighted in italics in the original version, along with a couple of introductory phrases, also in italics in the original draft. These serve no useful purpose in the revised draft. Finally, you will also notice in the revised version that the subject line (italicized in the orginal) has expanded from a vague placeholder to something specific and information rich.

Bottom-Line-First Draft of Weekly Status Report Email to Elaine Mayfield

Subject: Recommended Housing/Poppet Material and Bellow's Fill Fluid for Control Valve Prototype

Dear Elaine:

I have completed the next step of the bellows-actuated valve project and recommend that we begin building a prototype with a 316L stainless steel body and poppet, C95410 bronze alloy bellows, and synthetic ammonia as a working fill fluid inside the bellows.[12]

My first action was to seek input on material selection from our team's stress analysis and materials expert, Van Nguyen.[1] Van recommended 316L stainless steel for the valve body and poppet and C95410 bronze alloy for the bellows.[2] He verified these choices by plugging specifications for 316L and C95410 into a SolidWorks 3D virtual model of my design, and the analysis indicates that these materials satisfy strength and fatigue requirements with an adequate safety margin.[3]

We originally talked about filling the bellows with liquid benzene, which has a 0.00125 (1/°C) volumetric coefficient of expansion.[4] I further evaluated the suitability of this choice using MATLAB; unfortunately, it did not fare well.[5] My analysis indicated that, for our specified temperature change of +20 °C, liquid benzene expanded and pushed the poppet 1.0 ± 0.05 mm.[6] We are targeting a delta-L stroke length of 1.5 ± 0.05 mm.[7]

I did further research on potential fill fluids and determined that we can double the stroke length and achieve our target with GenAm synthetic ammonia, which has an expansion coefficient of 0.00247.[8] Synthetic ammonia, unlike natural ammonia, is inert when it wets bronze alloy/steel.[9] I have attached PDF copies of principal calculations, Van's simulation results, and my MATLAB printouts.[11]

Sincerely,
Pete

Pete's revised draft aligns well with the bottom half of Fig. 16.1. Below is a sentence stream illustrating the final arrangement of sentences. In this case, after the bottom line, the pertinent details that support it sequence satisfactorily in their original order. However, remain aware that for other status reports, it is okay to arrange the support details nonlinearly—if you choose to, and, more importantly, if you believe doing so will benefit your reader. Notice also that the chain ends with sentence eleven (S^{11}). Pete could've written a new ending sentence such as, "I look forward to our next meeting and further discussing the next step of the project." Instead, Pete has decided the original S^{11} sentence adequately closes the email's message. He opts for concision.

$$S^{12}—S^{1}—S^{2}—S^{3}—S^{4}—S^{5}—S^{6}—S^{7}—S^{8}—S^{9}—S^{11}$$

Sentence stream: bottom-line-first email text (11 sentences, adjusted order)

Now that Pete has completed his final draft, all that remains for him to do is to ensure that the message is correctly addressed and that all of the referenced files are attached. Figure 16.3 displays what Pete's effective email looks like when he presses **send**. Figure 16.4 shows its antithesis, an example of an ineffective email—what Pete might have written if he were not a proficient document engineer.

Effective Email Example

from: Pete Knox <pknox@bui.aerospace.com>

to: Elaine Mayfield <emayfield@bui.aerospace.com>

date: *<month/day/year>*

subject: Recommended Housing/Poppet Material and Bellow's Fill Fluid for Prototype Valve

❖ *< calculation file attachment >*
< SolidWorks simulation printout >
< MATLAB data/analysis sheet >

Dear Elaine:

I have completed the next step of the bellows-actuated valve project and recommend that we begin building a prototype with a 316L stainless steel body and poppet, C95410 bronze alloy bellows, and synthetic ammonia as working fill fluid inside the bellows.

My first action was to seek input on material selection from our team's stress analysis and materials expert, Van Nguyen. Van recommended 316L stainless steel for the valve body and poppet, and C95410 bronze alloy for the bellows. He verified these choices by plugging specifications for 316L and C95410 into a SolidWorks 3D virtual model of my design, and the analysis of these build materials indicates that they satisfy strength and fatigue requirements with an adequate safety margin.

We originally talked about filling the bellows with liquid benzene, which has a 0.00125 (1/C) volumetric coefficient of expansion. I further evaluated the suitability of this choice using MATLAB; unfortunately, it did not fare well. My analysis indicated that for our specified temperature change of +20 Celsius, liquid benzene expanded and pushed the poppet 1.0 +/- 0.05 mm We are targeting a delta-L stroke length of 1.5 +/- 0.05 mm

I did further research on potential fill fluids, and determined we can double stroke length and achieve our target with GenAm synthetic ammonia, which has an expansion coefficient of 0.00247. Synthetic ammonia, unlike natural ammonia, is inert when it wets bronze alloy/steel. I have attached pdf copies of principal calculations, Van's simulation results, and my MATLAB printouts.

Sincerely,
Pete

o *The report is formatted as a standard email with a descriptive title for the subject line.*

o *The email begins with a bottom-line lead-in; thus, a busy manager could stop reading here and still know the overall status.*

o *The remainder is chunked into four short paragraphs made of clear, descriptive sentences.*

o *The text provides details (evidence) to back up the bottom-line (claim).*

o *The email closes effectively and succinctly.*

Fig. 16.3 Effective status report email

Ineffective Email Example

from: Pete Knox <pknox@bui.aerospace.com>

to: Elaine Mayfield <emayfield@bui.aerospace.com>

date: *<month/day/year>*

subject: Materials, Etc.

❖ *< empty: no attachment(s) >*

Elaine,

Another week of arduous work has passed for me as I move along my assigned project. Like you, I also had concerns about the material selection for the valve body and poppet, so I approached our expert in stress and materials issues, Van Nguyen. Though not identical, Van has some related experience with our materials. Nguyen recommended a stainless steel for the valve body as well as for the poppet. He verified this choice by plugging the new specs into our SolidWorks virtual model for the valve, and the material passed. I also evaluated the suitability of our initial choice of benzene for the fill liquid for the bellows. I did this myself, applying my own expertise with MATLAB. Since you were the one who suggested benzene in the first place, I am sorry to say it did not fare well regarding poppet stroke length. As a fill fluid, it was indicated by my analysis that liquid benzene only pushes the poppet 1.0 +/- 0.05 mm for our desired delta temperature change. So I had to do further research on materials. It turns out we can double the stroke of the poppet by using a synthetic ammonia produced by GenAm, which has an expansion coefficient of 0.00247 rather than 0.00125. Another superior capability of synthetic ammonia is that it is inert and therefore is not corrosive when mixed with bronze and steel materials.

I can supply you with several pages of calculations related to these finds, and the Solidworks and MATLAB printouts, provided that you are inclined to ask for these. In my opinion, we are now very ready to build a prototype and my project status is excellent.

It's all good, boss! //Pete

- *The report is formatted as a standard email (good); yet the title is a vague label (not good).*
- *The email begins by telling the linear story of last week's work, but it does not yet reveal any final outcomes/ results.*
- *The text is chunked into multiple paragraphs—but the first is too long and needs further chunking.*
- *Some sentences contain negative phrasings (which are unnecessary).*
- *The email closes with a glib, unprofessional last line.*

Fig. 16.4 Ineffective status report email

Document Geometry: How to Format an Email

In today's workplace, email is a prolific and highly effective communication medium for both *internal* and *external* short-form message documents. Regarding typography and layout, email is a fusion of two predecessor formats—the *hardcopy memo*, which used to be industry's go-to, internal short-form document prior to the digital age, and the *hardcopy business letter*, which used to be industry's go-to, external short-form document.

Like both a traditional memo and a business letter, an email begins with a text zone known as the *header*, which logs traveling (transmission) information for the document. Specifically, the email header is where the sender/writer types in the email address of the target receiver/reader (**To:**) as well as a descriptive title (**Subject:**) summarizing what the email is about. The writer's email software automatically records the source (**From:**) and adds a timestamp (**Date:**) indicating both the clock time and calendar date of transmission. See Fig. 16.5 below for a plan-view illustration of the email's typographical layout with key features labeled in the header and elsewhere.

In the header, the writer has the option to type in a list (**Cc**: *visible* to the recipient, or **Bcc:** *invisible* to the recipient) of secondary readers who will receive a copy of the email that is identical to the original. The writer can also attach one or more electronic files—including word-processed documents, PDF files, images,

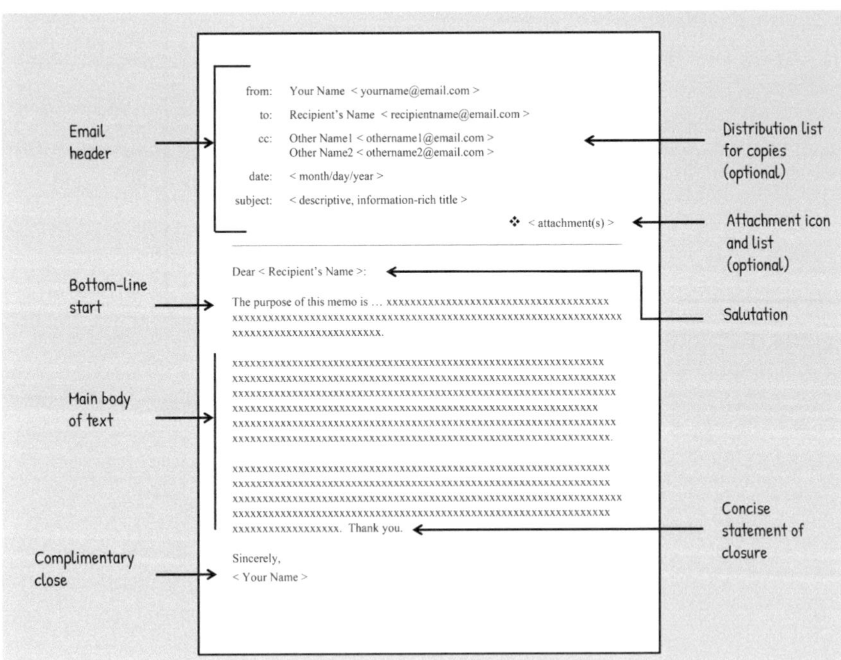

Fig. 16.5 Plan view of an email

spreadsheets, and video clips. Oftentimes, the icon for electronic file attachments is a *paperclip* or the email program's icon for a specific type of file. (In this book it's ❖.) Unlike memos and business letters, with an email, there is no signing or initialing of the original hardcopy. Authenticity of an email is established by its origin: the author/writer's email account and IP address.

As shown in Fig. 16.5, underneath the header is the email's body text. *Unlike* a memo and *like* a traditional business letter, the email's body text leads off with a salutation (Dear So-and-So:) and ends with a complimentary close (Sincerely/ Writer's Name), typed as two successive lines without a signature in between. The rest of the email's body text is typed in single-spaced blocks, flush left, ragged right, and with no tab at the beginning of paragraphs. The single-spaced text is double-spaced between paragraphs. Like almost all short-form documents, the text should incorporate a "chunked," quick-scan layout to enable quick, easy reading. Finally, as for the body text's length, *the shorter it is, the better*. Ideally, the reader's eyes will see the whole email as one static screenshot, which the reader can view, top to bottom, with no on-screen scrolling.

Document Engineering Tip

Unless corporate culture dictates otherwise, when deciding upon the level of formality for an email's salutation, err on the side of formality unless otherwise cued. For example, if you are contacting a manager with whom you have had no prior history or rapport and that person is listed as "Carla Simons, Vice President of Engineering," lead off with "Dear Ms. Simons:" rather than "Carla." The same is true for an email's complimentary close. Until protocol relaxes after a series of exchanges, use the standardized tag "Sincerely," instead of nothing or something less formal like "Best" or "Kind regards."

Action Items

α	While preparing a professional email, have you ever accidentally pressed **send** before you intended and then had that email go out to its recipient(s) unfinished or with glitches you meant to fix? Have you ever replied too quickly or rashly to a "bad news" or accusatory email because of your own emotionally charged urgency to issue a retaliatory or defensive response? In either of these two cases, and others like them, the email writer usually, in retrospect, wishes that he or she had delayed sending the email and that he or she had put more thought and time into the text to make its message effective and professional in its quality and tone.
α	The next time you have to write an email that invites restraint and diligence, before pressing **send**, try drafting the text of that email as a separate document—revising and editing the text there and then only loading it into your email text entry zone when you are 100% certain the message is best prepared for the occasion. Then, press **send**. Or, alternately, compose the draft email in the email program's text entry box, but, in this case, keep the recipient's name deleted until you are ready to press **send**. For reply emails to an individual, another caution to heed would be to check whether you are *replying to all* (all the parties on the email's distribution list) rather *replying solely to the writer/sender* of the message.

Recap

Key Learnings/Takeaways

- When it comes to short-form status reports, particularly emails, most managers prefer that the email writer place results first, details second. Managers are busy and are not so much interested in the beginning and middle of a workplace story that ultimately leads to a set of final results/answers/outcomes. Managers are interested in the story's ending, the bottom line.
- The document algorithm for status reports is built around *inversion* and has a two-phase structure:
  - *The first phase* directs the status report writer to write a *linear draft* that describes the work done, step-by-step, to complete an assigned task(s), and then finishes with a concluding statement that summarizes resultant progress made (interim results, findings, and so on). This first draft incorporates a chronological structure that tells a story—beginning, middle, and end.
  - *The second stage* guides the writer to take the content produced in the first draft and then reorganize ("flip") it so that the concluding statement moves from the end to the beginning and states the bottom line. Thereafter, the writer uses the remaining content as a source of supporting details that back up the leading "bottom line." This second draft ends with a brief statement of closure.
- The proper typographical conventions for formatting a short-form status report as an email are similar to those for formatting a memo. Both should feature a document geometry that places text on the page in quick-to-read chunks and uses shapes, symmetry, and lines that are easy on the eyes and easy for the reader to make sense of.

Chapter 17
Project Reports

Learning Objectives

- *Know the document algorithm for evaluative reports.*
- *Frontload an evaluative report with a statement of the bottom line, and then introduce critical components for argument setup.*
- *Strategically design and tactically advance a criterion-referenced, data-driven argument.*
- *Know and apply the "Rule of Three."*
- *Apply proper document geometry conventions when inserting graphics (such as tables and figures) into an engineering report.*

Project reports accomplish a variety of functions, including recording a project's purpose, scope, and metrics; showcasing a project's findings and results; and answering a project's objective with conclusions and recommendations. Project reports also take on the critical task of officially documenting the work done to complete an engineering activity from beginning to end. Rather than being a front-end document like a proposal or an interim document like a status report, this type of report is known as a back-end document because the engineering writer composes it after project completion and develops it from a retrospective point of view.

As with most engineering documents, it is important to remember that a project report's content must take aim at and be strategically tailored for a specific audience, purpose, and situation. And just like its front-end counterpart, the project proposal, a project report must develop a winning argument to persuade an engineering manager. However, in this case, rather than championing a proposed engineering activity, a project report must conclusively validate an already-achieved result.

B. Henderson, *A Math-Based Writing System for Engineers*,
https://doi.org/10.1007/978-3-030-10756-7_17

What Is the Algorithm for a Project Report and How Does a Writer Apply It?

Engineering operationalizes theoretical ideas into new products and processes. Engineering also optimizes existing products and processes by developing ways to improve their performance, cost-effectiveness, ease of operation, and so on. Because the final report on the development of a new product is usually a long-form document, this chapter focuses on writing short-form reports for engineering projects that evaluate solution alternatives for *improving technical products or processes*.

Most engineers who work in industry will, at some point, be called upon to write this type of evaluative report (or something similar). The evaluative report's basic structure is also an excellent form to learn and apply to reports for front-end design and technology research projects, feasibility studies, and for procurement/purchasing of best-choice equipment or components. These and numerous other engineering projects call upon the engineer to consider options; evaluate those options using objective, data-driven decision-making; and then recommend a best solution—a winner.

Figure 17.1, along with Figs. 17.2 and 17.3, displays this book's third essential document algorithm. Figure 17.1 illustrates the main algorithm's six operational blocks:

1. Introduction
2. Alternative Solutions
3. Evaluation Procedures
4. Results
5. Analysis
6. Conclusions and Recommendations.

Two dotted-line rectangles on the left side of the principal flowchart indicate links to two subroutines displayed in Figs. 17.2 and 17.3, each offering a further layer of detail for guiding the effective writing of an evaluative report's Results and Analysis sections.

You might also note that the main algorithm (Fig. 17.1) displays variables in counts of three: for example, three decision-making criteria (C1, C2, C3) and three viable alternatives (A1, A2, A3) because Fig. 17.1 illustrates a 3×3 evaluation—three Cs and three As. An acceptable variation upon this algorithm could be four criteria and four alternatives, three criteria and five alternatives, or even four criteria and six alternatives and any number of combinations. The Fig. 17.1 algorithm showcases a 3×3 scenario because (as we will see) evaluative arguments often develop best when considering three or more alternatives and, if possible, three or more decision-making criteria as well.

Here are further guidelines for engineering writers to use when applying this book's project report algorithm.

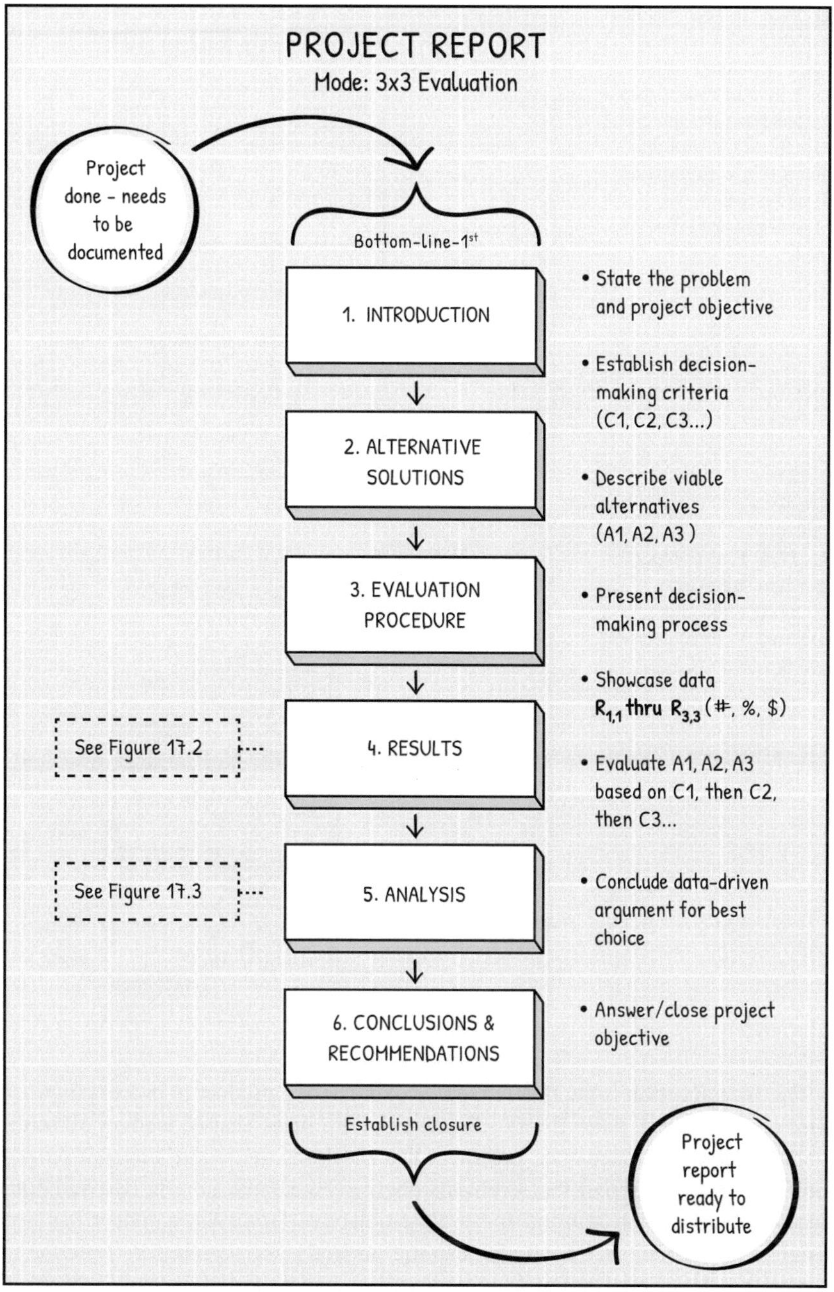

Fig. 17.1 Document algorithm for an evaluative project report, 3 × 3 scenario

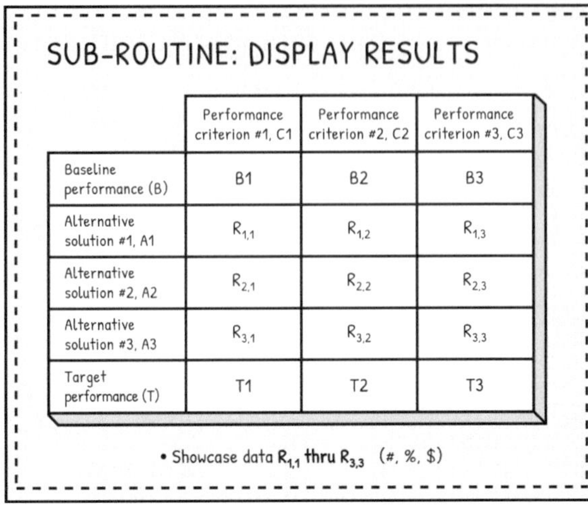

Fig. 17.2 Project report algorithm subroutine, a matrix populated with a 3 × 3 field of results data (R), generated by three alternatives (A) and three criteria (C), with baseline (B) and target (T) performance references noted

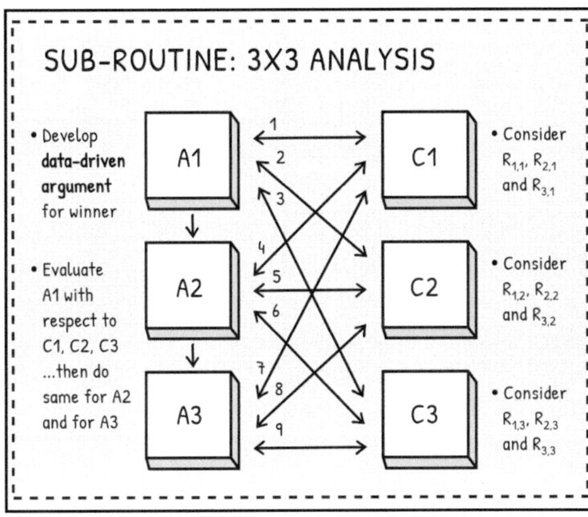

Fig. 17.3 Project report algorithm subroutine, evaluating three alternatives (A) based on three decision-making criteria (C)

The **Introduction** is the first functional element of a product or process improvement project report. After frontloading the bottom line—that is, leading off with a concise, one- to two-sentence statement that discloses the report's overall purpose and final result—a report that evaluates alternatives should present two additional pieces of information upfront: the project objective (*what* the engineer has been assigned to improve and *how*) and also the project's *decision-making criteria* (the metrics that define how and by what measures and means the engineer will determine a best solution—or pick a winner).

Almost never is an engineer given a project that's completely open-ended—no budget, no time frame, and no end-goal requirements. Real-world engineering projects are governed by concrete constraints—such as costs ($), project timelines (t), and target improvement specifications (#, %). In a report, these constraints should be recorded upfront, sooner rather than later. They define the "rules of the game" or judgment parameters that the engineer will use later to select a best option among alternatives. When a writer delays presenting key decision-making criteria (or neglects to present them entirely), a reader may become frustrated or confused. When decision-making criteria appear suddenly "out of the blue" during the decision-making process, from a reader's perspective, they can seem suspect or even contrived after the fact, to best fit the writer's opinions.

A report introduction can also contain a discussion of pertinent background information as well as a statement of project significance and projected impact. In a short-form report, add these if or when applicable. Always, however, include the project's bottom line, objective, and decision-making criteria in the introduction.

The second functional element is a section that presents **Alternative Solutions**. Real-world problems rarely have one tidy textbook answer. Engineering report readers know this. Therefore, in an evaluative report, presenting readers with a group of three or more viable choices allows the readers to feel that the writer is presenting a complete, reasonable set of choices and is engaging in fair play.

People have a tendency to process a group of one or two things differently from the way they process groups of three or more things. When given just one or two viable alternatives to choose from in a report, the reader is likely to feel a sense of incompleteness—an impulse to think things such as "why didn't you consider this?" or "how come you didn't include that?" Thus, presenting three or more viable improvement options upfront is an effective writer tactic to thwart readers from second guessing the writer's initial choices and assumptions. When presenting a field of options, I advise you to apply this Rule of Three.

The third functional element is the **Evaluation Procedure**. Just as the Introduction and Alternative Solutions sections set up the argument for a winner later on, this section further advances argument setup by describing how you, the writer/engineer, gathered or produced the performance data that you will display in the Results section. In a scientific report, this would be called the methodology section. In your project, if you collected performance data for alternatives through research or benchmarking, you document that process here. If you built and tested prototypes, you document that process here. If you used math modeling and simulations to generate results data, you document that process here as well.

The fourth functional element is the report's **Results** section. Here you showcase the data you will soon analyze to determine which of your improvement alternatives is best. In an engineering report, as a general rule of thumb, you choose the Results section's graphic devices (tables, graphs, or illustrations) on a case-by-case basis. The best graphic is the one that best displays the highs, lows, and trends of a given data set. The final segment in this chapter reviews some best practices for the typography or document geometry of graphic devices that showcase data sets.

In engineering reports aimed at math-based thinkers, results-driven arguments advance best when they're built around objective, quantitative measurements, rather than subjective, qualitative terms. For instance, test data expressed as numbers (#), percentages (%), and costs ($) are typically easier to discuss and analyze conclusively than word data (stories).

Consider the following outcome from a product trial: "We observed that a majority of subjects in our user test group stated that they liked how the improved widget performed as compared to the original model." Most engineers would probably prefer next version, which references a less vague, more rigorous set of findings: "We recorded that 83% of the subjects in our user test group scored functionality of the improved widget as 4 or higher on a 5-point scale, whereas only 58% of the test subjects rated the original (baseline) model to be 4 or 5."

Document Engineering Tip

Your project may call upon you to collect "soft data," that is, data that involves human perceptions or feelings, as is often the case in consumer/user testing of a product or process. For example, subject A said, "I think the device is very easy to use;" subject B said, "I, personally, feel the device is confusing to operate;" subject C said, "It was easy to turn on but I didn't like how heavy it was;" and so on. In cases like these, instead of collecting words, objectify/quantify the data by using a testing process that generates a scaled response. For example, rate a consumer preference for a feature using a 1–5 scale, where 1 is "strongly dislike," 2 is "dislike," 3 is "undecided," 4 is "like," and 5 is "strongly like."

To exemplify generic best practices, our algorithm references an idealized decision matrix (see Fig. 17.2) for the Results subroutine operation. Although not every project yields results that best display on a graphic device like this, the matrix represents a useful array of design features that can (and should, whenever possible) be incorporated into the graphic display of results—figure, illustration, or table.

Specifically, Fig. 17.2 shows a Results subroutine for a generic 3 × 3 decision matrix. The matrix catalogues data in three rows, A1 thru A3, with three columns for three decision-making criteria, C1 through C3. The criterion-referenced data ($R_{x,y}$) populates nine (3 × 3) cells, $R_{1,1}$ through $R_{3,3}$. A first row inserted above rows A1 through A3 records a log of baseline (B) or initial conditions—that is, how the product or process performs relative to performance criteria *before* improvement efforts.

Figure 17.2 also shows a fifth row inserted immediately below A1 through A3. This last row displays a log of objective, quantitative targets (project improvement specifications) associated with the project's performance improvement criteria (the same three Cs stated upfront in the report's introduction). Evaluating the performance of a product or process relative to a discrete set of baseline conditions and performance targets is typical engineering business. Unless the engineer can show appreciable and significant change in new data relative to the baseline or targets, the project results are usually of little or no use.

When applicable, I recommend logging (or noting) baseline conditions and targets in a report's Results section because baseline and target numbers enable a results graphic (table or figure) to communicate in an autonomous, self-contained fashion, without requiring readers to remember or look elsewhere (i.e., flip pages) for reference points for comparison and contrast.

The fifth functional element is the **Analysis** section. Figure 17.3 shows an analysis process subroutine for making sense of the Results data presented in Fig. 17.2. Since the data is a 3 × 3 set, the analysis process goes through a 9-step subroutine that methodically considers each of the three improvement alternatives (three As), one at a time, with respect to each of the three performance criteria (three Cs), and analyzes each corresponding subset of results (three Rs for each A). This step-by-step process reveals how each improvement alternative performs relative to the performance parameters (targets) stated upfront (in the report's Introduction).

To begin analysis, as shown by the first three numbered arrows on Fig. 17.3, you match alternative one (A1) with C1, C2, and C3 and consider (and discuss in your report's Analysis segment) the corresponding subset of results values, $R_{1,1}$, $R_{1,2}$, and $R_{1,3}$, and how they compare with your performance standards, both generally and as indexed to specific baseline (B) and target (T) numbers. Next, you step through the routine for alternative two (A2), considering all three criteria again and the corresponding subset of values, $R_{2,1}$, $R_{2,2}$, and $R_{2,3}$. Finally, you follow the routine as it moves through analysis for the results subset associated alternative three (A3)—$R_{3,1}$, $R_{3,2}$, and $R_{3,3}$.

As you cycle through the subroutine—examining the results for A1, A2, and A3, you will also be able to observe variances between the alternatives relative to individual criteria—that is, variances in a progression of results data that display on the Fig. 17.2 results matrix as a column rather than row—for example, $R_{1,1}$, $R_{2,1}$, and $R_{3,1}$. A key component of engineering data analysis, these criterion-referenced variances distinguish the significant highs, lows, and trends in specific data sets.

Figure 17.3 is a simplified block diagram. It does not explicitly illustrate all of the compare and contrast maneuvers described above. The figure does, however, illustrate a useful, rudimentary scaffolding for you, the engineering writer, to use when you are composing a report's Analysis section. If used as a conceptual model, and not as an explicit recipe, Fig. 17.3 (and its accompanying instructions) can guide you in developing effective (and persuasive) criterion-referenced, data-driven arguments for a variety of evaluation scenarios.

The final functional element, **Conclusions and Recommendations**, moves from analysis to judgment. The writer sums up the analysis and establishes which alternative shows the best performance numbers overall. Then, you complete your data-driven argument by recommending a winner and ending the report with a brief statement of closure.

To review, if at the beginning of a project report, you, the engineering writer, clearly and convincingly present the project's purpose and scope, and pertinent background, and establish an explicit project objective and set of decision-making criteria (the "rules of the game"), the report launches with a solid foundation for consensus. Immediately after the introduction, if you spell out a set of viable alternatives and then an evaluation procedure, what crystallizes is a contract between writer and reader.

From that point forward, if you stay on course and follow this contract and, in the end, if you satisfy the original project objective and criteria and recommend a solution based on terms and conditions already established up front, your report's conclusions and recommendations are not likely to be refuted. Data-driven arguments, when done well, are difficult to contest and are well received by engineers who like arguments that are supported by numbers and facts. Now, let's move on to this chapter's application example.

Application Example

Rock Logic Drum Company specializes in designing and manufacturing studio-grade percussion instruments, mainly drum sets. The company's product line is reputable and successful in the national and international marketplace. At present, the company's engineering team is developing a new five-piece drum set that will carry a signature endorsement from Rock & Roll Hall of Fame drummer, Jack Norem (*Bad Boys N' Roses* and *Velvet Gun*). The company anticipates first-year sales of 170–500 kits at $2.4K retail pricing per kit (production cost = $800 per kit, with a 3× markup).

Thus far, some of the engineering challenges presented by the Norem Signature Kit project have included, first, having to develop a new fast-track shell fabrication process to decrease manufacturing time and, next, having to scrap and remanufacture 1000 out-of-specification castings of tension-rod anchor mounts. The third and latest challenge now faced by Rock Logic's engineering design team is diagnosing and fixing an unsatisfactory acoustic response for the drum kit's smallest tom-tom shell (13″ wide × 13″ deep). The 13 × 13 drum is a ride tom mounted on the bass drum. The two other toms, 16 × 16 and 18 × 16, are floor toms.

The development engineer assigned to tackle the tom-tom acoustic response problem is Jeff Lee. His engineering manager, Linda Thomas, has asked Jeff to evaluate an optimal drum shell material (in this case, type of wood) to fine-tune the shell's frequency response, so it best complements the frequency of resonating drum heads. Linda has given Jeff a specific set of project parameters. She wants him to report back results in 3 weeks.

The Resultant Project Report

Given the above (hypothetical) project scenario, we will time travel forward and assume that engineer Jeff Lee has applied best practices, met his deadline, and arrived at a suitable result. To close the loop on the project, Jeff now must submit a short project report to his boss, Linda. Below is an example of what, in this situation, an effective report might look like.

Formatted as a memo, the *effective (good) example* (see Figs. 17.4a, 17.4b and 17.4c) shows the whole text of Jeff's final report, with annotated remarks on the right side of the document. In this effective example, you will see that Jeff has built his document around this chapter's featured algorithm. He leads off with a bottom-line answer to his boss's request and then immediately establishes a project objective and set of decision-making criteria. Thereafter, he methodically develops a data-driven argument for why his recommended solution is credible and worthy. Last, he closes his document with a single-sentence ending.

After the effective example, this segment includes an *ineffective (bad) example* (see Figs. 17.5a and 17.5b). This example displays a mock-up version of what Jeff's project report might have looked like if he had written it using an ineffective selection of content and structure. In this second example, the report writer focuses mostly on saying what he feels like saying about the project—what's significant to him. He is self-focused rather than audience-focused.

He also doesn't frontload the document with a bottom-line-first statement. In fact, he doesn't mention the project objective until the very end of the report. The bad example fails to incorporate the recommended algorithm and its subroutines to set up and advance a data-driven argument. Even though the second example is built out of sentences that are grammatically correct, overall it is not well engineered and most likely would *not* be well received and considered "winning" by an engineering manager.

Effective Report Example

TO: Linda Thomas,
Senior Engineer

DATE: *<month-day-year>*

RE: Recommend Maple Wood for Norem 13x13 Tom Shells

FROM: Jeff Lee, JL
Design Engineer

Introduction

As requested, I investigated three alternative wood types—maple, birch, and California oak—to enhance the acoustic profile of our new Norem Series 13-inch diameter by 13-inch deep (13x13) ride tom-tom. I recommend we replace the existing mahogany shell material with maple.

When I began work on the project, we agreed on the following targets and constraints:

- Increase the tom-tom's resonant volume and richness when the drum is played with top and bottom drum heads tensioned equally to output a 131 Hz (3rd octave C) percussive note.
- Consider wood type to be the design variable, with viable alternatives being maple, birch, and California oak.
- Control cost impact.
- Do prototype testing using our R&D Lab's acoustic response device, with C tuning and standard strike-test force.

Our tom-tom shell design features glued and pressure-treated 6-ply construction. See Figures 1a & 1b below. In the event that a new material's response might correlate differently regarding shell thickness and dynamic output, I decided to collect prototype data for two other common laminations, 4-ply and 8-ply, as well as for the existing 6-ply shell.

Figure 1a & 1b: Norem Series 13x13 Tom & 6-ply Tom Shell (Cutaway View)

I also consulted Patrick McElroy, lead manufacturing engineer assigned to the Norem project, to get his input and determine whether developing and building 4-, 6-, and 8-ply prototypes, as well as mahogany

Page 1 of 3

- *Begins page with a standard memo header.*
- *Commands the reader's attention with a descriptive, results-centered subject line.*
- *Features a bottom-line lead sentence, which states both the project objective (problem) and overall engineering response (solution).*
- *Establishes the decision-making criteria and decision-making process (rules of the game).*
- *Establishes a set of viable alternatives.*
- *Inserts a graphic properly, between split vertical text and with a complete label.*
- *Provides clarifying details that support the solution alternatives.*

Fig. 17.4a Effective project report, page one

baseline shells, would be feasible. Given the predicted high-profile and high-volume sales of the end product Norem Series drum set, Patrick concurred that additional data would be valuable and that his development schedule and budget could cover expanding the testing protocol.

Evaluation Procedure

Over a period of two weeks, we built and tested prototypes. Patrick's team used the fast-track shell fabrication process, also currently in the test phase. Quality Control verified the construction integrity of the prototypes, and, in fact, commented that construction quality was "uniformly excellent." I obtained cost estimates from our Manufacturing Supplies and Logistics Department for all versions of our test prototypes.

Patrick's team and I conducted the testing. For each of the twelve tom-tom prototypes (the new woods and the baseline wood), we followed a similar procedure. We installed a new pair of clear, 9-mm-thick Aqueous Rock Batter drum heads. We adjusted the tension rods on the top and bottom drum heads, using equal top and bottom increments, until we achieved a 131 +/- 0.1 Hz (or 3rd octave C) percussive note, measured with a hand-held digital tuner. We mounted the tom-tom on the R&D Lab's acoustic response tester and subjected the top head to a center-aimed, standard robotic stick strike, calibrated to a 2.3 +/- 0.01 pound impact. We took a dB reading at a distance of 1 +/- 0.01 meters directly perpendicular from the side edge of the top drum head.

To determine "richness," we used a qualitative process. We asked Mr. Norem to lend us his drum technician, Vinnie Boz. During the strike testing, we invited Mr. Boz to rate response "richness" for each prototype on a 1 to 5 scale, with 1 being least and 5 being most. The raw data indicated that the 6-ply construction was optimal for all three materials. Therefore, we used data for the 6-ply tom-toms for decision-making. Table 1 (below) showcases the key test data.

Results

Table 1: Results from 6-ply Shell Prototype Testing

	Volume @ 1 meter dB | Δv	Cost per unit $ | Δc	Richness, 1-5 (least to most)
Prototype 1, Birch Wood	100.12 | +0.97	$4.23 | -0.07	5
Prototype 2, Maple Wood	100.83 | +1.68	$4.37 | +0.07	5
Prototype 3, California Oak	98.4 | -0.75	$3.88 | -0.42	3
Current, Mohagany (baseline)	99.15	$4.30	4+

Note: Δv = (dB prototype - dB baseline)
Δc = (Cost per unit, prototype - Cost per unit, baseline)

Page 2 of 3

o *Uses frequent "journalistic" paragraphing to make the text a fast read.*

o *Describes key action steps.*

o *Provides additional details that explain the evaluation procedure for the several prototypes.*

o *Properly inserts another graphic between split vertical text and with a complete label.*

o *The graphic presents a complete set of decision-making data, as well as baseline data for easy compare and contrast.*

Fig. 17.4b Effective project report, page two

Analysis

The results indicate that the birch wood shell outperforms the baseline mahogany shell. It is more punchy with nearly a 1 dB (0.97) increase in volume relative to baseline. It costs 7 cents less per unit, and scored a perfect "5" for richness, exceeding baseline richness by almost 1 unit. As the table indicates, Norem's drum technician felt more comfortable assigning the mahogany shell a "4+" in richness, rather than a straight "4." Relative to the type 2 and type 3 prototypes, our type 1 prototype made of birch was even with maple for improvement relative to one criterion, richness, and superior to the California oak for two out of three criteria, volume and richness.

The maple shell also tested impressively overall. Like Birch, it increased volume, in this case to 1.68 dBs, 73% beyond the Birch prototype's amplification. It also scored a perfect "5" for richness. Although it carries a per-unit cost of 7 cents above baseline, this is a small increment.

The test results for the California oak shell were uneven. The cost analysis assigned the oak a per-unit price tag that undercut the cost of both maple and birch shells, as well as the original mahogany by around one half dollar. However, the oak prototype shell displayed decreased punchiness, - 0.75 dB, and a drop to "3" for richness. Birch, maple, and mahogany are drum industry standards. Oak is a new (and we thought promising) material choice. We anticipated better results than achieved. Oak is clearly the most cost-effective material. Yet I cannot recommend a significantly compromised acoustic performance on behalf of thrift.

Conclusions & Recommendations

Both the birch and maple wood prototypes demonstrated to be excellent choices for the new Norem Series 13x13 tom-tom. Scrutinized side-by-side, the trade-off was more punchiness with maple and less cost per unit with birch. I view the cost advantage of birch to be less significant than the resonance advantage of maple. I recommend maple for the small tom on our new high-performance Norem Signature drum kit. I look forward to continuing my work on your design team and am eager to be assigned my next project.

###

Page 3 of 3

- *The text presents a 3 x 3 data-driven argument anchored to already-established criteria.*
- *The argument progresses methodically, covering all nine steps of the algorithm subroutine for analysis.*
- *The text highlights specific results data that complement the final conclusions.*
- *To end, the argument revisits the original project objective and then recommends a winning solution that answers it.*
- *The final sentence establishes closure.*

Fig. 17.4c Effective project report, page three

Ineffective Report Example

TO: Linda Thomas,
Senior Engineer

DATE: *<month-day-year>*

RE: Project Report

FROM: Jeff Lee,
Design Engineer

My initial thinking regarding the Norem Series tom-tom project was originally that a new material's response might correlate differently regarding shell thickness and dynamic output; therefore, I decided to collect prototype data for the two other common laminations, 4-ply and 8-ply, as well as for 6-ply. I talked with Patrick McElroy (the manufacturing engineer) and got his input, which agreed with my position. He would develop and produce 4- and 8-ply prototypes in the three new wood types, as well as for the 6-ply scenario that you supplied me with. We of course produced mahogany shell prototypes to get adequate reference data. It took two weeks for us to build and test the prototypes, employing our fast-track shell fabrication process, in the test phase like our tom-tom project. The construction of the prototypes was verified to be in specification by Quality Control. The inspector said the construction was excellent. Cost estimates for alternative materials as compared to existing were obtained. Testing ensued.

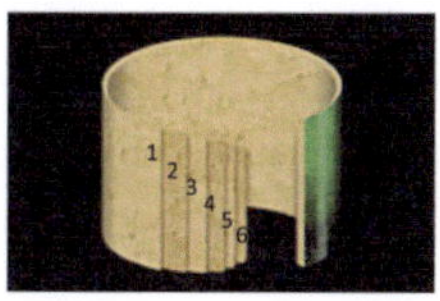

A standardized procedure was followed for all twelve of the tom-tom prototypes (the new wood and mahogany):

- a new pair of clear, 9-mm-thick Aqueous Rock Batter drum heads were installed;

- tension rods on the top and bottom drum heads, using equal top and bottom increments, were adjusted until a 131 +/- 0.1 Hz (or 3rd octave C) percussive note was achieved, and was measured with the hand-held digital tuner;

- the tom-tom was mounted on the R&D Lab's acoustic response tester and the top head was subjected to a center-aimed, standard robotic stick strike, calibrated to a 2.3-pound +/- 0.01 impact; and then finally

- a perpendicular dB reading at a distance of 1 +/- 0.01 meter from the top drum head was taken.

We obtained a "richness" rating with a more qualitative process. I asked Mr. Norem to lend us his drum technician, Vinnie Boz, and Mr. Boz said yes. So we brought Vinnie in as the expert in drum

Page 1 of 2

o *Begins page with a standard memo header.*

o *Yet labels the memo with a vague, broadly scoped, topic-centered subject line.*

o *Begins the memo's text with a generalized discussion about the report's topic.*

o *Further discusses the story-line of a series of project activities.*

o *Features a large block of single-spaced text, with the text wrapped around an unlabeled graphic.*

o *Implies the project's objective but does not state it explicitly.*

o *Adequately describes the decision-making process (test procedure), but does so in a monotonous passive voice.*

Fig. 17.5a Ineffective project report, page one

sounds and had him judge "richness" for each prototypical tom on a 1 to 5 scale, with 1 being least and 5 being most. See the collected data (below) for verification.

	Volume @ 1 meter dB | Δv	Cost per unit $ | Δc	Richness, 1-5 (least to most)
Prototype 1, Birch Wood	100.12 | +0.97	$4.23 | -0.07	5
Prototype 2, Maple Wood	100.83 | +1.68	$4.37 | +0.07	5
Prototype 3, California Oak	98.4 | -0.75	$3.88 | -0.42	3
Current, Mohagany (baseline)	99.15	$4.30	4+

Note: Δv = (dB prototype - dB baseline)
Δc = (Cost per unit, prototype - Cost per unit, baseline)

Please note: we did not need to look at the 4-ply and 8-ply data to make a decision, as the 6-ply was clearly optimal.

The California Oak was the cheapest as can be seen, but it didn't perform well. The birch wood shell was loud and on the data table one can discern it has positive cost factor. Norem's drum technician gave it a "5" for richness, which is a perfect score. We needed to look to the maple option. It exhibited very good numbers for resonance and richness, and maple does not appreciably increase cost, especially when compared to the birch wood option's opposite 7 cent reduction. All in all, maple wood is clearly the best choice for obvious reasons so that is what I recommend.

Sincerely,

Jeff Lee // JL

Page 2 of 2

- *Uses an unlabeled table to present the decision-making data.*
- *Reveals a complete snapshot of the project's decision-making criteria here in the table (rather than establishing the criteria upfront in the introduction).*
- *Rushes through an abridged, disorganized analysis of the data, without making reference to specific results numbers, except for one "5" and one "7".*
- *Finally, provides what the reader has been waiting for—a recommended solution.*
- *The report's concluding argument uses subjective bravado (rather than objective reasoning) to assert that the best choice is "obvious."*

Fig. 17.5b Ineffective project report, page two

Document Geometry: How to Design and Insert Tables and Figures into Text [1–3]

Aside from precise and accurate text streams (composed of words as well as numbers, equations, and unit-of-measure symbols) and effective and efficient layouts for placing text on the page, engineering documents also regularly include *tables* and *figures*. This segment focuses on straightforward applications and best practices associated with designing and inserting them effectively.

Why do engineers include tables and figures in documents? First, in engineering communication, the old adage holds true—*a picture is worth a thousand words*. Moreover, figures such as illustrations and images are especially good at capturing details that are otherwise difficult, tedious, and less efficient when described in sentences.

Engineers also use graphics to help them describe physical designs (e.g., a mechanical drawing), conceptual systems (e.g., a flowchart), and captured moments in time (e.g., a digital photo of first-, second-, and third-place auto racers crossing a finish line). In addition, and perhaps most important of all, engineering problem-solving and decision-making center around facts and data and *the interpretation of highs, lows, and trends within data fields*. Tables and figures provide effective visual means for displaying highs, lows, and trends and advancing data analysis.

I consider a *table* to be a graphic device used for tabulating and displaying data in rows and columns. A table is particularly useful for showcasing data precisely, since each data point appears individually in a cell, for instance, "4.387 (cell address: row 1, column 2, with amperes as the prescribed unit of measure)." I use the term *figure* to refer to a broad spectrum of graphics that includes both graphs and illustrations. Examples of graphs include line graphs, pie charts, and bar charts. Graphs are two- or sometimes three-dimensional grids, ideal for displaying data comprehensively and continuously so that the data fields reveal important characteristics such as maximum and minimum values, deviation, and slope. Illustrations, on the other hand, are visual aids that display physical structures and design concepts, including sketches, drawings, schematics, scans, and photos.

Universal Best Practices for Inserting Both Tables and Figures

Here are some universal best practices that almost invariably apply when preparing and inserting *both* tables *and* figures into professional-quality engineering documents:

- Avoid *free-floating* tables and figures that are not assigned a clear purpose and context by a statement preceding them within the document's text. Free-floating graphics confuse readers. Thus, make explicit reference ("See Figure *X*" or ". . . as shown next in Table *Y*, the *Z*-test test failed . . .") to any table or figure that you include in your document's text *before* the reader encounters it.

- Although one approach is to locate tables and figures as attachments in a short-form document and in the appendix of a long-form document, as a matter of convenience to readers, I advise engineering writers (unless, of course, directed otherwise by your workplace supervisor) to locate graphics *inside the text body* of a document and *as soon as possible* after the graphic is mentioned. This practice spares readers from having to flip back and forth between the document's text body and its end matter, while the reader is actively trying to understand the relationship between a segment of text and its complementary graphic.
- Center tables and figures in a vertical break (blank paragraph) in the document's text. This technique is a document geometry best practice that applies (and achieves benefit from) the principles of chunking and blank-space optical relief.
- Just like sentences, tables and figures do not need to be ornate and fancy to be effective—simple, clear, and correct is okay, if not preferred. Professional engineering readers do, however, expect tables and figures to be prepared using sharp, neat, and tidy production techniques.
- When the data (facts and numbers) in a graphic or an entire graphic itself (e.g., a copied-and-pasted figure copied from a technical article) originate from a source other than you and your company, *always give credit where credit is due*, and *identify the source*. In industry, citation practices are often more streamlined than those used in academe. One straightforward method for acknowledging facts and numbers you borrow from an outside source is to place an *attribution note directly underneath the table or figure*. This note should consist of the word "source" followed by a brief description of the source's origin or a URL linking to it (or even better, both). For instance, write "Source: National Science Foundation (NSF), Plutonium Research Project, 2016," "Source: https://. . .," etc.
- On both tables and figures, whenever applicable, record data field size(s) using the standard convention "n = [number of data points]," and locate this labeling so it's easily visible to readers. Also, clearly identify names of statistical methods, when applicable—for example, "Note: Column 2 lists p-values for each test run's data set . . . ."
- Finally, it is always best to avoid breaking tables and figures across pages. Thus, if a segment of text pushes a table or figure to spill over from one page to the next, it is better to "short-sheet" the text preceding the graphic—that is, insert a page break before the text reaches the end of the page's text-entry zone and allow a zone of blank space. Then, place the complete table or figure on the next page, and continue the document's text immediately after the graphic.

Figure 17.6 presents a side-by-side example of an effective insertion of a basic graphic (a computer-generated illustration) into a page of document text versus an ineffective placement of the same graphic. The two figures' labeling arrows point out the proper use (and improper nonuse) of several universal best practices mentioned above.

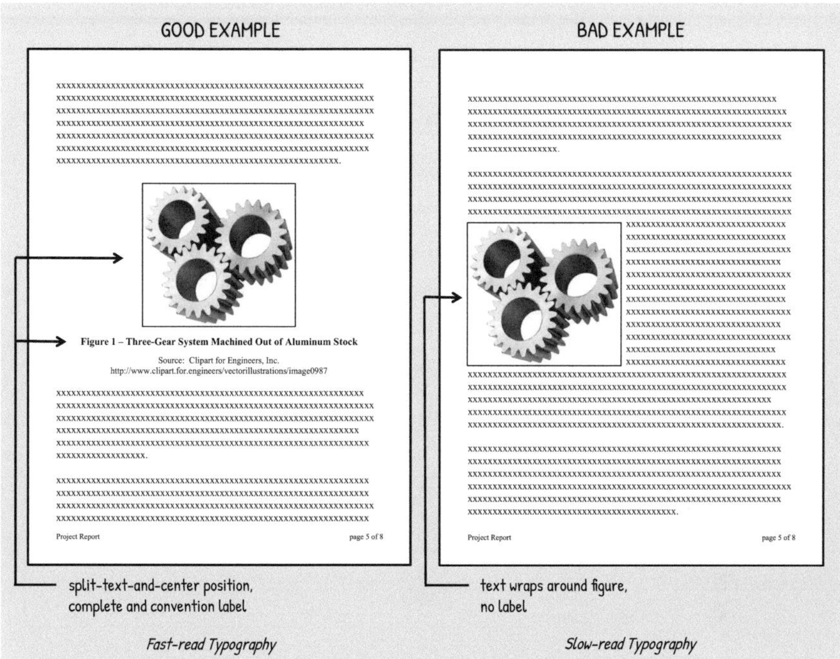

Fig. 17.6 Effective versus ineffective insertion of a basic graphic into a page of document text

Tables

Next, here are some best practices specific to preparing tables for placement in a document. Figure 17.7 illustrates an effective versus ineffective table design.

- Tables should always be labeled as such and assigned a designated number. If there's one table, it's "Table 1," and if more, they are "Table 1," "Table 2," "Table 3," etc.
- Added to the word "table" and its number using a short "n"-type dash ("–"), a table's label should also include a concise description of what the table represents (e.g., "Table 1–Stress and Strain Values for Casting Prototype Tests").
- The standard convention is to place a table's label immediately above it.
- The table's top-side horizontal header defines the type of data listed in each column, as well as (whenever applicable) associated units of measure. If units apply, *do not* forget to include them.
- The table's left-hand-side vertical header names each item to be tabulated, row by row.
- Finally, the standard convention that determines *what goes in rows versus what goes in columns* is as follows: arrange like data in columns, and list individual items being tabulated and the various data collected for each item in rows.

Table #–Characteristics of antibiotic-producing Streptomyces

Organism	Optimal growth temp (deg C) *	Color of mycelium	Antibiotic produced	Yield of antibiotic (mg/ml) *
S. Fluoricolor	-10	Tan	Fluoricillinmycin	4108
S. griseus	24	Gray	Streptomycin	78
S. coelicolor	28	Red	Rholmondelay	2
S. nocolor	92	Purple	Nomycin	0

GOOD EXAMPLE

yes, like data arranged in columns, with units defined*

Table #–Characteristics of antibiotic-producing Streptomyces

Organism	S. Fluoricolor	S. griseus	S. coelicolor	S. nocolor
Optimal growth temp	-10	24	28	92
Color of mycelium	Tan	Gray	Red	Purple
Antibiotic produced	Fluoricillinmycin	Streptomycin	Rholmondelay	Nomycin
Yield of antibiotic	4108	78	2	0

BAD EXAMPLE

no, like data arranged in rows, with no units defined

Fig. 17.7 Effective versus ineffective table design

Figures (Graphs and Illustrations)

Finally, here are some best practices to apply when preparing figures and placing them in a document. Below the bullet points of best practices, Figs. 17.8, 17.9, and 17.10 illustrate effective (left) versus ineffective (right) examples of a line graph, bar chart, and digital image, respectively.

- Figures should always be labeled as such and assigned a designated number. If there's one figure, it's "Figure 1," and if more, they are "Figure 1," "Figure 2," "Figure 3," etc.
- Added to the word "Figure" and its number using a short "n"-type dash ("–"), a figure's label should also include a concise description of what the figure represents (e.g., "Figure 1–Acceleration of Lug-X Projectile Over Time").
- For graphs that present *X*–*Y* or *X*–*Y*–*Z* plots of data, choose axis scales and increment spacing such that the data are presented ethically (highs, lows, and trends are not distorted).
- Also for *X*–*Y* and *X*–*Y*–*Z* plots, construct axis intersections where $X = 0$, $Y = 0$, and $Z = 0$; and use axis break lines ("—//—") in cases where data maps only to the higher values of an axis scale and the lower values, "0" to whatever, are not applicable in the data interpretation.

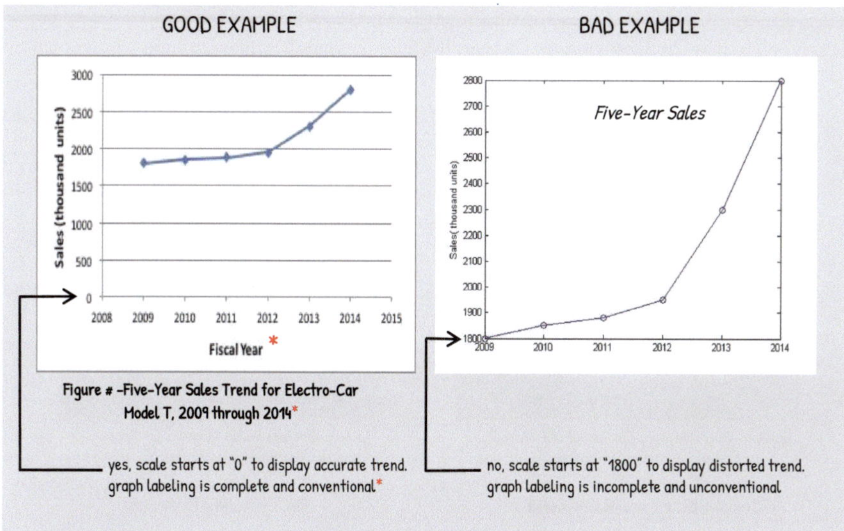

Fig. 17.8 Effective versus ineffective line graph design

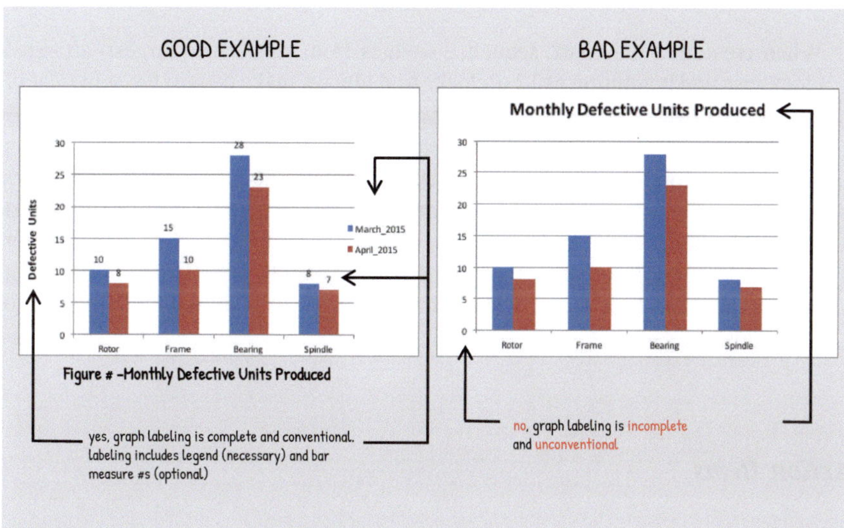

Fig. 17.9 Effective versus ineffective bar chart design

- Whenever applicable, on the axes of figures, be sure to clearly and explicitly assign and label data sets' appropriate units of measure (e.g., meters, kilograms, ohms, etc.).
- When necessary/applicable, provide a key or legend that defines symbols and colors featured on figures.

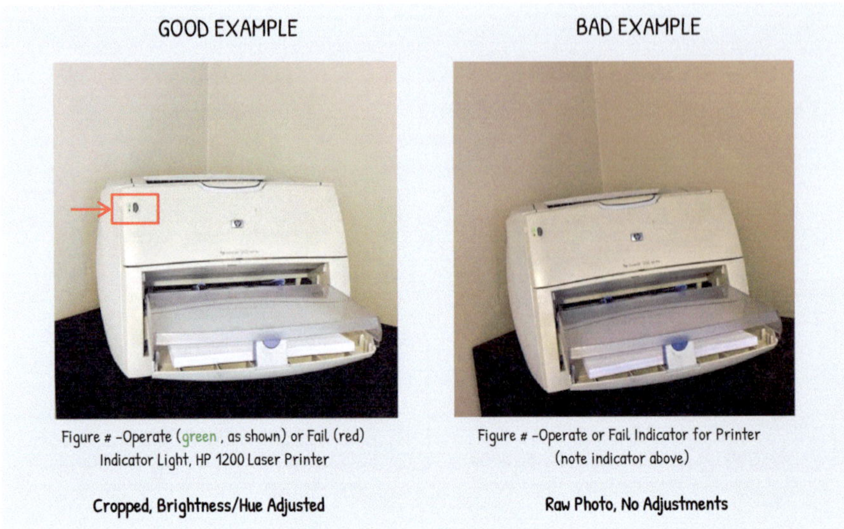

Fig. 17.10 Effective versus ineffective digital photo preparation

- When creating a pie chart, sequence wedges from smallest to largest, arranged clockwise and beginning at 12 o'clock. And always make certain the sum total of wedges is 100%.
- When using digital photos for illustrations, if applicable, use photo editing software to crop the photo so that it clearly displays the most important features and areas of interest. Also (if necessary), use photo editing software to adjust the lightness/darkness and exposure of an image to optimize visibility for readers. Finally (if necessary), add arrows and labeling to an image to point out features and areas of interest that would otherwise be difficult for readers to distinguish.

Action Items

α	To practice evaluative writing, the next time you decide to purchase something for work, home, or school that involves selecting a best choice among a field of options (e.g., choosing which new cell phone to buy), first, make a list of your top three choices (e.g., brand x, brand y, and brand z). Next, consider possible decision-making criteria, and list what you think are the three most important criteria (e.g., cost, battery life, and screen size). Then, in a paragraph or so, write your way through a "3 × 3" decision-making process, wherein you methodically evaluate all three options, one criterion at a time. Finally, after you've completed your criterion-referenced analysis of options, state which option you conclude to be the best overall, and briefly summarize why.

(continued)

α	Consider how managers evaluate their employees' on-the-job performance, how team leaders gauge the performance of individuals on a team, as well as how teachers evaluate the performance of students taking a class. In each of these situations, what decision-making criteria does the decision-maker (manager, team leader, or teacher) apply when doing the evaluations? List what you believe are reasonable judgment criteria for each situation.
α	Also, can you think of a time when you observed a manager, team leader, or teacher do an evaluation of an employee, team member, or student that was subjective and unfair. If yes, do you think that the evaluation might have been more objective and fair if the decision-maker used a predetermined, criterion-referenced method to guide the evaluation process? Could you, perhaps, illustrate that process with a document algorithm? If you drew a rough sketch of the algorithm, what would it look like?

Recap

Key Learnings/Takeaways

- Project reports document the work done by an engineer (or engineering team) to complete an engineering activity, from start to finish. This type of report is considered to be a *back-end document* because the engineer writes it *upon project completion* and develops the report's content from a retrospective point of view.
- Like a proposal, a project report must advance a winning argument, and therefore, organizing the report's parts in the correct order is key to making it persuasive. This chapter discusses project reports that document a product or process improvement project.
- Product or process improvement reports are built around this book's third essential document algorithm. The algorithm is evaluative, and its six main operational blocks are as follows:

  - *Introduction:* after stating the bottom line, the introduction presents the *project objective* (what the engineer is assigned to improve) and also the *project's decision-making criteria* (the metrics for how the engineer will pick a winner). The Introduction can also contain a discussion of pertinent background information and a statement of project significance and impact.
  - *Alternative Solutions:* presents viable options for solving the problem. And because real-world engineering problems rarely have one tidy textbook answer (and because engineering report readers know this fact), the engineer is advised to present a field of three or more viable options. This strategy is called "The Rule of Three."
  - *Evaluation Procedure:* describes how the writer gathered or produced the report's results data (presented in the following section). In a scientific report, this would be called the "Methodology" section.
  - *Results:* showcases the data that will be analyzed to determine which alternative solution is best. For projects that consider three alternatives, A1 through A3, with three decision-making criteria, C1 through C3, a handy vehicle for presenting data is a 3 × 3 decision matrix, in which criterion-referenced data ($R_{x,y}$) populate nine (3 × 3) cells, $R_{1,1}$ through $R_{3,3}$.

  - *Analysis:* methodically works through a subroutine that evaluates the results data (presented in the previous section) for all of the As with respect to all Cs. The analysis process yields a *data-driven decision*. Whenever applicable, the analysis process also considers baseline conditions (Bs) and target performance goals (Ts).
  - *Conclusions and Recommendations*: sums up the analysis and establishes which alternative shows the best performance numbers overall. Finally, the report writer recommends answer(s) to the project objective, which was stated upfront in the Introduction. The report's body ends with a brief statement of closure.

- Several important typographical conventions (among others also covered in this chapter) that you should know and apply when inserting tables and figures into engineering documents are as follows:

  - *Split text vertically and center tables and figures* when you are placing them on the page rather than wrapping text around them.
  - *Completely label tables and figures* by naming each table "Table [#]–[descriptive title]" and locating the title immediately above table and naming each figure "Figure [#]–[descriptive title]" and locating the title immediately below figure.
  - *Completely detail tables and figures* by including, whenever applicable, units of measure associated with data and also noting the data source when data (or a graphic displaying data) is not your own.

References

1. Alred, G.J., Brusaw, C.T., Oliu, W.E.: Handbook of Technical Writing, 11th edn. Bedford/St. Martin's, Boston (2015)
2. Beer, D.F., McMurray, D.A.: A Guide to Writing as an Engineer, 4th edn. Wiley, New York (2014)
3. Tebeaux, E., Dragga, S.: The Essentials of Technical Communications, 4th edn. Oxford University Press, Oxford (2017)

Chapter 18
Tech-to-Nontech Briefs

Learning Objectives

- *Know why tech-to-nontech communications are important in industry.*
- *Know how to strategically design and tactically advance a conceptual explanation that is centered around an analogy.*
- *Know the document algorithm for tech-to-nontech documents.*
- *Know how to use proper typographical conventions for formatting a tech-to-nontech document as an internally circulated informational brief.*

Engineers must routinely write technical documents for other engineers. However, since engineers are the ones who design a company's products and processes—and since the technical expertise and responsibility associated with those products and processes link back to engineering—engineers are also called upon to explain, review, and disseminate technical information to a variety of *non*-engineering audiences. As Fig. 18.1 reveals, engineers are positioned at the center of a communication network with message lines (channels) extending to a multifaceted array of personnel types and areas. A number of these lines fall into the category of technical to nontechnical or *tech-to-nontech* exchanges.

Nontechnical audiences can include factory floor laborers, marketing and sales staff, upper management, and more. Persons categorized as "nontechnical" ("nontech" for short) possess educations and areas of expertise that are different (though not less important) from the calculus-based, technical educations and technical competencies of engineers.

Nontechnical job roles support engineering in a variety of ways. In industry, engineering designs (no matter how brilliant) cannot provide widespread benefit to consumers until these designs are commercialized, marketed, and distributed—with product quality warrantied and safe usage assured. To support product design and manufacturing and do their jobs effectively, nontechnical personnel often do *not* need to know all of the technical details and theories that go into a technical product's inner workings. On the other hand, nontech audiences usually *do* need to know, in general, how a technical product or process works—its macroscopic

B. Henderson, *A Math-Based Writing System for Engineers*,
https://doi.org/10.1007/978-3-030-10756-7_18

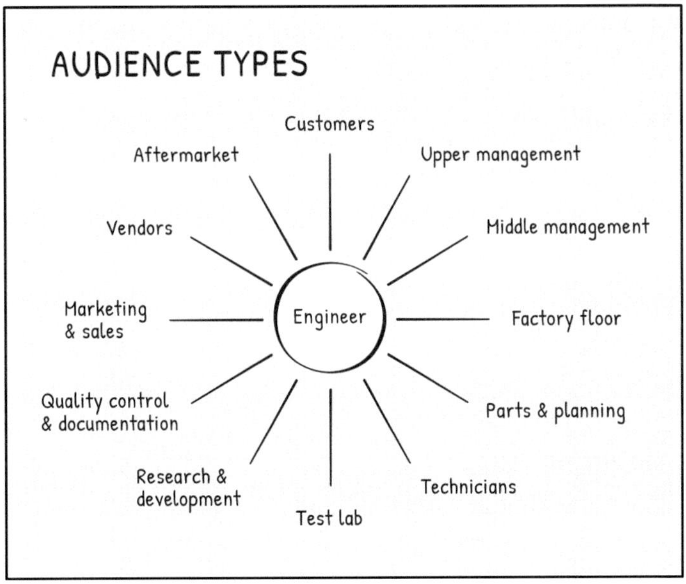

Fig. 18.1 Engineers communicate to a variety of different audiences

function, its input and output, and its key, top-level features that distinguish this product from those offered by competitors.

For example, an employee working in sales certainly needs to know how a product will benefit customers as well as how to operate the product, so the salesperson can demonstrate and sell the product with credibility and confidence. That same salesperson probably does not need (or perhaps want) to know how, in deep-level detail, to design and manufacture the product. In general, understanding the *gist* of how a technology works aids nontechnical employees in their job roles. Having a sufficient overall understanding of (and consequent intellectual connection with) a technology can also instill in employees a *vested interest* in related projects or programs. This can lead to extra measures of responsiveness, brand loyalty, and even work-team pride.

Nevertheless, despite all of the preceding, when engineers are assigned to write tech-to-nontech documents, it is not unusual for them to grouse about having to "dumb down" a product or process description to a degree (or so they allege) that is so stripped-down, even trivial, that it isn't worth communicating at all. Unfortunately, this response reflects "dumb-downed" thinking. As already established, tech-to-nontech communications are necessary and important. What's more, making a technical concept intellectually accessible to a nontechnical audience is no easy task. Instead of being humdrum and dull, transforming technical information into a nontechnical alternative that meets the audience's needs is a supremely challenging task for most engineering writers.

This job demands clever workarounds and out-of-the-box thinking. For instance, to the typical engineer, the higher math and applied physics that go into a design are a given. Yet, to many non-engineers, seemingly routine equations like $F = M \times A$ are neither basic nor common knowledge. For a nontechnical audience, a complex design's function can remain a mystery *until* the engineer who designed it associates that design's inner workings with concrete images and principles that come from outside the realm of engineering and exist in the mainstream, not-always-math-based world.

Using Analogies to Help Nontechnical Audiences Understand Technical Concepts

To translate technical concepts into nontechnical explanations, the engineering writer must often paint a picture of how a product or process works, from the outside looking in. A good way to accomplish this task is with an *analogy*—that is, likening something within the technical domain to something universally understandable in the mainstream world. The more concrete, straightforward, and vivid the analogy is, the better it will connect with a nontechnical audience. A favorite technical analogy of mine is one that was devised by one of Hewlett-Packard's "Fathers of Inkjet Technology," Niels Nielsen.

To help a multifaceted audience visualize contamination particles inside the micron-scale ink reservoir of an inkjet printer cartridge, Nielsen once said, "Think of two ping pong balls bouncing around in the back of an empty garbage truck. Next, scale this image down from something measured in feet to something measured in microns—specifically, an inkjet cartridge's pre-firing chamber. Now, continue to envision those two relatively tiny objects inside the chamber. These are the culprits that can clog a firing nozzle and stop a jet of ink from hitting its target on a page. Ultra-small, ping-pong-ball-like particles—just one or two of them in a space that, relative to them, is the size of a garbage truck—can present a mammoth problem to the operation of an inkjet firing system."

With an analogy like this one, anybody—either nontechnical or technical—can get the gist of the contamination threat being described, as well as why contamination control in production areas is so important. My hope here also is that our examination of Mr. Nielsen's clever analogy will help to dispel a myth: namely, that for one engineer to describe in simplified fashion some facet of engineering to another engineer—an audience for whom the subject matter is already "obvious"—is a waste of time or, worse, an insult to the second engineer's intelligence. If the analogy is creative and engaging, quite the opposite often occurs.

Even when an engineer already understands something technically, if that something has to do with engineering, gaining a new perspective on it helps that technical content stay fresh. Sometimes, a particularly clever analogy will become cherished, even iconic, among technical audiences. Consider the water and pipe analogy often used to teach introductory lessons on how an electric circuit works. Here, electron flow (current) driven by voltage in conductive wire is likened to water flow and pressure inside a pipe.

Document Engineering Tip

Here's a suggestion: on those occasions in your engineering work when you yourself think up a particularly clever tech-to-nontech analogy, consider integrating that analogy into your engineering writing—not only when you are composing a tech-to-nontech document for a nontechnical audience but also when you are preparing a technical to technical (*tech-to-tech*) report or proposal for your engineering peers.

To better understand this chapter's recommended method for writing a tech-to-nontech document, let's follow suit and use an analogy to further explore the task. In Fig. 18.2, we can see the task likened to a filtration process. On the left, unfiltered technical content feeds into a filtration chamber. The unfiltered mixture contains three types of content particles—large particles representing units of advanced concepts and details, medium-sized particles representing intermediate units, and, finally, small-sized particles representing basic concepts and details.

As you can see in the figure, once the feed flow of unfiltered technical content enters the chamber, the system strains out a filtered nontechnical mixture, and this filtered content dispenses out of the chamber's base.

The remainder, a mixture of residual technical content (intermediate and advanced), exits the chamber on the right. In Fig. 18.2, you may have also noticed

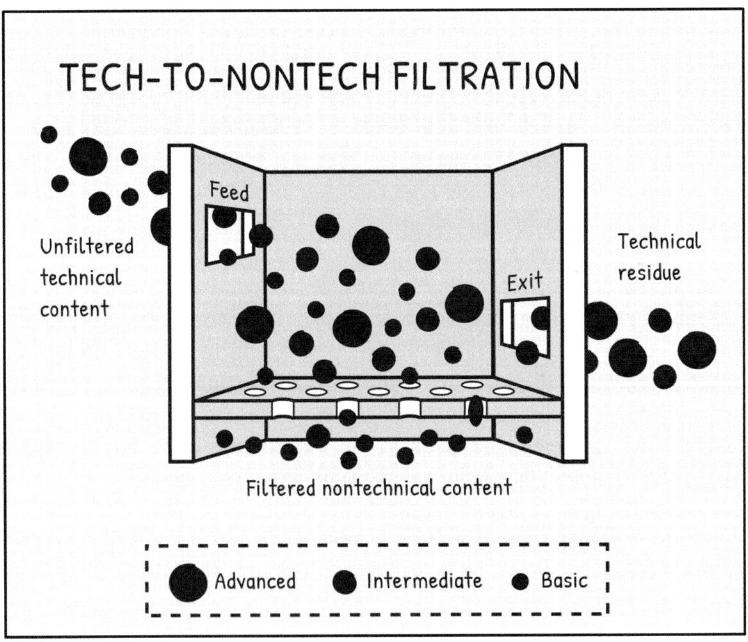

Fig. 18.2 An analogy for filtering technical content into a nontechnical extract

that the filtered content exiting out of the base contains a couple of intermediate-sized particles that have managed to squeeze through. This is consistent with reality. As an engineering writer, even if you are highly skilled and armed with an excellent analogy, seldom can you translate technical subject matter into something that is 100% nontechnical. But this reality is okay. The goal of tech-to-nontech translation is to provide nontechnical audiences with reasonable and sufficient access to technical information—that is, the gist—not to accomplish a perfect split of content into precisely 100% technical and 100% nontechnical streams.

Writing a Tech-to-Nontech Brief Using the Appropriate Document Algorithm

We will now examine how to write a specific type of tech-to-nontech engineering document—a tech-to-nontech "brief," which incorporates one or more analogies. The name "brief" comes from the document's purpose, *to brief* or provide a concise account. In this case, the document explains to a nontechnical or *layperson* audience how a complex product, process, or technology works. To guide document structure, the brief has a tech-to-nontech algorithm at its core. See Fig. 18.3 below.

Like its predecessors, the tech-to-nontech algorithmic flowchart, which is developed around the *mode of translation*, can be used to effectively guide the structure of short-form as well as long-form documents. Here, as usual, however, we will focus on the short-form document, one to four pages in length.

Here, specifically, is how the tech-to-nontech document algorithm works:

Start: The tech-to-nontech writing process begins with the engineering writer possessing an *unfiltered technical message* (e.g., a conceptual model of an engineering design and the way that it functions, including all of the details associated with higher-order math, physics, and so on). This technical message needs to be communicated to a nontechnical audience in a way that is simplified and understandable (yet still, at the same time, meaningful and adequately complete) for that nontech audience's application purpose(s).

First, before doing anything else, the engineering writer needs to carefully assess the nontechnical audience to accurately target the message at that audience. This, of course, is an important prewriting step for all document-writing tasks. Yet here, it is especially important, and the Fig. 18.3 algorithm explicitly notes audience assessment and targeting with a flagged sidebar operation on the flowchart.

Aside from standard assessment of *who* the audience is, *what* the audience *wants* to know, as well as what the audience *needs* to know, with a tech-to-nontech communication, it is particularly important for the writer to clearly identify the target audience's *level of technical knowledge*. This means assessing both the audience's *experience* and *education*. For example, if you determine that your nontechnical audience has not previously calibrated any piece of equipment in your work area, then you might need to review the difference between precision

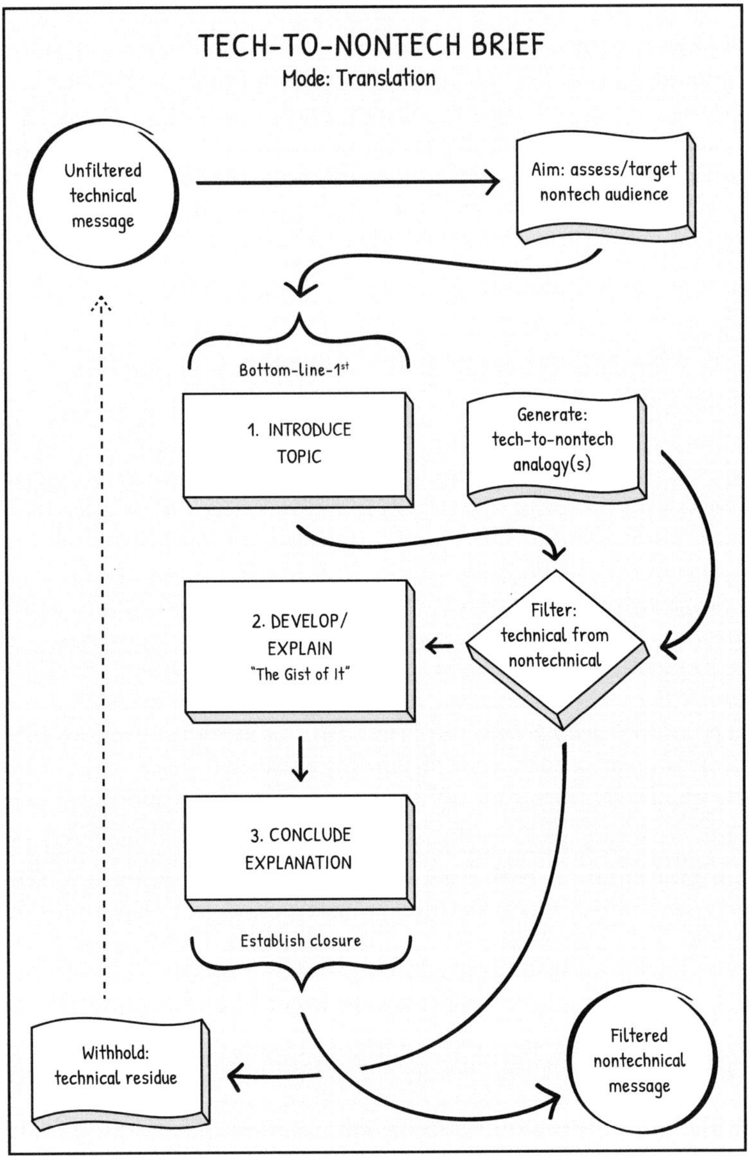

Fig. 18.3 Document algorithm for a tech-to-nontech brief

and accuracy before you explain how a new, self-calibrating laser detection station works. Or, if you establish that your target audience members have competency in math up to the level of algebra and basic geometry, yet no coursework beyond that, then it would probably be best for you not to refer to differential equations and integrals in briefs that address this audience.

Next, the algorithm identifies a sequence of three main structural elements to guide the writer in composing an effective tech-to-nontech brief's main text. These elements are displayed as flowchart operation blocks 1, 2, and 3:

1. *Introduce the topic:* Along with remembering to lead off with a one- to two-sentence statement of the "bottom line," which reveals to the reader right away what the document will go on to say and be about, the initial text should (as necessary) further expand on and complete the topic introduction. (Note: As usual, Fig. 18.3 signals the "bottom line" with the bottom-line-first bracket symbol.)

   It is often useful for the introduction to establish *topical relevance/significance* (an item or two to motivate the reader to read on, though not an exhausting catalogue of every argument *pro*). When useful and applicable, the introductory text can cover pertinent background information (key highlights rather than an exhaustive history). In addition, an effective brief's introduction may set boundary conditions for the topic—that is, establish *scope*. For instance, in the applied example that follows, we will examine a tech-to-nontech brief that discusses a specific type of device, a *pulse oximeter* (narrow *scope*) that belongs to a broader category of devices, *instruments measuring active features within human regulatory systems* (broader *topic*).

   Whichever of the preceding items your brief's introduction contains, all process block (1) content should strive to be concise. One short introductory paragraph is ideal for launching this type of document.

2. *Develop/explain the "gist of it":* With topic, scope, purpose, and significance and so on now revealed, the middle and bulk of a tech-to-nontech brief should focus on fully developing and completing the discussion launched by the introduction. This is accomplished in the main structural element, process block (2).

   Prior to this operation, however, Fig. 18.3's progression shows another flagged sidebar operation, calling for the writer to *generate* one or more tech-to-nontech analogies. There is a diamond-shaped filtration operation immediately below the sidebar. The diamond receives (top) incoming flow of unfiltered technical content (the original message) and filters it, separating unnecessary technical details from those details that are necessary for communicating an audience-appropriate version of content to the identified nontech audience.

   The analogies also enter (right) the diamond-shaped filter and combine with the filtered stream. Then, the mixture exits (left) the diamond-shaped filter and flows into the main process block (2).

   Note: The technical residue that has been filtered out exits in another stream (bottom of filter) and is transported to a third flagged sidebar block labeled

"Withhold." I added this sidebar to the algorithm mainly as a comfort feature for engineering writers. It serves to remind and assure you that any technical content you remove from your tech-to-nontech briefs—content which, most likely, is of high value to you—is stored, rather than destroyed. Later on, if you wish, you can add all of the technical content you have set aside back into the original mix—for other documents addressing a technical audience.

Let's move to the main element, process block (2). Now that you are prepared to explain the document's message using refined and adjusted materials, here are some additional recommended best practices that will help you successfully connect with your target audience's wants, needs, and level of technical knowledge:

- Avoid unnecessary technical terms, and instead use terminology that is as common and universal as possible. For example, instead of "*electromagnetic radiation illuminates* our days on earth" write "*sunlight lights* our days on earth."
- When you need to use a technical term and no simplified alternative word is available or relevant, *define or explain* the term immediately after you introduce it. Better yet, explain the term immediately before you use it. For example, instead of "... two cryogenic substances, liquid methane and liquid oxygen, fuel the rocket engine ...," write "... *when gases like methane and oxygen are cooled hundreds of degrees below the freezing point of a regular liquid like water, they become a special type of liquid known as a cryogenic substance*. For the rocket engine we are describing, cryogenic methane and oxygen fuel the engine ...." Alternatively, you could define "cryogenic substance" immediately after it is mentioned.
- Introduce your analogy sooner rather than later. If you wait too long to introduce a simplification into your brief, the technically dense and possibly inaccessible text leading up to it will put off and frustrate the reader, and your analogy will not be appreciated when the reader finally encounters it (if your reader has not already given up reading).

3. *Conclude the explanation:* The algorithm's final process block (3) guides the writer to complete the task of explaining a technical subject to a nontechnical audience. Just like the introduction, the brief's conclusion should be concise. In fact, the conclusion could be just a sentence or two. However, if useful and applicable, prior to the conclusion's last line(s), an effective brief's conclusion might, just as the introduction zoomed in from broad topic to tighter scope, *zoom out* to leave the reader with a point of general interest that relates to the brief's specific subject.

   For instance, if the brief's focus was on liquid rocket fuel and the engineering challenge of designing a container for its cryogenic ingredients, a point of general interest to readers might be to mention a new formula for more easily contained, all-solid rocket fuel, which is currently in development and predicted for release soon. Or, as in the applied example in the next segment, your conclusion might feature a *call to action*.

Regardless of what else the conclusion includes, its final sentence (or two) must establish closure. Figure 18.3 reminds the writer to do this with a *statement of closure bracket.*

End: As shown, the end outcome that results from following the Fig. 18.3 algorithm is an appropriately aimed, filtered, and structured message—now ready to deliver to and be understood by the document's target nontech audience.

Application Example

The Project Scenario

Timms Medical Equipment and Technologies is a biomedical applications company whose chief business objective is researching, developing, and commercializing products that monitor and control human cardiovascular and endocrinal systems. The company's product line includes a range of devices such as glucose monitors, insulin pumps, hemoglobin monitors, and pulse meters.

In response to market demand for a competitively priced, yet accurate and robust handheld pulse and blood oxygen meter, Timms developed and released the Model 3 Pulse Oximeter (M3PO). Projected first-year sales were 40,000 units. Actual sales, however, were 5000 units. Timms wants second-year sales to significantly improve and meet or exceed the first-year 40K target, and the management wants *all personnel* to be aware of this goal and to actively support it.

As a part of the year-2 sales push, management has asked the product's lead designer, Project Engineer Kali Pearson, to prepare a short informational brief that explains the general technology and operation of the M3PO to nontechnical employees who work in marketing and sales.

Timms' management knows from past experience that keeping employees at all levels technically informed helps to foster a company culture that optimizes internal support services and boosts end-unit sales. The company's Product Brief Series (PBS) is a well-established communication vehicle at Timms for disseminating in-house product information. Ms. Pearson's brief will be the 24th in this series. The management expects subject-matter experts like Ms. Pearson to write these documents clearly, concisely, and correctly.

The following example presents the result of Kali Pearson's tech-to-nontech brief developed using this chapter's algorithm. The two-page brief (see Figs. 18.4a and 18.4b) is structured around the recommended tech-to-nontech document algorithm. The example also illustrates a generic document geometry for preparing a short-form brief. This chapter's final segment further explains this format. After the effective example, you will find an ineffective example (see Figs. 18.5a and 18.5b), written for the same engineering scenario.

Effective Brief Example

Kali Pearson, Project Engineer
Timms Medical Equipment and Technologies
Product Brief Series, document #24, revision a
release date: <month-day-year>

Operational Overview of the Model 3 Pulse Oximeter (M3PO)

Timms introduced the Model 3 Pulse Oximeter (M3PO) into its commercial product line to address high-volume demand from healthcare professionals for a lightweight, battery-operated, ultra-compact device to measure patients' blood oxygen levels and heart rates. Until the current technology was perfected, testing a patient's blood oxygen level required a blood draw and subsequent lab work. Now, instead of an invasive process with time-delayed results, health professionals can perform the test immediately and also record heart rate using a noninvasive handheld unit—about the size of a tube of lipstick.

The health professional places the patient's finger into the M3PO device, and the device does the rest with tiny beams of light. The procedure is quick and painless. Collecting vitals takes less than 15 seconds.

How does the M3PO work? We all know from basking in the sunshine and looking at rainbows that light can present as an invisible source of sunburn (this type of light is *infrared* light) as well as in a spectrum of visible colors (*red* light, *blue* light, *yellow* light, and so on). Our pulse oximeter device uses the basic properties of light to operate. Inside each unit, the equivalent of one tiny high-tech light bulb creates a beam of red light, and another bulb creates a very small beam of infrared light. The high-tech light bulbs inside the device are actually low-energy Light Emitting Diodes (LEDs), the same type of lighting technology (though on a much smaller scale) we now see available for energy-efficient lighting in homes. Little light beams produced by the pulse oximeter

for internal circulation only page 1 of 2

- *At the top, the brief features a simple, left-justified header that records author, company, document category, and date.*
- *A descriptive title announces the document's topic and purpose; then, the text leads off with a "bottom-line" introduction of scope.*
- *Hereafter, the text proceeds to describe, step-by-step, how the product works.*
- *The text is double-spaced and "easy to read".*
- *To make the descriptions accessible, the author uses several analogies to connect the technical subject matter to nontechnical things—a tube of lipstick, sunlight, and rainbows.*

Fig. 18.4a Effective tech-to-nontech brief, page one

shine onto and into a patient's finger. Some of the light passes through and some gets absorbed. If you have ever held a flashlight to the palm of your hand in a dark room or at night, you have observed that your hand will glow on the backside because a portion of the light goes through.

Inside the M3PO, sensors detect the amount of light that passes through the patient's finger tissue. We know that to the human eye, oxygen-rich blood is red. This is because the blood reflects red light and absorbs other colors in the spectrum and other types of light. On the other hand, oxygen-depleted blood is not red; it is blue. The point is, in a human finger, there are physical properties that key to colors.

When installed on a patient's finger, the M3PO is able to keep a running tally of how much invisible infrared light gets absorbed and how much red light passes though the finger. The M3PO sorts out what this running tally of measurements means by using the properties of light reflection and absorption; and then the M3PO determines oxygen level in a patient's blood by the blood's redness. Since the M3PO measures redness to a fine degree, the device is also able to measure the patient's pulse by tracking the amount of blood inside the finger at any given time as the heart pumps it in and out.

Timms' competitively priced and durable blood oxygen and heart rate meter is an example of how our excellent engineering designs can benefit healthcare providers and patients alike, with our easy-to-use, vital-sign meters. Initial sales have proven the M3PO to be durable and relatively free from returns and warranty claims. The M3PO is a product that invites sustained product commitment and promotion from the Timms employee team.

for internal circulation only page 2 of 2

- *All paragraphs are "chunked" and composed mostly of eight vertical lines or fewer.*
- *As the discussion progresses, the level of technical complexity rises an increment or two, because the reader is now up-to-speed regarding the basics, fully engaged, and confident.*
- *The brief closes with some broader insights and a call to action.*
- *The footer contains a confidentiality tag and page number.*

Fig. 18.4b Effective tech-to-nontech brief, page two

Ineffective Brief Example

Kali Pearson, Project Engineer
Timms Medical Equipment and Technologies
Product Brief Series document #24, revision a
release date: <month-day-year>

PBS #24: M3PO Functional Prospectus

Testing blood O_2 levels of patients used to be a somewhat arduous process that required taking blood draws and laboratory analysis. Now, since the introduction of pulse oximeter technology, an individual can painlessly insert his or her finger into a device that incorporates usage of LEDs to calibrate the patient's blood oxygenation level and heart bps, or pulse rate. This process takes less than 15 seconds. The Timms Medical product response to this technology is our Model 3 Pulse Oximeter, a dual function device indicating two common vital signs monitored by health professionals relative to patients. The device functions using two LEDs. One source emits visible red light, and the other emits transparent light in the infrared spectrum. In addition, the device senses electro-magnetic radiation that passes through the finger's epidermis and underlying soft tissue and utilizes this to compute and read-out the patient's blood oxygen level. In other words, the light that is transmitted through the finger is simply the differential between the light emitted by the LEDs and that which is absorbed.

It is important to note that the pulse oximeter's high degree of accuracy and precision allows it to distinguish between the light absorbed by the actual blood versus that absorbed by the other surrounding tissue over time—that is, it takes a dynamic reading. Of course, the surrounding tissue does absorb some of the same wavelengths that the blood absorbs, but again, the M3PO measures the delta between the highest and lowest areas' absorbance of light. Light absorbed by static tissue is essentially a constant. The dynamic readings record the actual heart beating—that is, that patient's bps or pulse. We could say the blood oxygen meter is a mechanism for tracking cardio saturation and cardio pump cycle beyond the heart in the vascular system.

- *The brief's header is straightforward and presents pertinent info, yet its right-justified position is not as smooth and quick-to-read as a left-justified header.*
- *The document's title is technically dense and cryptic.*
- *The text itself begins in an esoteric, conversational manner, and as it progresses no analogies appear.*
- *The descriptions are riddled with unnecessary technical terms and jargon— "O_2," "arduous," "epidermis," electro-magnetic radiation," etc.*
- *The typography and layout are acceptable, but should be more chunked.*

Fig. 18.5a Ineffective tech-to-nontech brief, page one

The technology of the blood oxygen meter is an example of how Timms engineering advances serve to benefit medicine and improve human wellness. For nontechnical audiences outside of engineering, probably the most simplified way to look at the M3PO is to see it as an imaging device—not a camera but in the same category as an X-ray machine. One difference is that the M3PO's imaging field is very small and low-energy, and therefore the M3PO exposes patients to virtually no harmful radiation. Thus, our M3PO provides healthcare practitioners with a technology that is like specialized X-ray glasses, which "see" inside a human finger and detect an exact reading for 0_2% and bpm.

The Timms Model 3 pulse oximeter is relatively low-cost, less than one ounce in weight, and about the size of a business card, if you were to fold it lengthwise into the shape of a cylinder. This is a product that all Timms employees should know about, appreciate, and promote. Prior to designing, prototyping, and qualifying the M3PO, the company developed two other predecessors, the Model 1 and the Model 2. The Model 1 solely measured blood oxygenation within a patient's bloodstream, with a finger cusp device unit pair. The Model 2 only took a reading for heartbeats per minute. Yet management originally wanted a device that could perform both functions.

- *As the description advances, it continues in a manner that's unnecessarily technical for achieving its communication purpose.*
- *A couple of analogies (not particularly clever ones) finally do appear, but do so too late; most nontechnical readers, by this point, would be frustrated and confused.*
- *Finally, the text presents some interesting though inessential background material, and then ends without a solid feeling of closure.*
- *The pages are not numbered or labeled with a confidentiality notice.*

Fig. 18.5b Ineffective tech-to-nontech brief, page two

Document Geometry: How to Format an Informational Brief

This segment examines how to format an informational brief on the page or screen to establish a reader-friendly geometrical layout. Since there are many effective, company-specific approaches for preparing an internally circulated brief—which is in the same document category as a *bulletin* or *release*—this segment covers the typographical elements for preparing an effective *generic* brief.

Figure 18.6 below displays, in plan view, an annotated template for a short-form, two-page *internally circulated informational brief*. The template's scheme, from top to bottom, can easily be adapted for a one-page version, as well as for a document

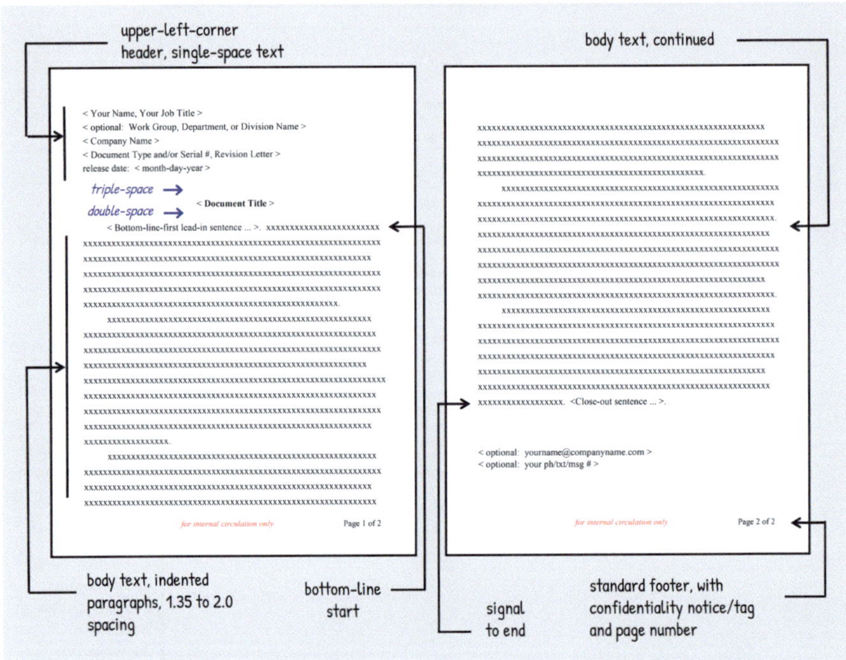

Fig. 18.6 Plan view of a generic brief's typography

consisting of three or more pages. Like a memo, the text of a brief begins with a *header*—though unlike a memo, the generic brief's header is placed in the upper-left-hand corner rather than all the way across the top. Single-spaced, the header lists the following key source information:

- The writer's name and job title
- An optional entry of the writer's work group, department, or division
- The company name
- A record of the document type and/or serial number, plus revision letter—"A" (first), "B" (second), etc.
- As with most all professional documents, a timestamp, in this case, the document's release date

Placing the header in the upper-right-hand corner would be acceptable too, except that the left location is generally easier to type and also generally easier for the reader to scan and absorb and understand.

Continuing with single-spaced text (word-processor setting 1.0–1.15), the writer spaces down three returns, centers the text, and types an engaging and information-rich *document title*, not just an empty label. Thereafter, the writer double-spaces vertically and begins typing the brief's *body text*.

Several necessary features guide the layout of the body text. The plan-view template displayed in Fig. 18.6 assumes that the writer generates and structures content for the body text using an *appropriate document algorithm* to produce an effective tech-to-nontech message (see Fig. 18.3). As noted in the plan view figure, the body text leads off with a *statement of the bottom line*, signaling the message's "start" and informing the reader immediately what the document is about. Also noted, the brief ends on page two with a concise, clear *close-out sentence* that signals the *end of the message* and releases the reader.

You will notice that the example template shows a *distinctive typography*—different from the left-justified block style this book recommends for composing short-doc emails and memos. Although using left-justified block-style type is not wrong or unprofessional, the generic template here displays (and recommends) paragraphs typed using *standard article layout* (used in newspapers, magazines, and journals). In this layout, paragraphs are announced with tab-once indents. Note: A tab-once indent measures ½ in., horizontally, for type on a standard 8½ × 11 in. sheet of paper. For other media, the best practice would be to strive for something proportionally equivalent.

With standard article layout, the text blocks feature a more open *multi-spaced text*, typically a line spacing setting of 1.35–2.0 on a word processor. You will notice that standard-article-layout text blocks place no additional spacing between paragraphs, just a return. This style should continue to follow the best practice of "chunking," keeping paragraphs to a target length of eight or fewer vertical lines.

After the close-out sentence that signals the end of the body text, Fig. 18.6 template shows a quadruple space and then two optional items, which can provide readers additional sources of information—that is, the *writer's email* and *phone number for calls/voicemails/text messages*. These are considered *optional* because inside an organization, this sort of information for an employee on the payroll is likely to be readily available in the company personnel directory.

Finally, you will see that the sample plan-view illustration displays a footer containing a middle and right-side item. As already established, page number on the right is a standard feature for documents of two pages or more in length. Since an informational brief, bulletin, or release document is likely to circulate beyond individuals in an engineering writer's engineering work group, the middle footer item is also essential to include. It reminds readers that the information in the document is "company business" and proprietary (to some degree ranging from standard to extreme). How, specifically, the confidentiality tag is phrased varies—"for internal circulation only," "for internal use only," or "private and confidential," and so on. Your company will define specific phrasing based on your company's security protocols.

Document Engineering Tip

Along with creating information-rich document titles, making a document's electronic file name sufficiently descriptive is also an engineering writing best practice. This enables the file name to reveal the document's source, purpose, and scope to the reader, even before the reader opens the file. Descriptive

(continued)

document file names also make it easier for writers themselves to organize and store files, as well as find and retrieve them later. For instance, an information-rich file name for this chapter's tech-to-nontech brief example might be something like "BriefSeries24_m3_pulse oximeter_rev a.pdf". In contrast, a vague, ineffective file name might be "oximeter info.pdf".

Action Items

α	In Chap. 15's first document-writing example on proposal memos, we considered a real-world scenario where an industrial team with a variety of education types and prior experiences—process operators, maintenance technicians, manufacturing engineers, and managers—needed to understand the work flow of a production line to troubleshoot and repair problems and speed up a new line start-up. This scenario and its communication device and those of the Timms Medical pulse oximeter example presented in this chapter are similar. Both examples feature an engineer using a type of show-and-tell device to make technical content accessible through concrete messaging. The former uses film (and soundtrack); the latter uses analogies and written text.
	Brainstorm a list of five to ten instances where you remember observing a technical concept being described via concrete text and images, with or without analogies, to make the concept accessible to a multifaceted (tech as well as nontech) audience.
α	Repeat the above action item, only this time, brainstorm a list of five to ten technical concepts that you yourself might wish to describe to a multifaceted audience. Now determine what sorts of concrete text, images, and/or analogies you might use to make each of the concepts accessible.

Recap

Key Learnings/Takeaways

- Because engineers design a company's products and processes and possess technical expertise associated with those products and processes, like it or not, technical communication responsibilities link back to engineering. These communication responsibilities include writing documents that describe technical content to nontechnical audiences (tech-to-nontech documents).
- Tech-to-nontech documents are important because they enable nontechnical employees to understand how technical products, processes, and technologies work. This understanding empowers those nontechnical employees with a *vested interest* in the engineering projects and programs they support. Technical understanding can also instill within nontechnical employees an extra measure of responsiveness, brand loyalty, and even pride related to their work.
- Use of analogies is an effective method for making documents on technical topics accessible to a nontechnical audience. As applied here, an analogy takes a technical concept within the sphere and rigor of an engineering discipline and

likens it to something similar (a nontechnical touchstone) in the mainstream world. An iconic example of an analogy that makes a technical topic accessible to a variety of people is the water and pipe analogy often used to teach introductory lessons on how an electric circuit works: In a basic circuit, electron flow (current) driven by voltage in a conductive wire is likened to water flow and pressure inside a pipe.

- The document algorithm for tech-to-nontech documents, such as informational briefs, is structured around the *mode of translation*—meaning its chief purpose is to translate an unfiltered technical message into a message in which unnecessary technical details have been filtered out to make the message accessible by a nontechnical audience. Given an unfiltered technical message, the writer's first task is to assess and take prudent aim at the document's target audience (determine the audience's needs, wants, and level of technical knowledge).
- Thereafter, the algorithm's three main elements are:
  - ***Introduce the topic***: *the writer establishes the document's purpose, topic, and scope.*
  - ***Develop/explain the topic***: *using at least one analogy, avoiding unnecessary technical language and clearly explaining technical terminology that must be retained, the writer communicates the gist of how something technical works.*
  - ***Conclude the discussion***: *the writer establishes closure to release the reader and optionally (and very briefly) discusses follow-on activities.*
- A common document type for tech-to-nontech communications is the *informational brief*. To prepare a generic version of one of these documents, the key typographical features that establish document geometry are as follows:
  - Begin with a left-justified, single-spaced header that contains key source information, and then write a title that is information rich.
  - Use *standard article layout*: use a line spacing setting of 1.35–2.0 for the body text and also announce paragraphs with tab-once indents.
  - Use a tech-to-nontech document algorithm guide, such as the one in Fig. 18.3, to structure the document's message.
  - Include a footer that records a confidentiality label (middle) and also a page number (right) if the document is more than one page long.

Chapter 19
Instructional Job Aids

Learning Objectives

- *Know how to use basic instructional systems design (ISD) techniques to analyze a human performance problem associated with a job task, and then determine a viable solution that might consist of or include a skill-transfer document.*
- *Know the document algorithm for an instructional (skill-transfer) document developed around the mode of transference.*
- *Know how to design and prepare a set of step-by-step instructions that begins with a three-part (Mager-style) performance objective (P.O.) and ends with a criterion-referenced check step.*
- *Know how to use proper typographical conventions for formatting an instructional job aid.*

What Is a Skill-Transfer Document?

In industry, people get paid to *do* things, not *know* things. College grade point averages are superseded by employee performance reviews as metrics for success. The essential workplace document that supports *skill training* or teaching an employee *how to do something* (typically an assigned job task) is called a *skill-transfer document*. The term "skill transfer" describes the document's chief purpose of capturing and recording a skill recipe for how to perform a particular task competently and then passing along (transferring) that recipe to a skill learner (or trainee).

Skill-transfer documents include procedures, manuals, instructions, and training guides. In the twenty-first century, a growing percentage of skill-transfer documents incorporates multimedia delivery platforms—online text, digital images, audio, and video. When the principal mode of instruction is a lesson delivered by a teacher, sometimes a skill-transfer document, such as a training guide or handout, complements the lesson. On the other hand, many skill-transfer documents must be stand-

B. Henderson, *A Math-Based Writing System for Engineers*,
https://doi.org/10.1007/978-3-030-10756-7_19

alone resources for self-directed learning. These documents are particularly challenging to write because, in the absence of a teacher to explain content and answer questions, the document *itself* must clearly and completely coach the learner through the learning process.

This chapter focuses on how to design and produce short, stand-alone instructional (skill-transfer) documents for employees in technical job roles who need to learn how to perform technical job tasks—*correctly, efficiently, and safely*. To begin, this chapter presents a rudimentary overview of instructional systems design (ISD) and a *process algorithm* for applying basic ISD techniques. Next, the chapter scopes down to present a *document algorithm* for creating a specialized type of instructional document known as a *job aid*. Thereafter, the chapter offers an application example of an engineer developing and writing a job aid document to address a human performance problem related to a technical job task. Last, the chapter presents a document geometry lesson on the recommended typography for step-by-step instructions that are formatted as a job aid.

Why Should Engineers Care About Instructions?

As technical experts who lead projects and manage technical personnel, in addition to their other responsibilities, engineers are regularly called upon to explain not only *how technical products and processes function* but also *how to perform technical job tasks* associated with the operation, installation, calibration, maintenance, and repair of production and test equipment. When a company hires new engineers, experienced engineers may be assigned to mentor and train the new hires. In your work with support personnel and other engineers, at some point in your career, your job will probably call upon you to do technical training and write instructional/procedural (skill-transfer) documents, and these documents will need to be well-written.

Poorly written instructional documents can produce a variety of repercussions—such as inefficiency, machine downtime, and inferior-quality manufacturing output—leading to scrap product, redos, and rework. Poorly written instructional documents can cause ineffective and improper use of expensive equipment, leading to excess wear and breakage. Worst of all, poorly written instructions can be the source of accidents—burns, cuts, and lost eyes, limbs, and even lives.

The good news is that well-written instructions—aside from helping to mitigate all of the preceding problems—provide a form of direct, self-contained learning and guidance to employees on your team. Rather than having to teach or supervise the same task over and over in-person, effective instructions (once written and in place) foster employee independence and allow you to assume a hands-off supervisory role—to do more engineering, less training.

Another bonus of learning the instructional writing method presented here is that it is excellent preparation for writing other important documents that involve assessment, such as position plans for employees in your work group (when you become an engineering manager). Like instructions, position plans must clearly and precisely define an employee's assigned job duties and metrics for how the employee's job performance will be measured.

The bad news is that, of all of the five essential documents covered in this book's chapters on document algorithms, instructional documents are, for most engineers, the most dry, meticulous, and tedious to write. This is why this chapter has a double agenda:

1. To teach you how to write instructional (skill-transfer) documents in the most efficient, effective way possible
2. To teach you how to avoid writing instructions when lack of skill is *not* the true root cause of a human performance problem

As we shall soon see, an employee's failure to perform an assigned job task does not necessarily mean that the employee lacks the ability to do it. There are three other factors besides *skill* that enable performance—*resources*, *motivation*, and *management*.

Instructional Systems Design (ISD) Basics

Education practices in industry differ from those in academe. Industry places chief emphasis on *skill training* and *skill transfer*. On the other hand, academic classes often focus more on knowledge acquisition. Most academic classes are designed around predetermined learning objectives, as in industry, yet these objectives are usually more abstract and theoretical than the narrow, prescriptive ones that guide industry training and documentation.

Academe's educational methodology is time-tested and has produced countless graduates who are well-prepared to enter jobs in the private sector, higher education, and technical industry. Once in industry, however, the training needs of technical employees become more pragmatic and applied to specific job roles in the workplace. These needs are best met with an alternative educational approach that has been actively used by industry trainers and nonacademic organizations for over a century [1]. That approach is called *instructional systems design (ISD)*, or *instructional design* for short.

Figure 19.1 below displays a systemized flowchart, or process algorithm, for applying basic ISD to address a *human performance problem* in the workplace—that is, a case where an employee is assigned to do a job task and fails to do the task satisfactorily. As an engineer in industry, when you are called upon to resolve a problem like this, the basic ISD process presented here provides you with a sound, 10-step method that will help you develop an effective training solution. And, as you can see in Fig. 19.1 immediately after step 1, there is a decision block that will help you avoid unproductive, unnecessary instruction writing if the human performance problem is not due to missing skill(s) but rather to a non-training issue requiring a non-training solution.

You should note that the ISD process used by professional training consultants in the industry is often more complex than the basic version displayed here, especially for large-scale employee development projects and comprehensive training programs [2]. However, for the training duties of a professional engineer, this abridged version will usually suffice. The remainder of this segment further explains the Fig. 19.1 process algorithm and each of its ten basic ISD steps.

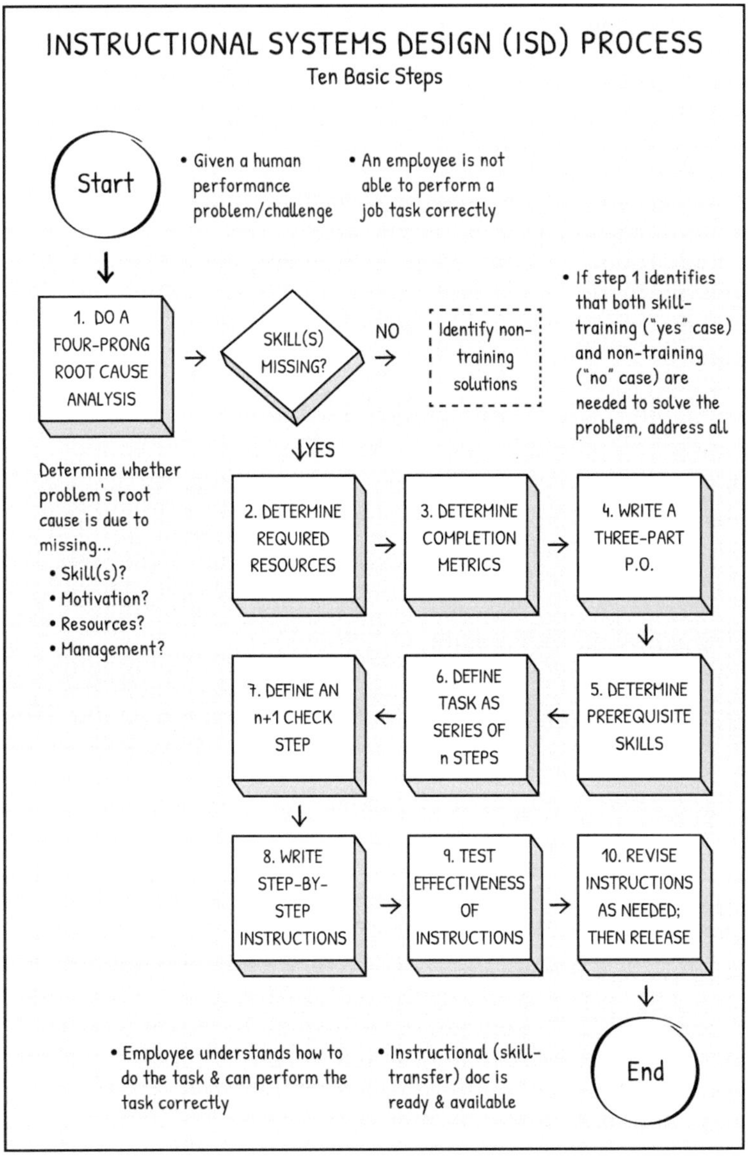

Fig. 19.1 Process algorithm for applying basic instructional systems design (ISD)

Here is a step-by-step explanation of how Fig. 19.1's process algorithm works and how to use it to guide basic ISD activities.

Start: You are confronted with a *human performance problem* in the workplace—a job task needs to be accomplished, and the employees assigned to the task are unable to do it correctly, efficiently, and safely. You must solve the problem.

Step One: Do a four-prong root-cause analysis to evaluate the problem with respect to four key factors that enable human performance:

- *Skills:* Does the employee (the task performer) possess the necessary skills to do the task?
- *Motivation:* Is the employee sufficiently motivated to do the task?
  - Even if all three of the other four factors support the subject's ability to perform the task, if the task doer *does not want to do* the assigned task, it will probably not get done (or if it does get done, the outcome will likely be unsatisfactory).
- *Resources:* Does the employee possess the necessary resources (materials and equipment) to do the job task?
  - When a person *does not possess the right tool(s)* for an assigned job, performance will likely be impaired—regardless of whether the doer knows how to use the tool properly, wants to use the tool, and is provided with sufficient work time and conditions.
- *Management:* Is the employee managed in a way that provides the employee adequate work time and favorable, safe work conditions for performing the task?
  - If a manager assigns an employee a task, and then that manager assigns the employee one or more other tasks to do before the employee has had adequate time to complete the first one, the employee's productivity is likely to diminish rather than increase, with the manager not the employee at fault for this problem.

Decision Junction: Are the Required Skill(s) for Doing the Task Missing, Yes or No? Based on the Step One root-cause analysis, determine whether the issue is that the task doer lacks the necessary skills for doing the assigned task. If *yes*, proceed to Step Two to develop a skill training solution. If *no*, explore non-training solutions to solve problems caused by lack of motivation, lack of resources, or ineffective management.

- If you determine that the performance problem is caused by both missing skills *and* a motivation, resource, or management issue, *address both* the skill-deficit problem and all other issues.

Step Two: After identifying and characterizing the missing skills necessary to do the task, determine all required resources (materials and equipment) necessary to do the task.

Step Three: Envision/define, as objectively and quantitatively (#, %, $) as possible, what metrics (criteria) define successful completion of the job task—that is, what does a "job well done" look like, and how do you determine/measure whether an employee has achieved this outcome?

Step Four: Write a three-part (Mager-style) *performance objective (P.O.)* for the task, which records the task *action*, necessary *conditions*, and performance *criteria* (performance metrics) associated with the desired performance outcome [3].

- Note: the next segment that presents the skill-transfer document algorithm will explain in detail how to prepare the three-part performance objective and will also provide an illustrative example.

Step Five: Determine and define, as applicable, the skill prerequisites—that is, the existing skills a learner must already possess to support and enable skill acquisition for learning how to do the task at hand.

Step Six: Break the task into a logical, orderly series of discrete instructional steps—*1, 2, 3* ... *n*, where *n* equals the final action step to complete the task—as characterized in the performance objective.

Step Seven: Determine and define an $n + 1$ check step to evaluate the instructional steps' final result—which should be *task complete and outcome correct* with respect to the P.O.'s performance criteria (performance metrics).

Step Eight: Write a set of step-by-step instructions with threaded notices (notes, cautions, warnings, etc.), where applicable [4].

- Note: the next segment that presents the skill-transfer document algorithm will explain in detail how to prepare step-by-step instructions and also how, when, and where to thread in notices.

Step Nine: User-test a draft of the instructional document on one or more trial users to determine the draft's instructional effectiveness. Are the instructions complete and clear? After consulting the instructions, was the trial user able to perform the task correctly? If yes, good. If no, revise the instructions and test again.

Step Ten: Continue to revise the draft if/as necessary to address opportunities for improvement as they are identified by trial user tests (Step Nine) until the document is clear of all known errors and potential user problems. Then, publish and release a finished instructional document.

End: The human performance problem has been resolved. A skill training solution (in this case, an instructional document) is now ready and available. The employee can use it to learn how to perform the job task and achieve a correct result.

- Note: If the ISD task analysis (Step One) has identified any non-training issues in addition to a missing skill(s), you will also need to address those with appropriate solutions—such as motivating the employee with positive feedback, supplying a missing resource, or adjusting your management style.

Before moving on to the next segment, here is a very brief application example that shows an engineer in industry using the first part of the Fig. 19.1 process algorithm (root-cause analysis) to troubleshoot a human performance problem that occurs in his work area. Later in this chapter, you will encounter a comprehensive application example that demonstrates, in much more detail, the basic ISD process from start to finish and see how it combines with this chapter's Skill-Transfer Document Algorithm to guide the production of a short, effective instructional document.

Basic ISD Example: Is Training Needed, and, If So, What Is the Right Solution?

John Quest works for a company that specializes in optics and lasers and is the engineer in charge of a manufacturing area that laser ablates 10- to 20-μm-diameter holes into a new type of integrated circuit (IC) that incorporates liquid cooling. John's team has recently installed a new laser apparatus manufactured by Sui Corporation, a high-tech Japanese firm.

John's company operates 7 days a week, 24 h per day, and there are four work shifts—weekday days, weekday nights, weekend days, and weekend nights. A team of 20 skilled process technicians (5 technicians per shift) operates the new Sui laser and its peripheral equipment. In the past, most of John's employees have learned how to operate factory equipment by "job shadowing." In other words, they have learned by "looking on" as an expert (in this case, usually John himself) performs the operation correctly. Until now, this type of informal training structure has achieved acceptable results. Yet last month, the work center's output was down 40%, there was an average of 10 h per week of throughput downtime, and the facility had to scrap a batch of 1200 out-of-specification ICs, costing $60K in lost product.

John's manager is upset and wants John to fix the problem by retraining his entire technician team on all aspects of their jobs. Instead of developing a training solution without first investigating the true root cause(s), John refers to the basic ISD process algorithm and performs a Step One task analysis. In doing so, he determines *several root causes*:

(a) Although all of the process technicians are responsive, motivated workers, John discovers that lately, when he is not present in the manufacturing area (especially night and weekend shifts), one of the factory-floor supervisors has been reassigning the technicians to do other tasks besides laser operation (management issue).

(b) John also discovers that the technicians have been having trouble logging on to the new laser's process control station because the station's software has coding flaws (resources issue).

John addresses non-training issues (a) and (b) immediately. They are easy to fix.

(c) The task analysis also determines that the technicians are having trouble calibrating a process control tracking mechanism that locates a pre-drill mark and checks X–Y coordinates prior to laser ablation of the holes (missing skill). During steady-state production, the device needs to be recalibrated every 8 h, and some of the process technicians on John's 20-person team have not been performing the procedure correctly.

(continued)

To address the missing skill, he follows Steps Two through Ten of the basic ISD algorithm and develops a short instructional document that correctly explains the calibration procedure. Then, he distributes this document to all of the techs and makes sure they read it, understand it, learn it, and apply it.

Writing a Short Instructional Document Called a Job Aid Using a Document Algorithm

When an engineer is dealing with a situation that involves a human performance problem, and if the engineer has already consulted the Fig. 19.1 process algorithm and used basic ISD to determine that there is, indeed, a valid training issue (a skill missing) that needs a training solution, a useful type of short instructional document for the engineer to be able to write is the *job aid*.

A job aid explains, step-by-step, how to perform a technical task efficiently, effectively, and safely. In high-tech work environments, job aids are often kept near work stations or in work areas where employees perform complex processes or operate high-tech precision equipment. Sometimes job aids are made available in hard copy format and are laminated with clear covers to keep them durable. Sometimes job aids are softcopy documents, designed to be viewed on a computer monitor.

Although job aids can record a wide variety of instructional sequences, excellent job aids embrace two key attributes that are present in all well-crafted skill-transfer documents:

1. The writing of the document is preceded by an ISD process to verify that a skill training need truly exists.
2. The document itself features a core set of ingredients that advance objective-driven, criterion-referenced instructions.

Figure 19.2 displays our fifth and final document algorithm. Centering on the *mode of transference*, this algorithm illustrates how to generate and structure text for an instructional job aid document.

Here is a detailed explanation of how Fig. 19.2's document algorithm works and how to use it to write an instructional job aid:

Start: The algorithm begins with an *identified task* that needs to be paired with a step-by-step, instructional document, in this case a *job aid*. Again, at this stage, we assume that the business need for the instructional document has already been verified with the ISD process (see Fig. 19.1).

Document Title: Since this type of document is written mostly as a series of steps rather than paragraphs, instead of its first sentence or two delivering the bottom-line

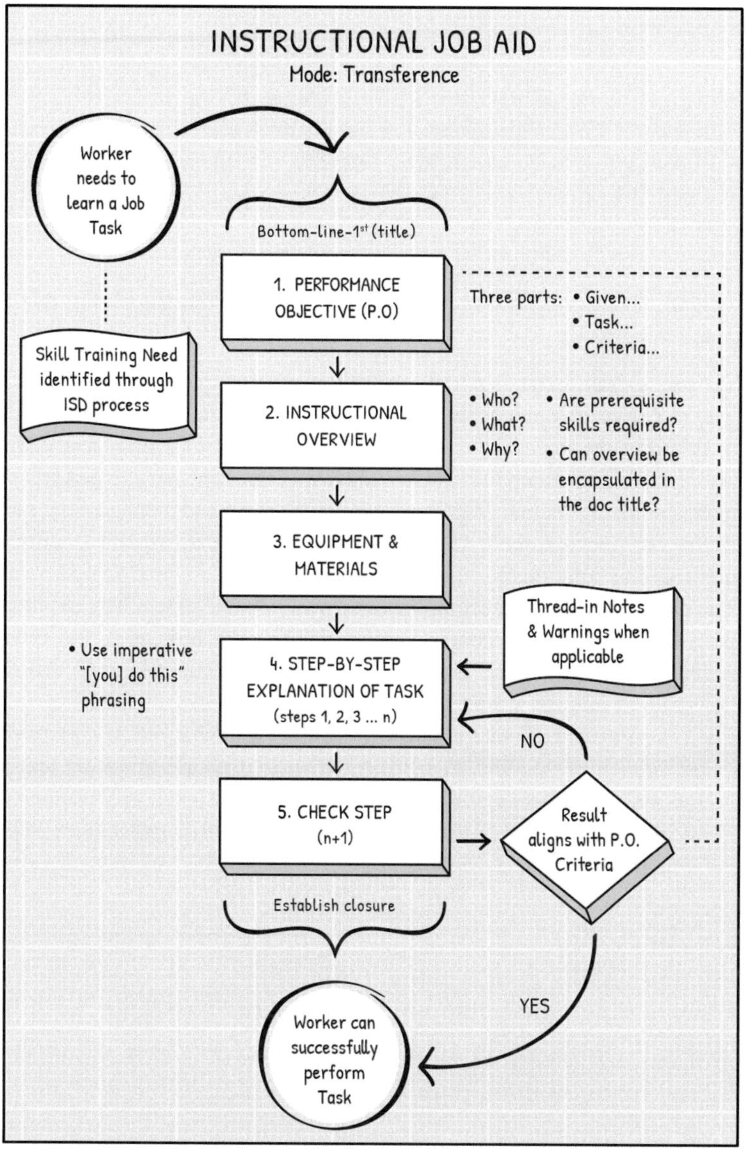

Fig. 19.2 Document algorithm for an instructional job aid (a skill-transfer doc)

(as in a short report or proposal), a job aid's title delivers the bottom-line at the top of the document's first page. Announcing the job aid's overall scope and purpose, job aid titles are often phrased, "How to [insert description of task]"—for example, "How to Calibrate the Process Control Tracking Mechanism" or "Calibrating the Process Control Tracking Mechanism."

Document Engineering Tip

Instead of using an empty, generic title, make your engineering document's title an active part of the content that the document intends to convey. You needn't wait until the first paragraph. Begin to tell your document's message right away in a focused, information-rich title. Make the title do some of the document's communication work.

Element One: The job aid's text begins by stating a performance objective (P.O.) that records three items:

- The key *resources* needed to do the task, such as required tools. A convenient way to phrase this part of the P.O. is to begin with "Given. . ." and then go on to describe the necessary resources. For example, "Given a test wafer and high-power microscope . . . ." Generally, it is *not* necessary to address performance enabling factors associated with motivation and management, because, if they are relevant, the ISD process has already identified and recommended non-training solutions for these.
- The overall *action* that accomplishes the task. A convenient way to phrase this part of the P.O. is to concisely describe the job task as a "do this"-type statement. For example, "operate the Model Z Excimer laser to drill a test-hole pattern into the wafer."
- The *criteria* (metrics) that define a job well done described as *concretely and quantitatively* as possible. A convenient way to phrase this final part of the P.O. is with a list that begins "such that . . ." and then spells out specific criteria. For example, "such that all test ablations are 20 ± 1 μm diameter holes with 5-μm maximum imprecision relative to target *X–Y* coordinates on the test wafer."

Sample P.O.

Resources:	Given a test wafer and high-power microscope
Action:	Operate the Model Z Excimer laser to drill a test-hole pattern into the wafer
Criteria:	Such that all test ablations are 20 ± 1 μm diameter holes with 5-μm maximum imprecision relative to target *X–Y* coordinates on the test wafer

Element Two: The next section of text on a job aid is a brief *instructional overview* defining the target audience, the purpose of the instruction set, as well as its significance. The overview also records any necessary prerequisite skills (if applicable).

- Note: If the job aid's overview section can be presented adequately, and more economically, in the document's bottom-line-first, descriptive title, then the title itself can serve the purpose of element two, and the job aid omits this section.

Element Three: The third element is a text section that expands on the key resources necessary for doing the job as defined in the P.O.'s statement of givens. In particular, if the job task *requires special equipment (tools) and/or materials* and these items number more than just a few, the job aid itemizes all necessary equipment and/or materials here. If each item self-identifies as either equipment or material, the items can be listed together in one bulleted (single- or double-column) list titled "Materials and Equipment." Otherwise, present two separate lists—one for required "Materials" and one for "Equipment."

- Note: If the job task does not require any special equipment or materials, or, again, if these items can be presented completely and more efficiently in the P. O., the job aid omits this section.

Element Four: The most lengthy text section of a job aid, almost always, is the step-by-step description of how to do the task as an incremental series of individual actions that lead to overall task completion. When preparing this central section of a job aid, you should keep in mind the following process advice:

- Strive to break down the job task into steps that can be fully described in one to two sentences, with one sentence being ideal.
- When you write the steps, *format them as a numbered list*, and present them in the exact order in which you intend the worker to do the steps. Sometimes you may find it useful, especially when instructing a job task that involves assembling pieces into a whole, to record a sequence of sub-steps underneath a main step. Sub-steps should also be written in one to two sentences, ideally one. Sub-steps might, for example, describe a sub-assembly of a component that must be constructed multiple times using the same sequence of actions. For example, "build four Z-type sub-assemblies by repeating the following sub-assembly steps four times in-a-row ...." Each step (or sub-step when you use them) should describe a single action that the worker should perform.
- Finally, use imperative phrasing for writing the steps—for instance, (1) *do this*, (2) next, *do that*, and so on. Recall that for imperative sentences, the subject is an invisible "you"—"[You] do this." Some examples of steps are "Turn the knob 45 degrees clockwise ..." and "Torque the bolt to 5.4 N·m ...."

Special Notices: When applicable, you can thread *special notices* between the instructional steps where they apply. Be aware of and use proper conventions, label types (signal words), and appropriate reasons when inserting "special notice" text [4]. Some common notice labels (with generic descriptions) are as follows:

- An insertion that begins "Note:" emphasizes or reiterates a particular point made by a step or a notable exception regarding the step.
- An insertion that begins "Attention:" alerts the task doer of the potential (at this point in the process) for botching up the task progression or damaging a piece of equipment.
- A "Caution:" notice alerts the task doer of a potential minor injury (burn, pinch, cut).
- A "Warning:" notice alerts the task doer of a potential major injury (loss of life, sight, or limb).
- A "Danger:" notice warns the task doer that here in the process, if precaution is not taken, major injury or death is very likely to occur.
  - Note: For all special notices associated with *safety threats/hazards* (any risk of injury to the task doer—minor or major), be absolutely certain that your text complies with your company or organization's safety regulations and required standards imposed by external regulation agencies such as the Occupational Safety and Health Administration (OSHA), American National Standards Institute (ANSI), and International Organization for Standardization (ISO).

Element Five: The final element is a *check step*. If the job aid explains how to do the task in a sequence of *1, 2, 3, . . . n* steps, the job aid does not end with *n*, the last step the doer performs to complete the job. Rather, the final step of a job aid is $n + 1$, an additional check step to ensure that the task stated in the performance objective has been completed fully and in a manner that satisfies all of the previously stated criteria. Another way to think of this step is as the action that "closes the loop." You will see on Fig. 19.2 that the check step connects with a decision block: *yes* means that the results aligns with the P.O. criteria, and *no* means that the criteria are not met and the doer is directed to try doing the task again. The check step for a job aid is often a self-directed test. Sometimes, as in the case of certification to operate production equipment in a manufacturing area, the check step directs the doer to take a test administered by a trainer or manager.

End: The desired outcome for a job aid is that the worker who uses it is now able to perform the task described in the job aid's P.O. while meeting the criteria stated in the P.O. and using the key resources catalogued in the P.O.

What about adding graphics to a job aid—line drawings, exploded views of assemblies, and digital photos showing where components are and how they fit together? Though the Fig. 19.2 algorithm flowchart does not explicitly reference graphics, certainly, with skill-transfer documents, the old adage is true that "a picture is worth a thousand words."

Therefore, between element one, the P.O., and element five, the check step, illustrations can be threaded into a job aid's text anywhere that they might serve to help a worker better understand an instructional step. Like any graphic in an engineering document, these images must be properly labeled—"Figure # Title"—and explicitly referred to in the job aid's text, prior to the figure. Occasionally, a set of instructions can be conveyed predominantly (sometimes entirely) using pictorial graphics rather than written steps. That type of document, however, is not the focus of discussion here and requires expertise in designing illustrations that most of us do not possess.

Application Example

Custom Track Fabrications (CTF) is a small-scale track-and-field equipment company that repairs and custom builds a variety of track-and-field equipment including javelins, shot puts, and hammers. Their specialty, however, is building high-performance pole vaulting poles. CTF supports a wide range of clients from high school athletes to aspiring Olympians and designs and produces approximately 250 customized vaulting poles per year. Each pole is created to match an individual client's height, weight, and vaulting technique, as well as any other special performance needs.

The ideal pole design vaults the athlete the highest from the ground and steers them successfully over the crossbar. During prototype testing, CTF actively involves the client-athlete, who often tries a selection of test poles that vary in construction material, pole length, and diameter. During the test vaults, a CTF equipment customization specialist observes each pole's performance and collects test data using a two-piece wireless strain gauge assembly attached to the pole.

The test poles are manufactured with axial guidelines on the pole's exterior that mark the athlete's grip area at the top of the pole and also a test zone for the strain gauge halfway down from the pole's top. The top grip area line and middle test zone line are linearly matched so that if the athlete grips the pole with their knuckles directly above the grip guideline, the test zone guidelines farther down the pole will "face out" when the athlete vaults. The middle-position test zone area is flexed and critically stressed during pole vaulting.

The CTF equipment design specialists assigned to work with individual client-athletes are responsible for installing wireless strain gauge assemblies in proper position with "face-out" orientation and then collecting data for engineering analysis. Recently, some of the flex data sets have been incomplete and imprecise. This led Jake Stellow, the engineering manager in charge of product production, to investigate the problem.

Through basic ISD analysis, Jake determined that no non-training issues need to be addressed and that the problem's cause is the equipment design specialists' lack of skill training in strain gauge installation. Therefore, Jake has developed a job aid document and has distributed it to the specialists to guide them to do the installations properly. Figures 19.3a, 19.3b, 19.4a, and 19.4b below display this application example's effective versus ineffective version of the job aid.

Effective Job Aid Example

JOB AID

How To Install a Strain Gauge Assembly in Face-Out Orientation on a Test Vaulting Pole

Performance Objective

Resources: Given a test pole with face-out orientation lines marking the athlete grip area and the test zone location, a two-part Model 49a wireless strain gauge assembly (gauge and transmitter/battery), and appropriate application materials and equipment

Action: properly install the strain gauge assembly on the vaulting pole

Criteria: such that the gauge and transmitter/battery are positioned securely as per Figure 1, the gauge's output terminals produce 800 $\pm$ 5 Ohms resistance when tested manually, and the transmitter sends an equivalent digital signal to a Model 49a wireless receiver.

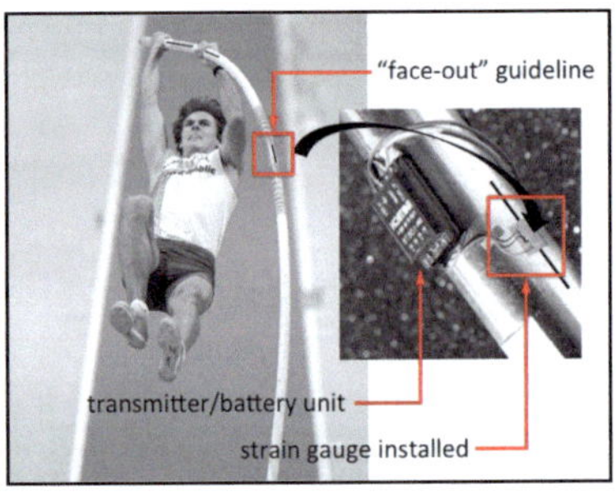

Figure 1 – Location of Gauge in Test Zone with Respect to Face-out Orientation Lines

Materials and Equipment

- emery paper
- epoxy & toothpick
- tissue paper & cellophane tape
- lacquer thinner
- soldering equipment
- tweezers
- ohmmeter
- Model 49b wireless receiver

Procedure

1. Clear off an adequately-sized, smooth, flat work surface; and situate the test pole, packaged strain gauge assembly, and necessary materials and equipment on the work surface.

Jake Stellow/031516/rev a Page 1 of 2

- *At the top, the job aid features a descriptive title.*
- *Next, there is a text block that presents a three-part performance objective (P.O.), with all necessary ingredients.*
- *Here, the writer decides that the doc's title is sufficiently descriptive and that no task overview is necessary; instead, the writer adds a properly labeled, descriptive image.*
- *This job aid includes a combined list of materials and equipment.*
- *Here is where the instructions begin.*
- *Author name, date, and revision letter are noted in the footer.*

Fig. 19.3a Effective job aid, page one [5, 6]

Note: if your work surface is raised, you may need stands to support portions of the pole that extend beyond work surface to keep the pole in a stable position.

2. Locate the black axial line (about two meters below the athlete grip zone at top of pole); this line indicates the face-out position of the surface test zone for strain measurements.

3. Using emery paper to remove the middle segment of the line and surface finish underneath, prepare a clear, roughened space for attachment of the gauge, as shown installed in Figure 1.

4. Also referring to Figure 1, use emery paper to prepare a clear, roughened space for attachment of the transmitter/battery unit. Relative to the test zone surface for the gauge, the other attachment surface is located radially about 45 degrees (counter-clockwise) and shifted up axially about 2.5 centimeters, in the direction of the pole's top and grip zone.

5. Apply cleaning agent to both application surfaces and wipe the surfaces clean with tissue paper.

Caution: Do not handle the strain gauges or their bondable terminal pads with your hands, as you may introduce oils and contaminants that will cause bonding problems.

6. Rotate the pole so the gauge installation surface is up and hold the pole in place.

7. Use tweezers to remove the strain gauge from the package and then place the gauge onto the prepared surface for attachment, making sure the lead wire terminals are on the left-side of the axis guide line as shown in Figure 1.

8. Using a strip of cellophane tape, placed gently on top of the strain gauge, lift the gauge from the application surface. Be sure to leave a hinge of tape extending off one side of the gauge to allow easy manipulation without touching the gauge's bottom.

9. Using a toothpick, apply a coating of epoxy adhesive to the entire underside of the gauge. Take care not to apply excessive adhesive.

10. Carefully re-apply the gauge onto its proper location on the vaulting pole, and press firmly down on the gauge to squeeze out any excess adhesive. Leave the tape on the gauge's top until the epoxy sets and the gauge's bottom is securely fastened to the pole.

[Steps 11 - 14, how to attach the transmitter/battery unit, not included in this example doc.]

15. After soldering, apply a small amount of epoxy to bond the two lead wires and one smaller leakage wire to the vaulting pole to finish securing the strain gauge assembly. The installed strain gauge assembly should look exactly like the installed assembly displayed in Figure 1.

16. Using an Ohmmeter, measure solder-pad to solder-pad to check the resistance of the strain gauge, now connected to the transmitter/battery unit. Resistance should be 800 ± 5 Ohms.

17. Last, check that the signal transmitted to the receiver from the fully installed strain gauge assembly reads 800 ± 5 Ohms on the receiver's digital display.

###

Jake Stellow/031516/rev a   Page 2 of 2

- *A notice is correctly inserted here, as well as farther down.*
- *All of the instructional steps are chunked into discrete, one to two sentence increments.*
- *The last action step (to complete the task) is n = 15.*
- *The final n + 1 check step (actually two steps) is n = 16 and 17.*

Fig. 19.3b Effective job aid, page two

Ineffective Job Aid Example

JOB AID

Vaulting Pole Gauge Application, Including Process and Wire Connection

Overall Description of the Task:

This procedure explains how to properly install a strain gauge assembly on one of our test vaulting poles. Specifically, this means attaching a Model 49a wireless strain gauge assembly (which includes the gauge itself and a transmitter/battery) to the correct place on the pole's surface—called the "test zone" and indicated by face-out orientation lines marking the athlete grip area and the test zone location. The accompanying images show what face-out orientation looks like and how it positions the assembly.

Necessary Items:

- emery paper
- epoxy & toothpick
- tissue paper & cellophane tape
- lacquer thinner
- soldering equipment
- tweezers
- ohmmeter
- Model 49b wireless receiver

Here Is the Recommended Installation Procedure (or Instructions):

Part A: Preparing the Surface

First, you will need to clear off an adequately-sized, smooth, flat work surface; and situate the test pole, the packaged strain gauge assembly, and the necessary materials and equipment on the work surface. *You should note that if your work surface is raised, you may need stands to support*

o *The job aid begins with a title that incompletely describes the job task.*

o *Next, there is a more detailed description of the job task and the document's purpose, but no performance objective (P.O.).*

o *Here, the writer inserts a descriptive image but does not properly label it.*

o *The list's title should specify "materials and equipment," rather than "items."*

o *When the instructions begin, they are parsed out in chunks that are too large.*

Fig. 19.4a Ineffective job aid, page one [5, 6]

portions of the pole that extend beyond work surface to keep the pole in a stable position. Next, locate the black axial line (about two meters below the athlete grip zone at top of pole); this line indicates a face-out position of the surface test zone for strain measurement devices.

Then, using emery paper to remove the middle segment of the line as well as the surface finish underneath, prepare a clear, roughened space for attachment of the gauge. Also use the emery paper to remove the pole's surface finish and prepare a clear, roughened space *for attachment of the transmitter/battery unit.* Relative to the test zone surface for the gauge itself, this other surface is located radially about 45 degrees (counter-clockwise) and shifted up axially about 2.5 centimeters, in the direction of the pole's top and grip zone. *You can refer to the images for actual location of the two parts of the strain gauge assembly, relative to a test vaulting pole.*

The next step is to apply cleaning agent to both application surfaces and wipe both surfaces clean with a clean tissue. *Please do not handle the strain gauges or their bondable terminal pads with your hands, as you may introduce oils and contaminants that will cause bonding problems.*

Part B: Attaching the Strain Gauge onto the Pole

Now that you have prepared the surface, you can rotate the pole so the gauge installation surface is up and hold the pole in place. Use tweezers to remove the strain gauge from the package and then place the gauge onto the prepared surface for attachment, making certain the lead wire terminals are on the left-side of the axial guide line. Thereafter, using a strip of cellophane tape, placed gently on top of the strain gauge, lift the gauge from the application surface. Be sure to leave a hinge of tape extending off one side of the gauge to allow easy manipulation without touching the gauge's bottom.

Using a toothpick, apply a coating of epoxy adhesive to the entire underside of the gauge. Take care not to apply excessive adhesive. Carefully re-apply the gauge onto its proper location on the vaulting pole, and press firmly down on the gauge to squeeze out any excess adhesive. Leave the tape on gauge's top *until the epoxy sets* and the gauge's bottom is securely fastened to the pole.

Part C: Attaching the Transmitter/Battery Unit onto the Pole

[note: Part C is not included in this example document.]

Part D: Connecting Part A with Part B

After soldering, apply a small amount of epoxy to bond the two lead wires and one smaller leakage wire to the vaulting pole to finish securing the strain gauge assembly. Your installed strain gauge assembly should look like the installed assembly in the images presented at the beginning. Congratulations. You have now completed the attachment process for a Model 49a Strain Gauge Assembly. When you are ready, turn on the Model 49b Wireless Receiver and begin using it.

###end###

- *The instructional steps are not numbered.*
- *As the instructions proceed, the text blocks continue to be too large, ranging from three to up to seven vertical lines.*
- *Once the last action has been described, the instructions stop.*
- *There is no check step.*

Fig. 19.4b Ineffective job aid, page two

Document Geometry: How to Format a Job Aid Instructional Document

Figure 19.5 displays a top-down, plan-view template for a one-page *generic* instructional *job aid* document. The template shows the arrangement of key elements in the form of text blocks with specific spacing schemes (document geometry). Each element is annotated to briefly explain the element's purpose. Although Fig. 19.5's template is one page, this job aid layout works well for longer documents, up to four pages and even beyond.

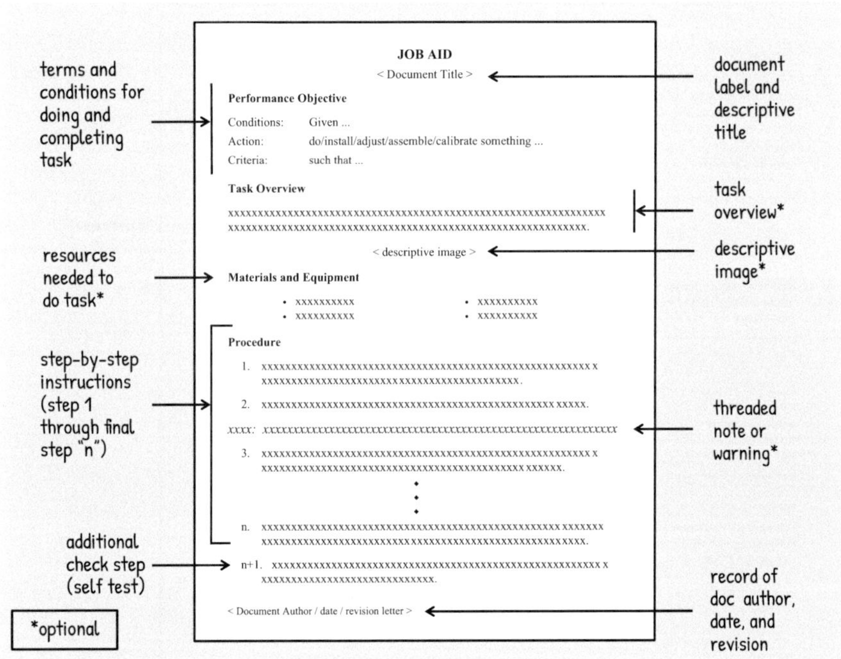

Fig. 19.5 Top-down view of a job aid

As you can see, the template's text is highly modularized, with most of the text blocks consisting of just one to two lines. Moreover, these textual chunks have bands of white space above, below, and in-between. For a generic job aid, recommended font and size is Times New Roman size 12, formatted with 1.0–1.15 spacing between lines within text blocks, and with 1.5–2.0 spacing between blocks. Headings are in **bold face**. Occasionally, the text can use *italic* for emphasis—for example, the Fig. 19.5 template uses italic text for the inserted notice between steps 2 and 3.

At the top of the page, the generic job aid begins, not surprisingly, with the label "JOB AID" centered beneath a top margin of about 0.5 in. Directly underneath the label is the document's descriptive title, with a recommended length of about 8–12 words. It is often useful to begin job aid titles with the phrasing "How To . . ."—for example, "How To Install a Servo Motor on the Lateral Transfer Robotic Arm"—or with the gerund form of the verb, "Installing a Servo Motor on the Lateral Transfer Robotic Arm." An effective job aid title should announce the document's overall purpose (to instruct) and its scope (what specific task will be explained).

The text block after the document label and title is a statement of the **Performance Objective**, which concisely defines what the task is (*action*), what resources the task doer needs to perform it successfully (*conditions*), and finally, for the task, what specific terms or metrics define successful performance (*criteria*). Although there are other acceptable phrasing strategies for performance objectives, the generic template suggests something along the lines of "Given these [things] … / do this [task] … / such that [this results] … ."

Next comes the **Task Overview**. Note that if the document's title by itself provides a sufficiently effective overview of the task, the task overview text zone can sometimes be omitted (as it was in this chapter's Fig. 19.3a and 19.3b effective job aid example).

Immediately beneath the task overview text zone, the template points out a space where an optional descriptive image might be added. The image should be centered horizontally and placed between vertically split text. Never wrap text around images. Like any document graphic, a descriptive image must be labeled "Figure # <title>" and be referred to explicitly in the job aid's text. As long as you follow these best practices, it is permissible (and sometimes very useful) to add descriptive images in other locations in a job aid—for example, threaded between instructional lines that invite use of a visual aid.

The next element shown in Fig. 19.5 is a text block with the heading **Materials and Equipment**. If the resources necessary to do the job task extend beyond those mentioned in the job aid's performance objective and the task doer will require additional materials or equipment, list them here as bullet points. The list may be in a double-column format (as shown) or single column if that format works better. It is also okay to split the text block into two separate lists—Materials first and then Equipment second or vice versa. Omit this text block when not applicable.

Thereafter comes the **Procedure** (or Instructions). Here, the job aid explains step-by-step how to do the task. As described earlier in this chapter's segment on the job aid's document algorithm (see Fig. 19.2), the step-by-step instructions are listed in the order you want the doer to perform them. Furthermore, the steps should be chunked into a series of discrete actions that can be described in one or two sentences, using imperative (verb-first, subject understood to be "you") phrasing.

When useful for clarifying an instructional step or identifying a safety risk, the job task writer can insert special notices between instructional lines. Depending on purpose and urgency, these insertions begin with different signal words: "Note: …, Attention: …, Caution: …, Warning: …, Danger: … ." The earlier segment on the job aid document algorithm discusses conventions for including notices in a generic job aid. However, your company or organization's safety policies and external safety regulations stipulated by agencies such as OSHA supersede the advice on notices given here.

Finally, the job aid template's annotation reminds writers *not* to skip or forget the "$n + 1$" step—that is, the check step(s) that must occur after the task doer completes the last action step. An effective job aid guides the job task learner to verify whether he or she has satisfactorily completed the task—*exactly* as described in the performance objective.

Additionally, in the job aid template's lower left-hand corner, you will see a place for the job task writer (sometimes called "document originator") to record authorship, date of document release, and document revision letter (a, b, c, and so on).

Action Items

α	Think of a technical task that you know how to do proficiently, something small in scope that takes you no more than about 3 min to do and requires no more than ten or so action steps. The task can be high-tech (e.g., modifying a webpage Javascript application) or low-tech (e.g., resetting a circuit breaker switch in a household breaker box). Think through how you would go about teaching the task to a new learner, step-by-step. Write down your step-by-step instructions. Do you think these instructions will work? Next, if possible, select someone to review your instructions, and then have them attempt the task. Were they successful? Why or why not?
α	Use the same technical task you selected for the above action item, or choose a new task with the same small-scope parameters. Again, think through how you would teach the task to a new learner. But this time, first develop a three-part performance objective (action, conditions, criteria) for the task; and then prepare and write down a set of step-by-step instructions for the task. The result is a mini job aid. Do you think it will work? Next, if possible, select someone to review the mini job aid, and then attempt the task. Were they successful? Why or why not? Did you remember to include a check step?
α	When's the last time you made a document title earnestly work for you? What was the document and what, specifically, was the title? Now ask yourself, why do you think this title was good—does it convey the document's purpose, scope, and significance? Does it convey something else? Next time you have the opportunity to create a title for an engineering document, make that title work even better than the example you just considered.

Recap

Key Learnings/Takeaways

- In industry, employees get paid *to do work*, **not** *to accumulate knowledge* like scholars in academe. Industry education uses a different approach than traditional school education, which centers around *knowledge transfer*. Instead, industry training centers around *skill transfer* and use of instructional systems design (ISD) to create learning that is driven by objective, measurable performance objectives.
- The basic ISD process outlined in this chapter is composed of the following *ten key elements* that can be illustrated graphically using an algorithmic flowchart, similar to a document algorithm:
  - *One:* Do a four-prong root-cause analysis to evaluate the problem relative to four key human performance factors: *skills*, *motivation*, *resources*, and *management*.
  - *Decision junction:* Ask whether the issue is a lack of required skill(s) for doing the task. If *yes*, proceed to Step Two. If *no*, explore non-training solutions to solve problems caused by lack of motivation, lack of resources, or ineffective management. If necessary, *address both* the skill-deficit problem and the other factors.

- *Two:* After identifying and characterizing the missing skill(s) necessary to do the task, determine all key resources (such as components and tools) necessary to complete the task satisfactorily.
- *Three:* Define concrete and objective performance criteria (metrics) to define successful completion of the job.
- *Four:* Prepare a three-part performance objective that records the task *action*, necessary *conditions* (key resources), and performance *criteria*.
- *Five:* Determine, as applicable, skill prerequisites.
- *Six:* Break the task into a logical, orderly series of action steps—1, 2, 3 ... *n*, where *n* equals the final action step to complete the task.
- *Seven:* Determine an $n + 1$ check step to evaluate the action steps' final outcome relative to the performance criteria.
- *Eight:* Write step-by-step instructions with threaded notices, where applicable.
- *Nine:* Test a draft of the instructional document with one or more trial users.
- *Ten:* Revise the draft if/as necessary and provide the final document to learners.

• The document algorithm for the skill-transfer document known as a *job aid* is composed of the following *five key elements*:

- *One:* After first front-loading the document with an information-rich, bottom-line-first title, the job aid's text begins by stating a performance objective (P.O.) that records three items: *resources* needed to do the task, overall *action* to accomplish task, and *criteria* to define job completion. A practical way to phrase a P.O. is "Given this ... , do that ... , so that this ... ."
- *Two:* The next section of a job aid is a brief instructional overview defining target audience, purpose of the instructions and their significance, and any skill prerequisites. This section can be eliminated if the document's title sufficiently communicates the equivalent.
- *Three:* The third element is a text section that lists additional resources (beyond those outlined in the P.O.) that are necessary for successfully accomplishing the job task—in particular, specialized *materials* and *equipment*. If not applicable, you can omit this text block.
- *Four:* This section presents a step-by-step description of how, specifically, to do the job task as an incremental series of actions (steps 1, 2, 3, ... *n*). Where useful, the writer may thread in special notices between steps that have to do with cautions, extra notes, etc.
- *Five:* The final element is a check step. If the job aid explains how to do the task in a sequence of 1, 2, 3, ... *n* steps, the final step of the job aid is $n + 1$, an additional step that confirms all criteria have been met.

• The generic typographical conventions presented in this chapter for formatting a job aid call upon the engineering writer to design the document so that it leads off with bottom-line-first title and immediately thereafter presents a three-part performance objective (P.O.). To follow, the template illustrates a series of "chunked" text blocks that include setup elements like a task overview (when applicable), the instructions themselves presented as an easy-to-read sequence of one- to two-line steps, and a concluding check step to verify proper task completion. Standard recommended type is Times New Roman size 12. Although including graphic images is optional, they can serve as highly effective devices in job aids. As usual, all graphics must be labeled and centered between split vertical text. You can insert graphics into a job aid's text anywhere readers will likely find them useful in supporting the task learning process.

References

1. Swanson, R.A., Torraco, R.J.: 1. The history of technical training. In: Kelly, L. (ed.) The ASTD Technical and Skills Training Handbook, pp. 1–47. McGraw-Hill, New York (1995)
2. Biebel, M.G.: 8. Instructional design basics. In: Kelly, L. (ed.) The ASTD Technical and Skills Training Handbook, pp. 243–265. McGraw-Hill, New York (1995)
3. Mager, R.F.: Making Instruction Work or Skill Bloomers, 2nd edn. The Center for Effective Performance, Atlanta (1988)
4. Beer, D.F., McMurray, D.A.: A Guide to Writing as an Engineer, 4th edn. Wiley, New York (2014)
5. Pole Vaulter, photo by Herbert Kratky, 123RF Stock Photo, standard license https://www.123rf.com/
6. Strain-Gage-Amplifier-Installation image, Raetech Corporation, online product brochure https://www.raetech.com/

Chapter 20
Expanding and Innovating Short-Form Documents into Long-Form Documents

Learning Objectives

- *Know a basic strategy for using the five essential document algorithms to guide the design and production of corresponding long-form (≥5 pp.) engineering documents.*
- *Know a basic strategy for using the five essential document algorithms as springboards that inform the design and production of other types of short-form (1–4 pp.) and long-form (≥5 pp.) engineering documents.*
- *Know how, in general, to expand the content, structure, and typography of a short-form document into a long-form document, as well as how, specifically, expansion manifests in an example long-form engineering report.*

In doing your job as a professional engineer, solving real-world engineering problems sometimes requires you to engage in creative, out-of-the-box thinking to *invent* new solutions or technologies. Yet invention is not always the best path for addressing engineering problems. In other cases, your education and applied experience already equip you with existing solutions, technologies, systems, or methods that match need satisfactorily or upon which you can *innovate* a new solution rather than create one from scratch.

For example, let's say you were assigned the task of designing and building a high-performance motorcycle with a state-of-the-art hydrogen fuel cell engine. Developing the motorcycle's power plant would most likely require some degree of original design. However, when deciding how to translate angular momentum into rolling motion on a roadway, you would probably not "reinvent the wheel." Instead, you would leverage/benchmark existing motorcycle wheel designs and determine a *best-fit innovation*.

Our discussion about invention versus innovation also extends to engineering writing. A good way to think of the five essential document algorithms you have learned thus far in Chaps. 15–19 is that they are core document structures that you can directly apply to short-form (1–4 pp.) proposals, status reports, final reports, instructional aids, and tech-to-nontech briefs. But they are also structures upon

B. Henderson, *A Math-Based Writing System for Engineers*,
https://doi.org/10.1007/978-3-030-10756-7_20

which you can *expand and innovate*. You can use these five algorithms, often with only minor modifications, to guide the preparation of similar long-form (≥5 pp.) documents. You can also use the algorithms as general, conceptual guides for producing a variety of other document types that you will need to write in the technical workplace.

For instance, this book's third algorithm (see Fig. 17.1 and its subroutines, Figs. 17.2 and 17.3, in Chap. 17) that defines the structure behind a product improvement project report also works well as a general guide for structuring a final design project report. In the latter case, instead of evaluating *several viable alternatives (A1, A2, A3, and so on)* with respect to *several decision-making criteria (C1, C2, C3, and so on)*, the engineering work has most likely already converged on *one selected final design concept* (D_{FINAL}) during a design review process.

Figure 20.1 below shows how you can adapt the 3 × 3 product improvement project report algorithm (shown left) into a 1 × 3 final design report algorithm (shown right). Here, we reduce the number of alternatives (A) from three (or more) for product improvement to one (D_{Final}) for final design. On the other hand, since most real-world design projects, at a minimum, tie progress to at least one design parameter (performance metric) such as target speed (#) or fuel efficiency (%), a cost/profit constraint ($), and a project time constraint (t), we keep the number of decision-making criteria (C) at three (or more) on both figures.

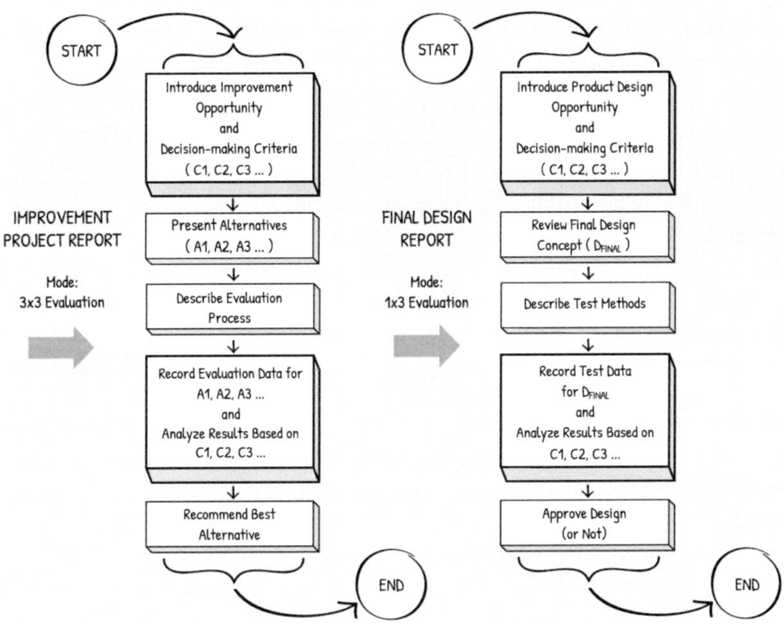

Fig. 20.1 Simplified document algorithm for an improvement project report (left) modified into a structure for a final design report (right)

Furthermore, the fifth essential algorithm, which shows how to structure a skill-transfer document or job aid (see Fig. 19.2 in Chap. 19), can inspire the document structure of an effective annual employee evaluation (performance review). Both structures, shown side by side below in Fig. 20.2, are similar because when an engineering manager writes an employee evaluation, it is critical that the review establish upfront—clearly, completely, and concretely—what the reviewed employee's assigned job duties are as well as what performance metrics (M1, M2, M3, and so on) or judgment criteria will be used to assess the employee's performance. For example, on an annual evaluation, a recorded job duty might be to *support and continuously improve manufacturing practices and processes for Product X*. A related performance metric, partnered with that duty, might be the following: *record a decreased annual accident rate and injury rate at the end of the fiscal year.*

As with most documents channeling an argument, the success or failure of an employee performance review relies heavily on front-loading and proper setup. In fact, an experienced manager knows that most of the argument setup for an employee review has (or should have) already taken place at the beginning (rather than at the end) of the review period, since this is when the manager records and activates the employee's position plan and explicitly spells out agreed-upon job duties, goals, and accountabilities for a given service period (see Fig. 20.2).

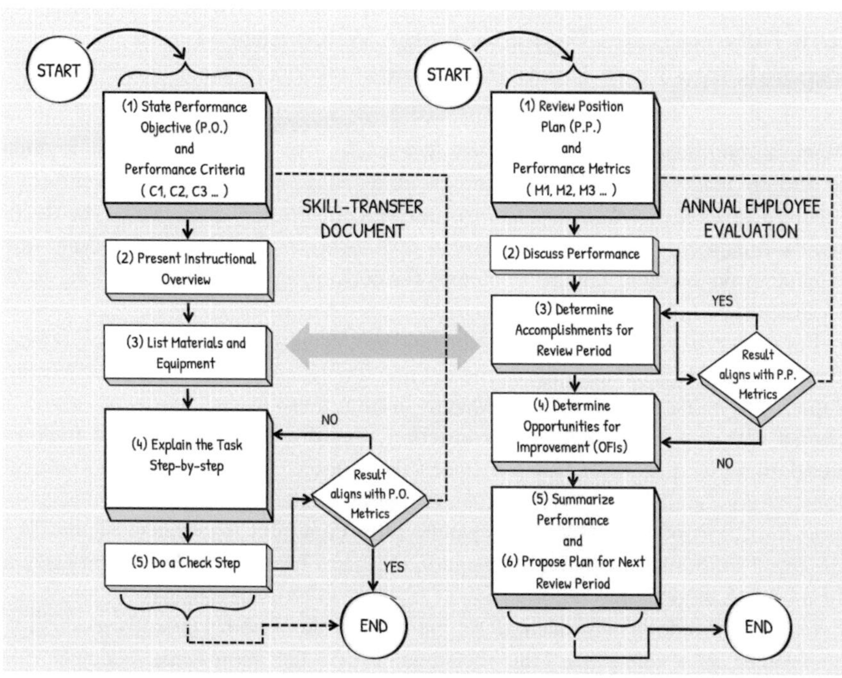

Fig. 20.2 Simplified document algorithm for a skill-transfer document (left) modified into a structure for an annual employee evaluation (right)

Routine engineering documents with predetermined algorithms and document geometries are, of course, generally easier to write than documents that are more customized and less formulaic. In some ways, however, writing any engineering document, whether routine or heavily inventive, must be done on a case-by-case basis, because each document is usually unique in two ways:

- It records *document-specific content*—with a specific level of significance, context, and necessary versus extraneous details required to precisely explain its particular content.
- It targets a *document-specific audience*—with its own wants, needs, biases, and level of technical knowledge.

Thus, the degree of uniqueness of content and uniqueness of audience must be accounted for in any engineering document we write, no matter how routine—or unconventional—that document is. It would be ideal if the five document algorithms covered in Chaps. 15–19 prescribed exact recipes for all writing situations and audiences that the engineer is likely to encounter in industry. But this is not reality. Therefore, use this book's five document algorithms with discretion, and when they don't apply, consider innovating upon them and using them as adaptable benchmarks/launching pads.

How Structure and Format Change When Short-Form Expands into Long-Form Documents

When an engineering document's length exceeds five or more pages, it is often helpful to readers, as well as common convention, to expand the document's structure to *include additional features* as well as formally modularize the document's functional elements into *labeled text sections*. Doing the preceding, and expanding page count, are the signatures for converting short-form into long-form documents.

For most types of engineering reports (and proposals and procedures, too), typical long-form structure expands the document into three main pieces—the Front Matter, Report Body, and (optional) Back Matter. In turn, these pieces divide into sections labeled with *headings* such as **Introduction**, **Results**, **Conclusions and Recommendations**, and so on. (And sometimes the sections further divide into subsections.) Additionally, long-form documents often contain features not included in short-form documents, such as a separate **Title Page** and **Appendix**. Figure 20.3 presents a simplified conceptual illustration of a short-form versus long-form version for a generic engineering report.

On the left side of the illustration, you can see the report's short-form version, symbolized as a one-piece gray rectangle. Inside the short-form rectangle, the illustration contains a sequence of typical short-form elements leading off with "the bottom-line" and then listing the key elements phrased as short prompts. For each prompt, a complete short-form report would develop a stream of sentences chunked into short paragraphs. Though not shown explicitly, we know that at the end of the short-form's text stream, there is a final sentence (or two) that establishes closure. If formatted as an email or memo, a short-form report would have a subject line title, date, and other circulation information above the body text.

On the right side of the illustration, you can see a graphic representation of the report's long-form version. Notice the three main pieces—*Front Matter*, *Report Body*, and *Back Matter (optional)*—with dotted lines around them. The *Front Matter* contains two essential elements—the **Title Page** and **Executive Summary** (first section). A significant difference between short-form and long-form reports is that long-form reports communicate "the bottom-line" (what the report is about and what it has to offer) in a summary paragraph (or sometimes multiple paragraphs) serving as the report's first section, rather than in a lead sentence (or two) at the start of the document's body text. The front matter piece, especially in reports longer than 10–12 pages, can also include additional features such as a **Table of Contents**, an **Acknowledgments** section, and **Lists of Tables, Figures, Abbreviations, and/or Symbols**.

The long-form report illustration on the right side of Fig. 20.3 also displays a separate Report Body broken into six separate text sections—beginning with the **Introduction** section and ending with the **Conclusions and Recommendations** section (note: the next segment in this chapter will describe in more detail what type

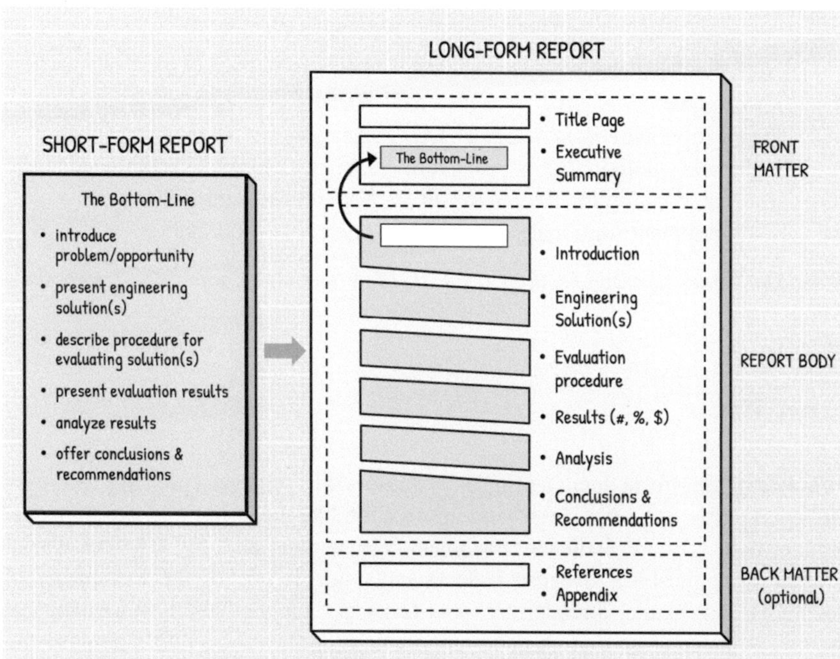

Fig. 20.3 Short-form versus long-form structure for a generic engineering report

of content each of the sections contains and will discuss additional sections that frequently appear in a long-form document's main Report Body).

Finally, Fig. 20.3's long-form illustration displays a **Back Matter** (optional) piece. Two features that regularly appear, as applicable, in a report's Back Matter are a **References** section and an **Appendix**. The References section contains a list of cited sources that have been quoted or paraphrased within the Report Body's text. An Appendix, or sometimes multiple Appendices A, B, C, and so on, contains supporting materials and attachments. Though not shown in the figure, a report's Back Matter can also contain a **Glossary** and/or **Index**.

More Detailed Section Descriptions for a Long-Form Generic Engineering Report

The previous segment provides a macroscopic overview of how short-form reports can be expanded into long-form reports using Fig. 20.3 as a visual aid. We will now examine some more detailed guidelines for developing content for the 7 standard sections of a generic engineering report, as well as 5 optional sections not displayed in the previous figure (12 sections total, besides the report's Title Page and *optional* Table of Contents):

Title Page
Table of Contents (optional)

I.	Executive Summary	VII.	Evaluation Procedure
II.	Introduction	VIII.	Results
III.	Theory (optional)	IX.	Analysis
IV.	Engineering Solution(s)	X.	Conclusions and Recommendations
V.	Test Materials (optional)	XI.	References (optional)
VI.	Test Equipment (optional)	XII.	Appendix (optional)

Note: The standard seven sections are I, II, IV, VII, VIII, IX, and X.

Front Matter

Title Page: Long-form documents usually begin with a separate Title Page recording the report's title and other classifying information like the writer's name, company/organization, primary audience, document release date, and any circulation constraints (if applicable). Because a document's title is one of the first pieces of document content that a reader sees and takes in, it is particularly important that the title is information-rich. A good report title embodies the report's purpose and scope and does so in a way that is compelling to read. Another way to think of a report title is as it being a concise phrase (target length about 8–12 words) that encapsulates or previews the document's overall message.

Table of Contents: A table of contents that lists a report's sections (and subsections) and those sections' corresponding page numbers can make a document that exceeds 10–12 pages more "user-friendly" for readers—especially those who are reading the report selectively, rather than start to finish. For relatively short reports, I recommend that the Table of Contents be omitted because it serves a less useful purpose.

I. Executive Summary: This first text section of a report, located just prior to the start of the report's main body, provides a brief, overall summary of the report's content in a paragraph (or two) consisting of about 100–250 words. Since this section is sometimes the only section read by high-level decision-makers who are chiefly interested in the project's Conclusions and Recommendations, not so much in supporting details, a good Executive Summary should emphasize bottom-line outcomes.

To be most effective, the Executive Summary should be ***autonomous***—in other words, it must make sense by itself without requiring the reader to reference material in the Report Body. A good Summary does not contain statements such as "See Table 7 on page 14," nor does it contain citations linked to References at the end of a report.

> *Note*: Since a report's Executive Summary must "summarize" a report's content from start to finish, it is generally easiest and best to *write this section last*, even though it comes first in the final document.

Report Body

II. Introduction: This first section of the report's main body introduces the report's topic and scope. The Introduction also presents pertinent background information, baseline conditions, key engineering assumptions, and a governing ***project objective(s)*** and set of ***decision-making criteria***. Ideally, the decision-making criteria (alternatively expressed as design parameters or solution criteria) are quantitative, measurable conditions and performance targets (numbers #, percentages %, costs $) that define and distinguish desired engineering outcomes.

In addition, an Introduction might also provide contextual information showing how the report relates to other in-house publications or why the report is important to the company's upper-level business agenda. Since a long-form report's Introduction is immediately preceded by an Executive Summary section, which provides an overview of the report's beginning, middle, and end, the Introduction ***does not*** necessarily need to reveal the report's final solution or outcome, nor a lot of details from the report's middle. Rather, the Introduction announces a course of action and establishes an initial reference point for engineering problem-solving/decision-making that will occur later on in the report.

> *Note:* One handy way to think of an engineering report's Introduction is to imagine it as a transcription of a tape recording that documents the initial conversation you have with your boss when he or she first assigns you a project

and gives you a set of guiding instructions and parameters. In a sense, the Introduction records an implicit work contract you enter into with your boss when you take on a new engineering project.

III. Theory (Optional): If your report's technical subject matter is particularly state of the art, conceptual, or obscure, and if your targeted technical readership will benefit from a dedicated tutorial on your subject's theoretical angle prior to learning more about the technical scope, you can include a separate Theory section. Place this section immediately after the Introduction section and prior to the subsequent sections detailing designs, problem-solving, data collection, and so on.

When you work on an engineering project and write a corresponding report, keep in mind that it will sometimes be quite adequate for you to weave any necessary theory into the standard sections of the report, such as within the Introduction and Evaluation Procedure sections, as opposed to having a separate Theory section. This is especially true when integration of pertinent theory requires only an occasional additional paragraph or two.

IV. Engineering Solution(s): Scientists explore and validate scientific hypotheses using the scientific method. Similarly, engineers develop and refine the designs of technical products (and processes) using criterion-referenced, data-driven decision-making. The next standard section of an engineering report presents one or more solution concepts that respond to the engineering project objective and decision-making criteria presented in the Introduction. If the project objective is to design a new product, this section will most likely describe one principal solution generated in a design review(s). If the project objective is to improve an existing design, the section will probably showcase a field of viable improvement options.

The Engineering Solution(s) section is descriptive, not analytical. Analysis and decision-making do not take place here. These occur later in the report after it has presented the facts and data that enable the writer to derive objective and quantitative—rather than subjective and qualitative—conclusions and recommendations.

Note: The same way that a scientific report parallels the scientific method and withholds judgment of a hypothesis until after the experimental methods and results sections, an engineering report withholds judgment of a candidate solution until after the report sections that describe the project's decision-making procedures and resulting facts and data.

V. Test Materials (Optional): If the evaluation procedure your project uses for decision-making (i.e., for evaluating one or more solutions) requires specific testing materials, the Test Materials section lists them here. The purpose behind the Test Materials section is simply good applied science. It records the ingredients (e.g., process chemicals, adhesives, test specimens, and so on) necessary to ensure that your evaluation methodology is "reproducible and repeatable."

Note: "Materials" here **are not** design build materials used to build your tested prototypes, products, or processes. If applicable, present those build materials earlier in section IV.

VI. Test Equipment (Optional): If the evaluation procedure your project uses for making decisions requires specific test equipment, the Test Materials section lists them here (e.g., an oscilloscope, ohmmeter, mounting clamps, and so on). Like the previous section, the purpose behind the Test Equipment section is simply good applied science.

Note: If the listings of Test Materials and Test Equipment are brief, it is okay to consolidate them into one combined section, "Test Materials and Equipment."

VII. Evaluation Procedure: This section records the procedures or methods that you used to evaluate how well your candidate engineering solutions fared relative to the decision-making criteria stated in the Introduction. For engineering projects, the decision-making process is usually one of the following:

- Build and test one or more prototypes.
- Do math modeling and simulations.
- Do field/application research.
- Perform a cost analysis.
- Engage in benchmarking.
- Or some combination of these.

When writing this section, be sure that it remains objective and descriptive. At this point, don't evaluate what your results mean. Here, simply explain *how you produced or arrived at the data and facts* that you will present in the *next* section, Results.

Note: If your project uses "benchmarking" to evaluate and/or judge the performance of your project's candidate solution(s), this means you have opted to characterize your solutions' performance using another similar project's findings and data. Thus, your benchmarking procedure will present how and why the selected benchmarks are viable indicators of your solution's performance and then present a set of corresponding benchmark data.

Though not as rigorous as prototyping and math modeling, benchmarking or "equivalency" is sometimes the most feasible and cost-effective source for facts and data. For instance, let's say you are a civil engineer in charge of building a mile-long, commuter-traffic bridge across a deep bay. The bridge must have a service life of 100 years. For this scenario, the projected degradation data and potential failure modes informing your bridge design might best come from real historical data from an existing 100-year-old bridge rather than from computations and simulations.

VIII. Results: This section presents the key decision-making data and facts generated by your report's solution-evaluation process (as described previously in the Evaluation Procedure section). The Results section should showcase refined, bottom-line data, not comprehensive raw data sets. Generally, large sets of raw data go into a report's Appendix. Since you have presented a project objective and set of decision-making criteria (or metrics or parameters) earlier in your report, now it is time for your report to answer its objective and criteria with criterion-referenced data.

In an engineering report, the Results section should consist chiefly of one or more table(s) and figure(s). Where necessary, you can further annotate graphics with notes, definitions, and sources. However, the principal communication vehicle in the Results section should be numbers and graphics generated from numerical data sets, not words. Be descriptive. Defer narrative analysis and discussion of "what the data means" to the Analysis and Conclusions and Recommendations sections.

Although not all engineers and technical writing experts would agree with the preceding advice to make your report's Results section "numbers rich and word poor," I strongly endorse this strategy as a best practice. The reasoning is twofold.

1. The Results section is often the message center for an engineering report, the crux of the report's argument. Your engineering readers will be eager to get to this section, because all that has come before has created in those readers an anticipation of and expectation for relevant data—data that readers want to look at and that readers expect will reveal answers to a previously stated report objective.
2. In general, engineering readers like numbers and are expert at finding meaning in numbers. Thus, design your Results section so that it gives your readers a chance to review your data's highs, lows, and trends themselves, before you play the role of tour guide (which you will do in the next section).

Note: For further discussion of how to make your Result section's tables and figures well-designed and user-friendly, refer to Chap. 17's section on document geometry and ways to insert tables and figures into engineering reports.

IX. Analysis: Now that your report has presented engineering solutions, criteria for evaluating those solutions' effectiveness, and a set of evaluation results indicating how well the solutions test relative to the criteria, you are ready to shift from description to analysis. In the Analysis section, you enter into and advance a criterion-referenced, data-driven argument, carefully and methodically reviewing the refined data displayed in the Results section and evaluating it with respect to your report's decision-making criteria.

Since the Results section precedes the Analysis section, your reader has already previewed your report's data and, most likely, has established an initial impression of what the data means. Now playing the role of tour guide, your analysis should guide the reader through each table and figure of the Results section from start to finish and point out the highs, lows, and trends that you think are significant. Don't rush here. Narrate your analysis at a steady, methodical pace. As you "walk through" each set of results, explicitly reference key numbers extracted from each table or figure—that is, partner your analysis claims with quantitative evidence.

If your report presents a single engineering solution (for instance, a final design report), be sure to analyze it relative to *all* of the decision-making criteria that you state in the report's Introduction. If your report presents a field of possible engineering solutions (for instance, a product improvement report considering viable alternatives), be sure to analyze *all* alternatives relative to *all* of the decision-making criteria. Be thorough.

X. Conclusions and Recommendations: Now that you have completed your data-driven argument, it is time for you to answer the problem or opportunity your report led with and encapsulated in a project objective. Based on the criterion-referenced observations that you presented in your Analysis section, you must synthesize your argument into conclusions and then recommend, or not recommend, the engineering solution(s) your report presents. Avoid being vague or abstract. Here again, reference key numbers to back up your claims.

In this section, you may also wish to restate key assumptions, limitations, etc. that further qualify your conclusions. You have spent the entire report up to this point staying objective and using sound engineering methodology, and, in doing so, you have established credible expertise regarding your topic. Therefore, this is one place in an engineering report where it is okay to interject your expert opinion. You have earned the right to do so, and your audience will most likely expect and appreciate it—if your report's content leading up to this point has been effectively and persuasively delivered.

Here, in addition to making one or more final recommendations, you might also consider giving your reader (especially if that reader is your manager) a bonus by extending your report's final section to include *insights and foresights*. Doing so takes the report beyond answering a stated objective with a set of conclusions and recommendations for "today" and into the realm of "tomorrow." Be visionary and suggest future actions that might leverage your project's results. For example, recommend a follow-up or follow-on activity that will multiply your Results' benefits or create spin-off projects or ideas for recycling or reusing purchased equipment or resources. Reflect on key findings that your work has generated that go beyond your project's original scope and that might lead to new opportunities for your company.

Back Matter

XI. References (Optional): In an engineering report—whether it is an internal-circulation-only report (to be seen exclusively by members of your work group, your division, or your company as a whole) or an external-circulation document (to be shared with customers, clients, or the public)—*professional engineering ethics obligate you to give credit where credit is due to outside sources*. Thus, in your report's text, if you quote, paraphrase, or summarize information, numbers, or statistics that are *not* a part of the technical "public knowledge domain," and if any one of these items comes from a source published (or owned) by someone else—an article, conference paper, brochure, book, and so on—whether in digital (online) or hardcopy format, you must acknowledge it by listing the source's author(s), title, and publication information at the end of your report in a References section. Even if you don't summarize, paraphrase, or quote directly from a source, if that source significantly informs your document's text, you should list it as a general reference.

If you do use formal citations in a workplace document—that is, tagging a paraphrase or direct quote with a marker that refers to one of the sources listed in the References section—I recommend using the IEEE citation system [1] or some similar *number-source system* [2]. With a number-source system, you tag in-text quotes and paraphrased and summarized material with a *number (between brackets, within parentheses, or in superscript)*, and then you provide a numbered list of corresponding sources at the end of your report in References. The source entries are formatted as *author (first), title (second), and so on*. You will notice that this book uses a number-source system to acknowledge sources and tags in-text references to them with bracketed numbers.

An example of a formal in-text quote using IEEE citation would be "the phenomenon of bearing friction was first explained by Pertroff using the assumption that the shaft is concentric" [1]. This citation would partner with a numbered source entry in References at the end of the report:

[1] J.E. Shigley, *Mechanical Engineering Design*, 3rd ed. New York: McGraw-Hill, 1977.

As a writer in industry, you must choose how to acknowledge references, and, most likely, your manager and your company's documentation guidelines will provide guidance.

Note: If you choose to tag source references in text, I suggest that you always tie them to a References list at the end of your report, rather than to bottom-of-page footnotes. In fact, in engineering reports, I recommend that you avoid footnotes of any kind. If you have something to note, integrate the note into the document's text. From a busy reader's point of view, footnotes are intrusive, because they ask the reader to pause and veer off the main message trail in search of auxiliary information. Typographically, footnotes also add clutter.

XII. Appendix (Optional): The last section contained in a report's back matter, if required, is an Appendix or set of Appendices A, B, C, and so on. If there is more than one appendix, they are named in the order that they are referenced in the main text. This section is meant to contain detailed information (either textual or graphic) that is not suitable for inclusion in the main body of the report text because the material is cumbersome, unpolished, or ancillary in nature (i.e., not all readers will feel compelled to look at it). Some examples of suitable Appendix material include the following:

- Large format blueprints
- Detailed calculations
- Raw data from experimental procedures
- Vendor brochures

All appendix items should be directly tied to text within the report's main body (Introduction section through the Conclusions and Recommendations section) via parenthetical references—e.g., "(see Appendix B)." Readers who seek either a top-level or abbreviated understanding of a report's content will often *not need to consult* items in the Appendix—and, in fact, view them as invisible. On the other hand, readers who seek a complete technical understanding of a report may need to consult items in the Appendix and then factor that information into their overall understanding of the report's content.

Document Engineering Tip

If your engineering report's Results section presents a series of tables or figures that report similar types of data sets, and these sets present incremental and individual quantities that produce averages or sum totals used for principal decision-making, consider consolidating the data averages or sums in one "snapshot" table or figure that displays the "bottom-line" numbers in a convenient "one-stop shopping" manner. Since a snapshot table or figure presents the bottom line, also consider placing it (front-loading it) at the beginning of the Results section.

Paragraph Typography

The general best practices for formatting the typography, or document geometry, of long-form documents are similar to those of the five short-form documents previously presented in the Part III chapters. *First*, type sentence streams (paragraphs) in long-form documents as flush-left, ragged-right blocks of text that are relatively short in length—about eight vertical lines or fewer. *Second*, surround the blocks of text with bands of blank space, before and after, as well as on the left and right. As established earlier, presenting text this way makes it easier and quicker for readers to scan, discern, and upload.

Also similar to short documents, the recommended width for top, bottom, left, and right page margins—that is, the band of blank space around the perimeter of a page—is 1.0 in. Unlike short documents, where recommended vertical spacing for lines within text paragraphs is 1.0–1.15, for long documents, I recommend 1.35–2.0. For spacing between paragraphs, both short and long documents share the same convention: make it double that of a document's vertical spacing for lines within text paragraphs. To accomplish this, at the end of a paragraph, either press the keyboard return key twice or set up your word processor's paragraph spacing (i.e., number of points inserted before and after) to produce the equivalent when you type one return. For instance, in Microsoft Word, for standard size 12 font, setting paragraph formatting to insert 0 points before and 12 points after each paragraph will achieve the "double-space" equivalent of two returns with standard spacing, 0 points before and 0 points after.

To ensure user-friendly reading for hardcopy documents, I recommend that you routinely type sentences and paragraphs using a size 12, *serif* font such as *Times New*

Roman. A serif font means typed characters incorporate wide variations within, as well as small curls and finishing strokes on the ends of, the main strokes of letters, numbers, and symbols—for example, "**T**," "**1**," and "**&**." For hardcopy pages, a serif font is easiest and quickest to read. For digital documents displaying on-screen text that resides on a backlit electronic page rather than on a top-lit paper page, the recommended font type is *sans serif*, wherein the main strokes forming characters are without (sans) embellishment—for example, "**T**," "**1**," and "**&**." A recommended example of sans-serif font is *Calibri*. Since screen size displaying digital documents can vary widely, so, too, does the best choice for online font size. For emails, generally the program's default text type and size are the best choices.

The above best practices apply to a generic engineering document's main body text. Text formatting for a report's front matter and back matter pages varies based on a specific document type. For further guidance, I recommend that you refer to your company's documentation standards and models. For a general glimpse at what a report's textual typography looks like for a generic engineering report (front matter, report body, and back matter), you can examine the page formatting displayed in this chapter's long-form document example.

Typing a Document's Title and Headings

For most documents, long and short, the same general recommendations apply for the typography of the document's title and its text body headings, as differentiated by a standard first-level, second-level, and third-level hierarchy. To distinguish the document's title, type it in bold face font and capitalize the first letter of the first word of the title and subtitle, and the first letter of all other words that follow, except for articles and short prepositions such as "in," "on," "of," etc. To position a title on the page, center it horizontally. Although opinion varies, I recommend that you do not type your documents' titles in all capital letters.

To distinguish a first-level (primary) heading, type it in **bold face font** style and use the same first letter capitalization scheme as described above for a main title. Position a first-level heading flush left, and double-space it above the text block it heads. To distinguish a second-level (sub) heading, type it in *italic font* style and use the same first letter capitalization scheme as described above for a main title. Position a second-level heading flush left, and double-space it above the text block it subheads. Finally, for a third-level heading, also position it flush left relative to the text it subheads. Type it in italic font with *only* the first letter of the first word capitalized (like in a sentence), and then end the title with a colon and a space. Start typing the paragraph's first sentence immediately after the space following the colon on the same line.

Here is an illustration of the above hierarchy displaying the appropriate type style and capitalization:

Title
First-Level Heading
Second-Level Heading
Third-level heading: …

Here is a mock-up example that demonstrates correct horizontal spacing, as well as type face and capitalization:

Final Design for the Falcon 20C Engine's Main Turbine Blade Profile

Failure Mode and Effects Analysis (FMEA)
Xxxxxxxxxxxxxxxxxxxxxxxxxxxxxxxxxxxx …

Test Procedure for Space Radiation Exposure
Xxxxxxxxxxxxxxxxxxxxxxxxxxxxxxxxxxxx …

Step one: Xxxxxxxxxxxxxxxxxxxxxxxxxxxxxxxxxxxx …

What follows next is an example of a basic long-form engineering report. The report incorporates many of the standard features, pieces, labels, and formatting conventions that have been discussed thus far in this chapter. See Figures 20.4a, 20.4b, 20.4c, 20.4d, 20.4e, 20.4f, 20.4g, 20.4h and 20.4i below. The algorithm guiding this long-form document's structure is the same one that guides the structure of the essential short document covered in Chap. 17—the algorithm for a product or process improvement project report including a 3 × 3 evaluation (three viable options judged against three decision-making criteria).

A Model Long-Form Engineering Report (3 × 3 Evaluation)

High Efficiency Solar Panel Design:

Increasing Efficiency through Nanotechnology

XYZ Solar

month/day/year

[Sample Engineering Report]

prepared by: Writer Name
Design Engineer

prepared for: Reader Name
Project Manager

o *This long-form report begins with a title page with all the standard items.*

o *Notice that the document geometry is sparse—with items arranged proportionally around the page's center, rather than top.*

Fig. 20.4a Model engineering report (3 × 3 evaluation), cover page

Executive Summary

In order for solar panels to be a feasible source of green energy, they need to be both affordable and efficient. XYZ Solar's new thin-film solar panel model has a significant advantage in the current market with respect to low cost and ease of installation, but unfortunately, it is less efficient than competing designs. We have addressed this problem by adding nanostructures to the outer surface of our product's thin-film panels in order to make its design superior to the competitions'.

Through the use of mathematical simulations and benchmarking, we determined that silver nanostructures significantly increased the efficiency in the lower wavelengths but absorbed too much light in higher wavelengths. To correct this problem, we evaluated different additions of dielectric silica for their effect on our thin-film solar panel's efficiency, price, and lifespan.

We found the best overall design was the one that incorporated alternating silver and silica nanostructures. While this design produces a slight increase in product price and decrease in sustainability, it has a substantial increase of 40% efficiency relative to our current baseline model. This design will give XYZ Solar a distinct advantage in the thin-film market because it offers a both economical and high-performing alternative energy option.

o *The Executive Summary is a concise, standalone (autonomous) overview of the whole report—revealing both the initial project objective and end result (bottom line).*

Introduction

Heightened awareness of the waste and toll on the environment caused by burning fossil fuels and the destructive bi-products of nuclear power has led to a wide-scale call for better energy options. Industry has been responding with a push to develop cleaner, more efficient, green energy solutions, and one promising alternative is solar energy. Solar energy is safer to manufacture than other alternatives, and has a much smaller negative impact on the environment. Solar energy is also sustainable, for the next billion years or so, and best of all, it is free.

o *The Introduction begins by establishing the topic, scope, and document significance.*

A significant obstacle to widespread use of solar panels is the cost of manufacturing them and the amount of time it takes for installed panels to recoup investment costs and generate steady-state savings. As a result, XYZ Solar has moved toward the use of thin-film solar cells. These cells can be manufactured quickly and with less waste. Since the panels are thinner, less material is needed and the manufacturing cost can be significantly lowered. Thin panels are also easier to

Writer Name – 1

Fig. 20.4b Model engineering report (3 × 3 evaluation), page one

install and can be installed onto a more diverse range of surfaces. This makes them more competitive with other green energy options.

However, at present, these features come with a significant decrease in efficiency. Our thin-film model, usually made with silicon, absorbs far fewer photons of solar energy than our regular panels, and therefore needs much more surface area to match the performance of our standard-sized panel models. Several different methods have been investigated to address this problem. One of the most promising new alternatives is adding nanostructures to the reflective surfaces of the thin-film panels.

The new method attaches nanostructures to the top of the panels in order to increase the reflectivity of incoming light. This allows the silicon base to absorb more photons and therefore make up the efficiency lost in the thin-film design. Nanostructures come in many different configurations and materials, each with their own benefits and disadvantages [1].

This project seeks to improve and optimize our new thin-film solar design by studying three different nanostructure models and evaluating them based on their *efficiency*, *price*, and *sustainability*. The efficiency measured will be the quantum efficiency of the transmitted energy into the base silicon. Price will be evaluated as the additional manufacturing costs of the different configurations of nanostructures. Sustainability will be determined by the number of years the panel can be expected to work reliably and without need of significant maintenance.

Thus, this project's overall objective is to produce a new thin-film solar panel design that is cheap, efficient, and long-lasting that will give XYZ Solar an edge over current market options from our competitors.

Product Improvement Alternatives

Option One

The first configuration focuses on using metal nanostructures to improve the efficiency of the thin film panels. Metal has high reflectivity properties and increases the scattering of light. If more light is reflected back onto the structures, the silicon is able to absorb more photons in a smaller area. Various modeling tests compared different metals, including aluminum, gold, silver, and copper, against the base performance of silicon. These tests determined that silver had the highest increase in transmitted energy efficiency.

Writer Name – 2

o *The Introduction further explains the problem at hand with technical details.*

o *Toward the end, the section explicitly states the project's decision-making criteria—efficiency, price, and sustainability—as well as the project objective.*

o *Here, the Product Improvement Alternatives section is the equivalent of a generic Engineering Solutions section.*

Fig. 20.4c Model engineering report (3 × 3 evaluation), page two

The first option embeds silver nanostructures onto the upper reflective surface of the solar panels. We ran mathematical simulations to determine the optimum placement and size of the silver nanostructures. The software measured the efficiency per period of several layouts with different spacing, and determined that the closed packing structure had the greatest increase in efficiency. A closed packing structure places one nanostructure followed by a space equal to its diameter. These spaces allow for better light trapping between the nanostructures and therefore the absorption of more photons. In addition, we compared different radii to balance the increased light reflected per unit area versus the light absorbed by the silver. In a range of 25 to 100 nanometer radii, 90 nm was recorded as being the most efficient. (See Figure 1.)

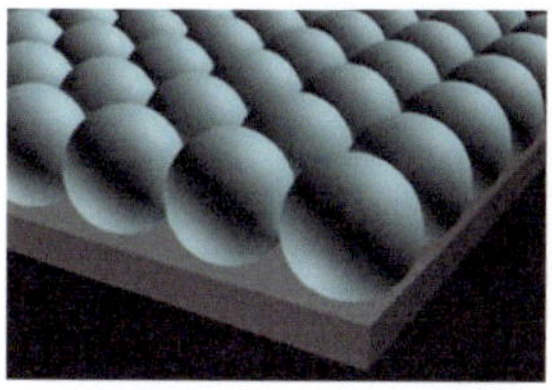

Figure 1: Only Silver Nanostructures Simulation

Source: *High-efficiency Solar Cells: Physics, Materials, and Devices* [2]

Option Two

The second option studied in this project retains the optimum closed packing configuration determined previously and embeds it within a layer of dielectric silica (SiO_2). Dielectric silica is a material that acts as an electrical insulator because of it high electrical resistance. Since silica has a low electrical conductivity, it retains excess energy and is capable of storing electrical energy via polarization in the form of an internal electric field. This layer also acts as a protective coating that shields the nanostructures from the environment and increases the lifespan of the solar panels. Various simulations modeled the best thickness of the silica layer and we determined that 120 nm led to the highest efficiency solar cell.

The second configuration takes into account a significant disadvantage of using silver configurations. Silver nanostructures have a strong reflectivity in the lower wavelengths but

Writer Name – 3

o Notice how each subsection, labeled "Option One," "Option Two," etc., describes, rather than evaluates/judges, each viable alternative.

o To establish an organization that is predictable and quick and easy to read, the text presents all three Options similarly, using a consistently patterned delivery and formatting style.

Fig. 20.4d Model engineering report (3 × 3 evaluation), page three

absorb light in the higher wavelengths. This decreases their overall efficiency though they are still more effective than the base silicon model. This option two design combats the loss of energy due to absorption by embedding the silver nanostructures in a layer of dielectric silica. Silica also has an increased reflectivity, though lower than that of metal, but silica absorbs far less light. With silver's increased reflectivity in low wavelengths and silica's reflectivity combating the loss of energy from absorption in high wavelengths, this design takes advantage of the benefits of these two different materials and has a more consistent and efficient capture of photons and transmission of energy. (See Figure 2.)

Figure 2: Silver Nanostructures in Silica layer

Source: *High-efficiency Solar Cells: Physics, Materials, and Devices* [2]

o The figure is properly labeled and also properly positioned using a "split text and center" approach.

Option Three

The third design option tested in this project was that of alternating nanostructures of silica and silver added to the reflective surface of the original silicon panel. This option attempts to limit costs and resource consumption by using nanostructures instead of an entire layer of silica.

This design retains the closed packing method of the first option but alternates the material that the nanostructures are made of. Mathematical simulations determined the optimum ratio of silica to silver radii to be 1:2 allowing for both increased light reflection and capture as well as decreased light absorption. The best radius was then chosen from a range of 40 nm to 100 nm to be 80 nm silica and 40 nm silver nanostructures. (See Figure 3.)

Writer Name – 4

Fig. 20.4e Model engineering report (3 × 3 evaluation), page four

Figure 3: Alternating Silver and Silica Nanostructures

Source: *High-efficiency Solar Cells: Physics, Materials, and Devices* [2]

This method of studding the panel allows the nanostructure to reflect more light with greater access to sunlight. With the added scattering from the dielectric structures, the radius of the silver nanostructures can be decreased to 40 nm without a significant decrease in the light scattered and energy transmitted. This decreases the amount of light absorbed by the silver and improves the efficiency of the overall design.

Evaluation Procedure

Since the materials used in this product are costly and complicated to make, it was more reasonable to use mathematical modeling than a prototype for the measurement of efficiency. Simulations also enabled testing of the solar panels under a diverse range of conditions and projected time periods that would not have been realistic to attempt to physically measure.

The mathematical software this project used was Lumerical FDTD Solutions. FDTD Solutions is a 3D solver that analyzes the interaction of ultraviolet, visible, and infrared radiation over a wide range of complicated structures and surfaces. This software is able to simulate the entire spectrum of sunlight as well as replicate the period and movements of individual photon particles.

We used the FDTD software to compare the differing quantum efficiencies. Quantum efficiency indicates the amount of current a solar cell produces when bombarded with photons from sunlight. Here, more specifically, the internal quantum efficiency characterized the ratio of electrons captured by the cell to the number of photons it absorbed. The software plotted

Writer Name – 5

o *The Evaluation Procedure remains objective and focuses on description, not judgment, as did the previous section.*

o *Here, the author chooses to use the active voice—"We used" vs. "The FDTD software was used." This is good.*

Fig. 20.4f Model engineering report (3 × 3 evaluation), page five

quantum efficiency as a function of wavelength allowing us to not only compare efficiencies but also see at what wavelengths designs functioned best and where there needed to be improvements.

We compared different types of materials against the base silicon to determine the most efficient choice. Our procedure also determined optimum configurations and the changes in performance that occurred when more than one material were combined. This significantly reduced the time and money needed to determine the best conditions for maximum efficiency.

We analyzed product cost using current market prices of silver and silica. We focused on the additional manufacturing costs of the nanostructures per panel and not on the money already required to build the base solar panels. Silver was currently valued at five times the price of silica and therefore had a larger impact on the overall cost of the nanostructure design. We compared cost per solar panel to determine the price range of the three options in order to determine which design would generate the most profit. It was important that the prices not significantly increase relative to the current silver only option, since the goal of this project is to increase the commercial feasibility of thin film solar panels, while maintaining low manufacturing cost.

We measured sustainability by benchmarking. This was necessary since, aside from prototypes being costly and difficult to make, we concluded that there were too many confounding variables for a realistic simulation of several years of service. We compared our first two design options with similar solar panel designs already tested and presented in studies by Blacklett Laboratory in London. Just as our designs feature metallic structures that protrude from a silicon base, the Blacklett panel designs had aluminum studs embedded into their surfaces. These types of textured surfaces present a similar degree of wear and exposure risk. Therefore, we approximated how long our design option one and design option two might last in service based on the Blacklett study's published data.

For the third option with the protective layer, we benchmarked the option against a panel currently on the market from XYZ Solar. This panel mitigates wear and exposure with a protective glass shield. To project relative lifespan, we assumed that our third option's materials would offer similar protective properties.

Writer Name – 6

o *The paragraphs explain how engineering produced the decision-making data—presented next, in the Results section—and not what the data "means."*

o *Notice that most sentences use past tense verbs—for example, "We measured" not "We measure" In engineering reports, verb tense can vary from past to present to future. However, past tense is usually the preferred choice for documenting procedures or methodology sections.*

Fig. 20.4g Model engineering report (3 × 3 evaluation), page six

Results

Table 1: Average Quantum Efficiency

Design	Average Quantum Efficiency
Current Silicon Only	0.5
Only Silver Nanostructures	0.70
Silver Nanostructures with Silica Layer	0.85
Alternating Silver and Silica Nanostructures	0.9

Table 2: Additional Price per 1000 cells

Design	Additional Price ($ per 1000 cells)
Only Silver Nanostructures	83.8×10^{-6}
Silver Nanostructures with Silica Layer	642.0×10^{-6}
Alternating Silver and Silica Nanostructures	8.29×10^{-6}

Table 3: Expected Lifespan under optimal conditions

Design	Expected Lifespan (years)
Current Silicon Only	20
Only Silver Nanostructures	15
Silver Nanostructures with Silica Layer	25
Alternating Silver and Silica Nanostructures	17

Analysis

The first option, only silver, has an efficiency of 70% compared to the baseline efficiency of silicon only at 50%. There is an increase in price because of the high cost of silver. However, since very small amounts of silver are used per panel, cost is within an acceptable range. This option's textured surface caused by the addition of the nanostructures leads to faster deterioration of the solar cell, and decreases the amount of time the solar panels can be expected to work reliably by about 5 years as compared to our silicon only baseline.

Writer Name – 7

o The Results section presents three complete data sets that align with the report's three decision-making criteria.

o The tables are properly labeled and positioned on the page.

o The section focuses on presenting decision-making numbers. There is no additional text discussing or narrating the data, since this occurs in the next section.

o The discussion of the data—its highs, lows, trends, and what these "mean"—begins here.

Fig. 20.4h Model engineering report (3 × 3 evaluation), page seven

The second option of embedding silver nanostructures in a silica layer has an average efficiency of 85%, which is a significant increase (35%) from the silicon only option and a 15% increase over the pure silver design. The addition of a silica layer does, however, substantially increase the price because of the large amount of extra material needed. This design costs almost 8 times more than using only silver nanostructures. This model uses a layer of dielectric silica that helps protect the nanostructures and solar cell and is therefore more sustainable because of the flatter and more durable surface. This allows the solar panel to last up to 25 years, giving it the longest lifespan of the three designs and making it the only design with an increase in sustainability from the original model.

o The evaluative discussion of the data continues, cycling through a repeating delivery pattern for all three alternatives relative to all three criteria (3 x 3).

The third design option contains a pattern of alternating material nanostructures that combines the strength of silver at low wavelengths with the silica's high refractive index at higher wavelengths. This results in a steady efficiency of around 90%. This is the most efficient design of the three and has a significant increase from that of the current silicon model. The smaller radius of the silver nanostructures and using silica nanostructures instead of a layer also make this the cheapest of the three configurations. There is a 10x decrease from the first design and nearly an 80x decrease from the second. The 17-year expected lifespan of this model is better than the first option by about 2 years, but is still 3 years less than our current design at 20 years.

o The Analysis section directly cites numbers from the tables in Results.

Conclusions & Recommendations

We recommend using the third design option of alternating silver and silica nanostructures. This model best meets the project criteria of increased efficiency at 90% quantum efficiency. It also has the smallest increase in manufacturing costs. Although this option has a 3-year decrease in lifespan compared with our original cell, we consider its sustainability rating to be acceptable. Overall, the alternating silver and silica nanostructures' combined high efficiency and low manufacturing cost make it our best option for releasing an innovative new thin-film technology and gaining the advantage in the solar panel market.

o Finally, the writer draws conclusions from the Analysis section's discussion and synthesizes these into a recommended best option.

References

[1] Y. A. Akimov, W. S. Koh, S. Y. Sian, and S. Ren, "Nanoparticle-enhanced thin film solar cells: Metallic or dielectric nanoparticles?", *Applied Physics Letters*, 96:7, 2010.

[2] X. Wang and Z. M. Wang, "Light Trapping for Solar Cells," *High-efficiency Solar Cells: Physics, Materials, and Devices*, Switzerland: Springer International, 2014.

o A References list acknowledges the two information sources cited in the text.

Writer Name – 8

Fig. 20.4i Model engineering report (3 × 3 evaluation), page eight

Action Items

α	Consider the various types of technical documents that you write at work (or in engineering school). Among these, pick a document that you think has a structure that is different (somewhat or even radically) from those connected with the five essential documents covered in Chaps. 15–19. For example, maybe it's a trip report for a professional trade show at which you presented on and answered questions about a new product and served as your company's representative subject matter expert. Keeping in mind this report's principal readership—most likely, a manager—think about what this document's key functional elements are and how each might be situated in a customized document algorithm.
	Then think about element sequencing—what needs to be communicated in the document's introduction, its middle, and its end? Now sketch and label a draft of your version of this algorithm on a blank sheet of paper. Do you see any elements in your sketch that are significantly different from those that make up other types of technical documents that you're familiar with? Do you think that it will be helpful for you to save your customized algorithm and use it as a future writing reference?
α	Continue to consider the same technical document that you identified for the above item—a document that you think has a structure that is different (somewhat or even radically) from those connected with the five essential documents covered in Chaps. 15–19. Now consider this document's document geometry or typographical format. Is there anything unique about this document's geometry—how you type it and place it on the page (or screen)? Conversely, what about this document's geometry is similar to the designs of this book's five essential documents?
α	One segment of this chapter explored what goes into the various sections of a generic long-form engineering report. For this segment, what would be three things about report sections that you already knew about, and what would be three things that you didn't previously know about (or haven't always considered) in report writing?

Recap

Key Learnings/Takeaways

- In engineering, oftentimes it is creative, out-of-the-box thinking, rather than standard formulaic thinking, that leads engineers to *invent* new products and technologies. Yet invention is not always the best path for solving engineering problems and doing engineering work. Sometimes it's best not to "reinvent the wheel" and instead to use or *innovate* upon what already exists. This philosophy also applies to the design and selection of the appropriate engineering document structure for a specific workplace writing scenario.
- A good way to think of the five essential document algorithms presented in Chaps. 15–19 is that they are core document forms that engineering writers can directly apply to short-form (1–4 pp.) proposals, status reports, final reports, instructional aids, and tech-to-nontech briefs, and they are also structures upon which to expand and innovate. You can use the five essential algorithms, often with only minor modifications, to guide the preparation of similar long-form

($\geq$ 5 pp.) documents. You can also use the algorithms as conceptual guides for producing a variety of other document types that appear in the technical workplace—such as a final design report or annual employee evaluation document.

- Aside from page length, there are other features that distinguish a short-form document from a long-form document. A short-form document typically begins with the author's name, document title, circulation information, and so on at the top of the first page and then immediately follows with a stream of text sentences that are chunked into small paragraphs for quick, easy reading. A long-form document usually is more formally modularized into sections with section headings and is divided into three main parts—the Front Matter, Report Body, and Back Matter. For long-form documents, author name, document title, circulation information, and so on often appear on a separate title page. Moreover, the "bottom line" is delivered in a front-loaded section, the Executive Summary.
- When a generic engineering report expands from short-form to long-form, typically the report begins with a **Title Page** and other front matter items, and then the report's middle and back matter pieces break out into the following sections, labeled with section headings (briefly described below):

  – *Front Matter*

    Executive Summary – presents a miniature version of the whole report.

  – *Report Body*

    Introduction – presents an engineering problem/opportunity and its solution criteria.
    Theory (optional) – explains the theory behind an applied engineering approach.
    Engineering Solution(s) – describes a candidate design or field of alternatives.
    Test Materials (optional) – records necessary items.
    Test Equipment (optional) – records necessary items.
    Evaluation Procedure – describes how results data were produced/generated.
    Results – showcases the results data using tables and figures and little narrative.
    Analysis – discusses what the data means in a comprehensive text narrative.
    Conclusions and Recommendations – answers the original problem/opportunity.

  – *Back Matter*

    References (optional) – credits outside sources.
    Appendix (optional) – provides additional materials that support the report's text.

References

1. IEEE Editorial Style Manual, IEEE Periodicals Transactions/Journals Department, version 8 (2014)
2. Beer, D.F., McMurray, D.A.: A Guide to Writing as an Engineer, 4th and 5th edns. Wiley, New York (2014)

Chapter 21
Twenty Universal Features of Excellent Engineering Documents

Learning Objective

- *Know 20 universal features that distinguish excellent engineering documents regardless of length.*

Aside from using the recommended document algorithms and geometries to guide the design and production of your engineering documents, I suggest that you also be aware of and incorporate (whenever useful and applicable) a set of essential features that often distinguish effective versus ineffective engineering documents. Below, I have put together a list of 20 of these hallmarks of excellence. The list is not meant to be absolute or comprehensive. I chose to highlight these 20 features based on my own practical experience as well as their alignment with and repeated appearance within the preceding chapters.

The list presents each feature as a "do statement" with a brief explanation of how to integrate it into a document. The features are organized not by degree of importance (they are all important) but rather by where they would usually be inserted into the body of an engineering document—beginning, middle, end, or throughout.

Document's Beginning

1. *Front-load the bottom line.*

    - In the first sentence (or two) or in an initial executive summary section, tell the reader in a nutshell what the document is about and what it requests or has to offer. Bear in mind that good document titles also work as front-loading devices. Titles are not meant to replace a document's initial bottom-line statement, but a good title can certainly enhance it.

B. Henderson, *A Math-Based Writing System for Engineers*,
https://doi.org/10.1007/978-3-030-10756-7_21

2. *Deliver a complete statement of the document's topic, scope, and significance* in the introductory segment.

    - Once your document delivers the bottom line, instead of lingering on preliminary top-level discussion and stage setting, most engineer readers will be eager for your content to scope down to specifics related to engineering action(s) and problem-solving. Typically, the reader will get the most out of these specifics and make the best sense out of them if he or she fully understands why the document's content is significant for the company's business and success.
    - The reader will also usually benefit from knowing how the document's content connects system components to bigger systems—as well as why a design is commercially viable and functionally relevant.
    - For example, if the report is about a gear design, the reader will probably need to understand how the gear's teeth connect with other gears, and what, in turn, the overall gear system does (What does it control? What does it transfer power to?).

3. *Present pertinent background information or existing baseline conditions* associated with the problem/opportunity being addressed in a document.

    - Especially for proposals that argue for the *need to do an engineering project* and for reports that argue for the *acceptance of a best design or solution*, establishing the "givens"—that is, what's present to begin with, what's missing, and how an existing problem or opportunity is being dealt with (imperfectly or not at all)—is a critical part of argument setup.
    - Baseline performance conditions often serve as useful reference points for ending arguments. If your project aims to improve an existing product or process, ultimately, your project's final report will need to show that your recommended solution's performance is superior and worthwhile (rather than marginal and unworthy) as compared to any existing solution's baseline performance.

4. *Record the project or activity's objective* established by you, your team, a secondary party (e.g., your manager), or a third party (e.g., the client/customer). And unless your connection to the objective and to the second or third party is implicitly obvious, record this connection too.

    - Project or activity objectives steer engineering work. Objectives also direct and anchor the successful development of project results and conclusions in project reports. A written project or activity objective serves as a contract—that is, *what I/we/my team will do*—and thus establishes a commitment that will be carried through, revisited, and, hopefully, fulfilled by the document's end.
    - Stating a project or activity's objective upfront also establishes boundary conditions. Clearly stated objectives contain reader expectations and

disengage the writer from having to report or discuss material that might be of interest to the reader but yet extends beyond the project's purpose and scope.

5. *Record the project or activity's target outcomes* as concretely and quantitatively as possible.

    - Target outcomes define a project or activity's tangible end goals and also set a precedent for how results will be assessed and measured. In engineering reports and proposals, target outcomes are often expressed as a set of *design parameters*, *design specifications*, or *performance specifications* that a new or improved product or process must meet and guarantee.
    - Strive to characterize target outcomes as concretely, objectively, and quantitatively as possible (using # numbers, % percentages, and/or $ costs). For instance, for a project designing a new hovercraft, an effective target outcome might be "the hovercraft's top speed must *meet or exceed 100 kilometers per hour*." A less effective target outcome (abstract and qualitative) would be "our new hovercraft must be *adequately fast*."

Document's Middle

6. *Strive to record and explain each critical thinking process you use and each assumption you make* (or, in the case of proposed future work, that you will have to make) as you advance the discussion of your project or activity.

    - Engineering work incorporates math and science tools and principles. Yet because engineering is an applied science, engineering work also relies in part upon the engineer's expert interpretations and well-reasoned assumptions. As a result, how real-world engineering work gets done can vary from project to project and from engineer to engineer based on individual problem-solving styles and choices.
    - When you are writing an engineering document, be sure to clearly and completely explain your critical thinking approach and the assumptions that support your text's development. It is never good to assume that your methods need not be stated explicitly because they are "obvious" or easily inferred by technically competent readers.

7. *Explain your project or activity's methodology for testing and data collection*—and also explain the data itself.

    - As with your critical thinking processes and assumptions, be sure to clearly and completely explain testing procedures and experimental setups, data collection and analysis, and the data itself. The impetus behind this best practice is the *scientific method*. For the facts and numbers that your report presents to be credible in the eyes of engineering readers, the processes generating those facts and numbers must be sound, logical, and (ideally) *reproducible and repeatable*.

8. *Insert best-choice tables and figures* to complement your document's textual messages, accent key points, and showcase outcome-referenced results whenever such visual devices will make the document easier and faster to read.

   - In general, graphics "show" and text "tells." Oftentimes, technical readers benefit from both working together simultaneously. To select a best-choice table or figure, consider the type of data you wish to showcase, the communication needs of the text segment into which you intend to insert the table or figure, and strategically, the highs, lows, or trends that the table or figure should reveal.

9. *Fulfill implicit contracts associated with all tables and figures* and their communication purposes.

   - If you choose to include a table or figure in a document, by doing so you establish both reader curiosity and reader expectations. You are then obliged to fulfill those expectations by adequately explaining to the reader the table or figure's purpose.
   - For graphs, charts, or tables, this often means textual narration—referring to and discussing the key highs, lows, or trends displayed by the data (e.g., tracing through the cycles of a sinusoidal wave). For images and drawings, this might require describing the subject's form and function (e.g., a three-dimensional robotic arm's anatomy and its range of movement and force output). Or for a system, this could mean explaining how the system's application pieces connect to theory (e.g., a "walk-through" of a system flowchart for a chemical processing plant).

Document's End

10. *Whenever possible, present conclusions and recommendations that are graceful and diplomatic*, in addition to being relevant and useful.

    - A report's final position comes from the report writer (the project doer or activity leader) and gets delivered to the report reader(s)—the primary audience, such as an engineering manager, as well as secondary audiences, such as upper management, project partners, clients, or product/process end users. A good report's final position should be tailored to be relevant and useful to multiple perspectives—the writer's and, more importantly, the readers'.
    - For positive project outcomes, establishing win-win relevance and usefulness for both the reader and writer is easy. However, when a project's final

outcome yields negative or unwelcomed impact for the reader, writing a report ending that readers will accept is more difficult.

- For example, if an engineering report presents the findings of a feasibility study to determine whether one of the company's favorite, long-standing product lines should be obsoleted and replaced with a new line, and the report's conclusion is yes, some of the report's readers are likely not to welcome this outcome—especially readers who may have a personal history attached to the old line and, perhaps, were even original inventors of it. In this case, the report writer must be graceful and diplomatic when delivering the bottom line. To accomplish this goal, the writer might choose to acknowledge and highlight a few of the old product line's successes out of respect. The writer might also communicate a few previously undisclosed advantages of the new line that demonstrate that it is a worthy successor and will make the loss of the old line somehow "worth it" to wary readers.

11. *Whenever possible, establish criterion-referenced, data-driven arguments*, rather than qualitative, abstract arguments.

    - A criterion-referenced, data-driven argument uses facts and data as its chief coinage of persuasion. Most of the time, an effective engineering report's argument links back to an initial project objective and set of target outcomes stated in the beginning, as well as a report middle that generates and presents an ample quantity of facts and data that relate to the report objective. Then, at the report's end, to derive conclusions and recommendations, instead of using philosophical rhetoric (subjective and qualitative) to substantiate final claims, the writer obtains reader agreement using factual/quantitative evidence.
    - A straightforward example of a closing argument might come from an annual report with a purpose of communicating a manufacturing company's profitability based on *production output* measured in units sold per year and associated *revenue* in dollars. In this case, to claim that the company achieved substantial sales during the previous fiscal year, the writer might refer to the following two results numbers as evidence: 515K units sold and $2.23M net revenue. These two numbers will likely impress the reader if their order of magnitude matches readers' general expectations. If, on the other hand, readers are hoping for bigger numbers, then they will be less convincing.

12. *Whenever possible, use concrete, quantitative baselines and targets to frame your data-driven arguments.*

    - To further strengthen a report's data-driven argument and ensure that it is difficult to refute or contest, I recommend that you evaluate your project's results and findings using terms and conditions established earlier in the document (ideally in the introduction). When you set up your argument, in addition to establishing the decision-making process, also establishing baseline and target values related to your decision-making process will anchor your argument to a concrete starting point and specific end-goal target(s).

- For instance, using Feature 11's example, the two numbers closing the argument (515K units/year and $2.23M revenue/year) would be even more persuasive if the writer had established earlier in the document that the previous year's annual sales for the same product were 400K units sold/year, with net annual revenue of $1.5M, and that this year's target goals were 500K total unit sales and $1.75M revenue.

13. *Show foresight and insight* in the document's conclusions and recommendations by going beyond simply answering the project or activity objective and offering more.

    - During your experience working on a project, your perspective probably expanded beyond the guidelines of the project or activity objective, and you made useful observations and gained additional insights that might be of interest to readers. These might include recommended future steps, bonus outcomes, projections of additional cost or time savings, equipment or process recycling, and possible spin-off projects. If applicable, offer any or all of the preceding to your reader as a bonus.

14. *Signal closure* to the reader when the document's message is complete.

    - After a document completes its communication purpose, a good engineering document ends clearly and quickly, so the reader experiences a sense of psychological closure and is released from the communication process to now move on to other tasks. In short-form documents, aside from announcing closure in a sentence or two, and sometimes also stating a call for subsequent action, there should be no unnecessary rehashing of points already made or ending quips, anecdotes, or meditations. The endings of long-form documents should be as concise as possible as well.

Throughout the Document

15. *Identify the document's target audience(s) and aim the document's message* at those readers.

    - When writing an engineering document, always begin by assessing your target audience's wants, needs, biases, and technical experience/education. Then aim your document's message at that audience, and tailor the document's text—its structure, format, level of detail, level of technical complexity, and tone—for that specific audience. Also, identify and assess probable secondary (and other) audiences. A simple and practical example of secondary readers for a short-form email would be the people you CC on the email. Be sure that your document aligns with these other readers' needs, too.

16. *Anticipate your target audience's objections and preemptively defuse them.*

    - Once you've identified a document's target audience, stay conscious of that audience as you develop your document's message. Try to predict places in the message where your audience is likely to object to a point you're making or be prompted to formulate a question or concern. In such places, be immediately proactive. Write a response—a counterargument or explanation—directly into the text. Doing this will likely defuse or at least soften an audience's objections, questions, and concerns.
    - On the other hand, if you anticipate that readers will argue that an idea that you have proposed has already been attempted and failed in the past, acknowledge that you already know about this failure and that you have modified your approach to avoid a repeat of nonsuccess.
    - Here is an example preemptive strike: "We understand that you might be concerned that adding a biodigestor to our production machinery in Beijing, China, will not be an effective solution for reducing organic contaminants, given that a similar installation failed at our São Paulo, Brazil, plant. However, we have determined that the cause of the São Paolo failure was insufficient process temperature and chamber size, and we have tested and verified a redesigned biodigestor for Beijing."

17. *Advance a coherent and cohesive discussion that partners claims (assertions) with evidence (facts and numbers).*

    - Partnering claims with evidence is a best practice that nearly always applies when writing an argument in a technical report or proposal, or in anything else involving persuasion. Consider attorneys in courtrooms and how they argue cases before a judge. Sometimes attorneys make claims and produce evidence to back them. Sometimes they produce evidence and deduce a claim. Attorneys win cases not by cataloging a list of free-standing claims or offering a sole display of evidence but rather by partnering these two items together. When you are writing an engineering document, and you assert a claim, remember that it will usually be more persuasive if presented with evidence.

18. *Deliver your document's message using an effective, professional tone.*

    - In general, the optimal tone for most engineering documents is one that is objective, confident, and solution-focused. To style sentences this way, avoid unnecessary hype and exaggeration, present your claims without tentativeness or undue qualification, and avoid negative phrasing.
    - For example, state that the project's status is "on schedule" (as opposed to "proceeding amazingly as planned"), conclude the report by recommending that "the best material choice is a composite" (instead of "I believe that we should probably try using a composite material"), and assert that "although the first test was inconclusive, we achieved success running the second trial" (not "the first test was inconclusive and thus it failed, so unfortunately, we had to run a second trial, which finally succeeded").

19. *Build your documents out of concise, clear, and correct sentences* (and paragraphs).
    - Add all 20 sentence optimization techniques presented in Chaps. 10–13 to your sentence engineering tool kit, and apply them generously when you write workplace documents. Concise, clear, and correct sentences (that are also appropriately styled) help make a document's text user-friendly. Excellent sentences are quick to read, easy to understand, and free from sentence-level glitches that cause readers to be confused, hesitant, and annoyed. They are a staple for building excellent documents.
20. *Build your documents using user-friendly modularization (chunking) and headings*, as well as "best-choice" typography.
    - In addition to sentence-level quality, how you place sentences on the page (or screen) also impacts the user-friendliness of text. As I have advised throughout this book, text that is delivered in small chunks with plenty of blank space before and after, and on the left and right, makes a document more user-friendly simply because its text is easy and quick to read.
    - As an engineering writer, it makes good sense to always apply feature 20 to your writing. Though both composing excellent sentences and optimally positioning sentence text can significantly improve document quality for readers, of the two, inserting frequent paragraph returns and adjusting tab stops is an easier task for the writer.

Action Items

α	Consider all 20 universal features covered by this chapter. Which features do you already incorporate as a "best practice" feature when you're writing engineering documents? What are some features that you do not regularly incorporate in your documents that you might want to add as a future "best practice(s)." Along with thinking through this action item, you might also find it useful to print out the Review Box added at the end of this chapter and put a check mark by the features you already incorporate and circle the number of those features that you would like to remember and incorporate when writing future engineering documents.
α	Front-loaded bottom-line-first statement and executive summary aside, the main body of many engineering reports structurally aligns with the scientific method (see basic definition below). Moving from introduction to end, the engineering report writer methodically develops an argument that: 1. States a problem to be solved or opportunity to be seized 2. Examines relevant facts and data first 3. Develops conclusions and recommendations based on those facts and data
	Conversely, academic essays often begin with a closed thesis stating a position toward some philosophical, political, or sociological issue; and then, from that point onward, the essay writer argues vigorously for or against that position until the writer aggregates argument points to persuade the reader to agree with (or at least accept/acknowledge) the writer's position. *Think about the similarities and differences between how you were taught to write an academic essay in school versus how this book advises you to write engineering reports.*

(continued)

	scientific method (definition) [1]: noun 1. A method of research in which a problem is identified, relevant data are gathered, a hypothesis is formulated from these data, and the hypothesis is empirically tested.
α	Continuing on with the preceding action item, make two side-by-side lists—one of the similarities between academic writing and engineering writing and one of the differences—and then put this observation in your engineering writing tool kit as a future reference.

Recap

Key Learnings/Takeaways

- This book presents a list of 20 universal features that often distinguish excellent engineering documents. All 20 are summarized in the Review Box below.

Review Box: 20 Universal Features That Distinguish Excellent Engineering Documents

Document's Beginning

1. Front-load the bottom line.
2. Deliver a complete statement of the topic, scope, and significance.
3. Present pertinent background information or existing "baseline" conditions.
4. Record the project or activity's objective.
5. Record the project or activity's target outcomes.

Document's Middle

6. Strive to record and explain each critical thinking process you use and each necessary assumption you make.
7. Explain your project or activity's methodology for testing and data collection.
8. Insert best-choice tables and figures.
9. Fulfill implicit contracts associated with all tables and figures.

Document's End

10. Whenever possible, present conclusions and recommendations that are graceful and diplomatic.
11. Whenever possible, establish criterion-referenced, data-driven arguments.
12. Whenever possible, use concrete, quantitative baselines and targets to frame your data-driven arguments.
13. Show foresight and insight.
14. Signal closure when the document's message is complete.

(continued)

Throughout the Document

15. Identify the document's target audience(s) and aim the document's message at them.
16. Anticipate your target audience's objections and preemptively defuse those objections.
17. Advance a coherent and cohesive discussion that partners claims (assertions) with evidence (facts and numbers).
18. Deliver your document's message using an effective, professional tone.
19. Build your documents out of concise, clear, and correct sentences.
20. Build your documents using user-friendly modularization (chunking) and headings.

Reference

1. Dictionary.com "scientific method," in *Dictionary.com Unabridged*. Source location: Random House, Inc. http://www.dictionary.com/browse/scientific-method. http://www.dictionary.com/. Accessed 26 Mar 2018

Appendix

Scan and Code Demo: Using Equations to Identify and Reconcile Imposter Sentences

To demonstrate sentence-level scanning using sentence algebra, we will use a text sample extracted from a short justification memo (first draft, unfinished). Although scanning can reveal a variety of sentence-level errors, here we will focus only on "top-level" scanning to determine whether the sentence equations under word streams that are typed as sentences properly align with one of five basic (B) sentence formulas or with an acceptable combination of the basic formulas making an advanced sentence (A).

The demonstration explicitly describes each step—(a) scanning the original paragraph to determine word streams typed as sentences, (b) examining each word stream *through the lens of math (sentence algebra)* and annotating (or "coding") the streams to reveal the core equations underneath (both correct and flawed), and, finally, (c) determining fixes for the faulty equations and applying these repairs to the text of a revised paragraph.

In practice, proficient engineering writers usually "do the math" of scanning in their heads. It takes place automatically while the writers write and filter out errors before they are made. When errors do inadvertently slip past a writer's *ongoing quality control process*, good writers identify and reconcile those errors in subsequent scans during editing/proofreading. Here, we slow down the process in order to illustrate how "seeing the grammar underneath sentences" can enable the engineering writer to identify and repair grammar errors using an objective as well as precise and accurate method.

To begin, I invite you to read the example text below in its original form (version 1). As you read, scan the text to identify and count the number of word streams that are typed as sentences. Next, using whatever grammar system you wish, try to envision the basic sentence structure(s), or basic sentence equation(s), underneath each stream typed as a sentence. Then, note which sentences you think are

B. Henderson, *A Math-Based Writing System for Engineers*,
https://doi.org/10.1007/978-3-030-10756-7

grammatical (legitimate) and which ones are ungrammatical (imposters). For now, don't worry about specific repair techniques for any imposters.

original text version 1

The purpose of this memo is to request $125K to build and test additional wing-profile prototypes for the RobinX4. At a standard 135 mph cruising speed and elevation, our new single-propeller craft must generate a lift force $\geq$ 2,583 pounds. Wind tunnel tests of wing prototypes A, B, and C failed. Even though our Bernoulli simulation program predicted sufficient margins and success. Given that pre-orders for the RobinX4 craft are accumulating rapidly, we need to perfect the wing design and all other aspects of the craft by next fiscal quarter, yet we will have ample qualification test time to ensure on-target product release. Our anticipated profit margin is $5K net per unit, thus the expense is justifiable. We can recoup additional development expenses in one fiscal cycle: predicted first-year sales are 500 planes.

If your initial scan of the version 1 text detected that there are seven word streams typed as sentences and that two of these streams are imposters rather than legitimate sentences, congratulations—you are correct. If you missed any of this, that's okay, too—the next version tags and counts (using superscript numbers) the streams typed as sentences, flags which streams are imposters (in *italics*), and also notes [in brackets] what type of error each imposter represents (run-on or fragment).

scanned and marked text version 2

[1]The purpose of this memo is to request $125K to build and test additional wing-profile prototypes for the RobinX4. [2]At a standard 135 mph cruising speed and elevation, our new single-propeller craft must generate a lift force $\geq$ 2,583 pounds. [3]Wind tunnel tests of wing prototypes A, B, and C failed. [4]*Even though our Bernoulli simulation program predicted sufficient margins and success* [sentence fragment]. [5]Given that pre-orders for the RobinX4 craft are accumulating rapidly, we need to perfect the wing design and all other aspects of the craft by next fiscal quarter, yet we will have ample qualification test time to ensure on-target product release. [6]*Our anticipated profit margin is $5K net per unit, thus the expense is justifiable* [run-on sentence]. [7]We can recoup additional development expenses in one fiscal cycle: predicted first-year sales are 500 planes.

Now, let's further investigate *why*, grammatically, the two marked structures—one fragment and one run-on—are imposters. Our next example paragraph, version 3, will use sentence algebra to code, stream by stream, the basic sentence equations that reside beneath the paragraph's text. First, however, to prepare all readers to follow along, no matter where they're at in their math-based writing studies, let's review the sentence algebra fundamentals that apply to our upcoming example.

Sentence Algebra Review

Since we're doing a simplified top-level scan for imposter sentences, the equations will express only core functional elements, that is, words (or word groups) represented by the variables "N" = noun, "X" = pronoun, "V" = verb, and (in one case) "M" = modifier, and the subscripts that further classify functional roles, "s" = subject, "o" = object, "i" = intransitive, and so on. The annotated scan omits sentence elements that add specificity (details) to a sentence's message; this omission does not impact the core framework that defines sentence type—basic, advanced, or imposter.

Three (of the five total) *basic sentence* types will appear in the upcoming scan's annotation (coding):

$\mathbf{B_1 = (N_s \text{ or } X_s) + V_i}$
Basic Sentence One = (Noun or Pronoun Subject) + Verb Intransitive

$\mathbf{B_2 = (N_s \text{ or } X_s) + V_t + (N_o \text{ or } X_o)}$
Basic Sentence Two = (Noun or Pronoun Subject) + Verb Transitive + (Noun or Pronoun Object)

$\mathbf{B_3 = (N_s \text{ or } X_s) + V_l + (N_p \text{ or } M_p)}$
Basic Sentence Three = (Noun or Pronoun Subject) + Verb Linking + (Predicate Noun or Predicate Adjective)

Additionally, four combinations of basic sentences forming advanced sentences will also appear in the coding:

$\mathbf{A = B_1 \,/\, B_2}$
Advanced Sentence = (B_1)—(Subordinating Conjunction)—(B_2), where "/" is the shorthand for subordinating conjunction (such as "although," "because," and so on)

$\mathbf{A = B_2 \mid B_2}$
Advanced Sentence = (B_2)—(Coordinating Conjunction)—(B_2), where "|" is the shorthand for coordinating conjunction (such as "and," "or," and so on)

$\mathbf{A = B_3 \,;\, c, B_3}$
*Advanced Sentence = (B_3)—(**;** conjunctive adverb **,**)—(B_2) where "c" is the shorthand for c = conjunctive adverb = "however," "in fact," and so on*

$\mathbf{A = B_2 : B_3}$
*Advanced Sentence = (Basic Sentence One)—(**:**)—(Basic Sentence Two)*

In version 3 below, you will also notice that the text annotation underlines words affiliated with variables in the sentence algebra equations, which are written immediately above each line of text. In addition, the annotation tags all imposters with an "✗" and legitimate constructions with a "✓." Finally, where applicable, the annotation recommends equation-level corrections in brackets [. . .].

Here is the annotated (coded) scan of our example paragraph that reveals the fundamental sentence equations underneath the text:

annotated (coded) scan showing equations version 3

$\mathbf{(N_s + V_1 + N_p) = B_3}$ ✓
The purpose of this memo is to request $125K to build and test additional wing-

profile prototypes for the RobinX4. At a standard 135 mph cruising speed and

$\mathbf{(N_s + V_t + N_o) = B_2}$ ✓
elevation, our new single-propeller craft must generate a lift force of $\geq$ 2,583

$\mathbf{(N_s + V_i) = B_1}$ ✓ /
pounds. Wind tunnel tests of wing prototypes A, B, and C failed. Even though

$(/\mathbf{N_s} + \mathbf{V_t} + \mathbf{N_o}\ [=/\mathbf{B_2}]) =$ **fragment** ✗ [correction: $\mathbf{A} = \mathbf{B_1} / \mathbf{B_2}$]

our Bernoulli simulation program predicted sufficient margins and success.

Given that pre-orders for the RobinX4 craft are accumulating rapidly, we need

$(\mathbf{X_s} + \mathbf{V_t} + \mathbf{N_o}) = \mathbf{B_2}$ ✓ |

to perfect the wing design and all other aspects of the craft by next fiscal quarter, yet

$(\mathbf{X_s} + \mathbf{V_t} + \mathbf{N_o}) = \mathbf{B_2}$ ✓ [okay as is: $\mathbf{A} = \mathbf{B_2} \mid \mathbf{B_2}$]

we will have ample qualification test time to ensure on-target product release.

$(\mathbf{N_s} + \mathbf{V_l} + \mathbf{N_p}\ [=\mathbf{B_3}])$ $,\mathbf{c}\ (\mathbf{N_s} + \mathbf{V_l} + \mathbf{M_p}\ [=\mathbf{B_3}]) =$ **run-on** ✗

Our anticipated profit margin is \$5K net per unit, thus the expense is justifiable.

[correction: $\mathbf{A} = \mathbf{B_3}\ ;\ \mathbf{c},\ \mathbf{B_3}$]

$(\mathbf{X_s} + \mathbf{V_t} + \mathbf{N_o}) = \mathbf{B_2}$ ✓ :

We can recoup additional development expenses in one fiscal cycle: predicted

$(N_s + V_l + N_p) = B_3$ ✓ [okay as is: $\mathbf{A} = \mathbf{B_2} : \mathbf{B_3}$]

first-year sales are 500 planes.

In the annotated scan shown above, the first imposter sentence to be revealed in both text and equation form is a sentence fragment, which begins with the subordinating conjunction "Even though." A legitimate basic sentence—specifically, a B_2—follows the conjunction, but since the B_2 is preceded by a subordinating element, the structure cannot stand alone. We can remedy the problem by attaching the subordinated B_2 to the basic B_1 sentence that stands before it. The result is an advanced sentence, $A = B_1 / B_2$, or, in text, "Wind tunnel tests ... failed, even though ... sufficient margins and success."

The second "imposter sentence" is a run-on structure. Here, the culprit is a "weak-link" connection created by an improperly used comma (comma splice). To connect two basic (B_3) sentences together with a word like "thus," a conjunctive adverb, punctuating with a lone comma does not work. The correct punctuation here is a *semicolon* [;] *before* and a *comma* [,] *after* the conjunctive adverb (represented in the equation as "c"). Thus, an easy fix here would be to upgrade the punctuation from "... per unit, thus the expense ..." to "... per unit; thus, the expense ..." with the resultant sentence algebra equation being $A = B_3\ ;c,\ B_3$.

If we incorporate all of the preceding, the repaired text looks like this:

repaired text version 4

The purpose of this memo is to request \$125K to build and test additional wing-profile prototypes for the RobinX4. At a standard 135 mph cruising speed and elevation, our new single-propeller craft must generate a lift force $\geq$ 2,583 pounds. Wind tunnel tests of wing prototypes A, B, and C failed, even though our Bernoulli simulation program predicted sufficient margins and success. Given that pre-orders for the RobinX4 craft are accumulating rapidly, we need to perfect the wing design and all other aspects of the craft by next fiscal quarter, yet we will have ample qualification test time to ensure on-target product release. Our anticipated profit margin is \$5K net per unit; thus, the expense is justifiable. We can recoup additional development expenses in one fiscal cycle: predicted first-year sales are 500 planes.

Index

B. Henderson, *A Math-Based Writing System for Engineers*,
https://doi.org/10.1007/978-3-030-10756-7